Current Challenges for the Aquatic Products Processing Industry

SCIENCES

Agronomy and Food Science,
Field Directors – Jack Legrand and Gilles Trystram

Food Chains, Subject Heads – Thomas Croguennec,
Catherine Renard, Véronique Verrez-Bagnis,
Sylvie Chevallier and Véronique Sante-Lhoutellier

Current Challenges for the Aquatic Products Processing Industry

Coordinated by
Véronique Verrez-Bagnis

iSTE

WILEY

First published 2023 in Great Britain and the United States by ISTE Ltd and John Wiley & Sons, Inc.

ISTE Ltd
27-37 St George's Road
London SW19 4EU
UK

www.iste.co.uk

John Wiley & Sons, Inc.
111 River Street
Hoboken, NJ 07030
USA

www.wiley.com

Library of Congress Control Number: 2023942087

British Library Cataloguing-in-Publication Data
A CIP record for this book is available from the British Library
ISBN 978-1-78945-149-8

ERC code:
LS8 Ecology, Evolution and Environmental Biology
LS8_11 Marine biology and ecology
LS9 Applied Life Sciences, Biotechnology, and Molecular and Biosystems Engineering
LS9_3 Applied animal sciences (including animal breeding, veterinary sciences, animal husbandry, animal welfare, aquaculture, fisheries, insect gene drive)
LS9_5 Food sciences (including food technology, food safety, nutrition)

Contents

Preface

Véronique VERREZ-BAGNIS
Ifremer, Nantes, France

In terms of food, fish is one of the most appreciated among consumers due to its image as a healthy product. In fact, fish is a food with a high nutritional value, being rich in proteins, vitamins and essential amino acids. Fatty fish also provide polyunsaturated fatty acids which cannot be synthesized by humans.

The latest FAO State of Fisheries and Aquaculture report shows the growing role of the aquatic product sector in the global economy and in human nutrition. Indeed, according to 2018 figures, the global production of marine and freshwater products (fish and shellfish) is estimated at 179 million tons, including 82 million tons of aquaculture products, worth USD 401 billion. Since the 1960s, global per capita seafood consumption has increased from 9 kg (live weight equivalent) to 20.5 kg in 2018.

Fish is a very perishable foodstuff, undergoing an alteration of its organoleptic properties due to chemical degradation linked to a significant bacterial development, and to the action of enzymes present in fish (proteases, TMOase, lipases, etc.). In order to extend the shelf life of fish, the development of efficient processing treatments has progressed on a rather large scale. Next to the methods traditionally used, such as salting and drying, fermentation, marinating or smoking, refrigeration and especially freezing have enabled the distribution and sales in areas farther from the coast and have upset the whole sector with the advent of factory ships and the construction of new supply, processing and marketing channels. Fish is currently one of the most globalized foodstuffs.

Current Challenges for the Aquatic Products Processing Industry,
coordinated by Véronique VERREZ-BAGNIS.

Like any agri-food sector, the fishery and aquaculture industry is constantly changing. Indeed, the management methods of fisheries are evolving and must be improved so that the exploited stocks can be reconstituted and support a sustainable management of fisheries and ecosystems. Consumption habits are also changing (e.g. the recent craze by Westerners for raw products such as sushi). Technological and physical treatments, as well as biological treatments, are evolving to ensure the safety of food products. Public policies and associated regulations are very often amended.

Although this book is part of the SCIENCES collection and is intended to provide an overview of the vast field of the aquatic product industry, our purpose is to focus on the main current challenges of the aquatic product processing industry. The issues addressed are somewhat more related to the problems of European consumers and processors, with Europe importing 40% of total world imports in fishery and aquaculture products. However, the information provided in this book can be widely extrapolated and useful to any reader interested in these subjects, whether a processor, a fishmonger, a consumer, a scientist or a student.

Thus, in this book, the first chapter is dedicated to the sociological definition of the "fish" as food and the perceptions of this product by consumers. The following chapters deal with the notions of quality sensu lato (Chapters 2–4), then with biological risks (Chapters 5 and 6) and those related to the pollution of oceans and ecosystems by microplastics (Chapter 7). Two types of processes and processed products which are very important for the European market (smoked products and surimi-based products – Chapters 8 and 9) are then addressed. Chapters 10 and 11 talk about valorization of by-products of the processing industry, including the "blue biorefinery", and the innovative pH-shift technology that plays a central role in splitting up poorly valorized resources into high value-added products.

April 2023

1

Consumer Perceptions of "Fish" Food

GERVAISE DEBUCQUET
Audencia Business School, Nantes, France

1.1. Introduction

Due to being omnivores (Fischler 1990), humans are both obliged to diversify food sources to satisfy nutritional needs and the willingness of choosing these sources to vary the pleasures provided by the act of eating. These choices are made according to the availability of food, individual expectations, as well as the more or less affirmed prescriptions of social and cultural groups. In all cultures, animal foods are the subject of the strongest prescriptions, which can lead to prohibitions, avoidance or the expression of simple aversions. Indeed, animal products have always been the object of beliefs and myths and have thus crystallized a certain number of fears (Vialles 1998; Ferrières 2002) due to a form of magical thinking inherent in the act of incorporation (Rozin et al. 1986). Thus, incorporating the meat of an animal implies, in addition to ingesting its nutritional components, ingesting, more or less consciously, its symbolic attributes (Douglas 2005). In the continuity of Hippocrates' thought, we can say that "one is what one eats" by virtue of these two mechanisms.

Incorporating fish into diets is therefore symbolically incorporating the real and supposed, imagined properties of the animal as well as of the sea (Fischler 1994); this gives rise to a symbolic equivalence between the sea, fish and those who eat them (Corbin 2005) and to a particular place for fish in collective representations. In fact, animals, depending on their nature and their origin – terrestrial, marine, lake or

Current Challenges for the Aquatic Products Processing Industry,
coordinated by Véronique VERREZ-BAGNIS.

river – have, from time immemorial and in various socio-cultural spaces, conveyed contrasting imaginations and representations, which determine what is edible and what is a food taboo (Leach 1980). In the order of food, the lexical oppositions between "fish" and "flesh (associated with meat)" reveal the distinction between what comes from the aquatic environment and what comes from other environments (Vialles 1998). As a result, land, animal, meat and fish, regardless of their nutritional profile, are not completely interchangeable for committed eaters.

Questioning the reasons for eating fish reveals, even today, some of the specificities of this singular food and, through it, of the fish animal, questioned since antiquity, endowed with a particular status in the Judeo–Christian era and finally, reconsidered more recently in the light of dietetic and ecological imperatives.

1.2. Thinking about fish

1.2.1. *Naming the unknown*

In the order of the living, fish has always occupied a singular place that should be placed in relation to the representations of its habitat as well as of its own characteristics such as they have been conveyed both by academic discourse and through religious or popular beliefs. For a very long time, fish and other aquatic creatures remained poorly known because, not immediately visible to "earthlings", they escaped any sensitive experience. Then, the rare occasions to observe them and to understand their habits, their ways of life and reproduction did not facilitate the work of species classification.

In antiquity, fish, their anatomy, their behaviors (olfaction, reproduction, sight, mobility, etc.) were thought of by analogy with terrestrial animals and by anthropomorphic projection and thus many errors of interpretation have been highlighted (Byl and Schouls 1990). Well before scientific nomenclatures, in his *History of Animals* (Barthélémy-Saint Hilaire 1883), Aristotle used vernacular names or names borrowed from terrestrial animals, such as mule, wolf or "sea toad", to designate fish, to which certain characteristics of very distant animals, terrestrial mammals, birds or batrachians were attributed (Louis 1971). At the end of the Middle Ages, only a dozen fishes were known.

The underwater imaginings deployed in Jules Verne's 1870 work, *Twenty Thousand Leagues Under the Sea,* then fed into collective representations of the sea and aquatic creatures, which have endured long despite the beginning of oceanography as a science in the 19th century. As Elkays (2018, p. 14) puts it, "from an epistemological perspective, the marine environment is only accessible through technological artifacts". And it is perhaps this that explains why, despite the advent of modern science, the sea remains a relatively alien world to terrestrial humans, still charged with imaginaries and beliefs that give aquatic foods a special place in the food order.

1.2.2. *The imaginary of the marine world: permanence and metamorphosis*

The marine world nourishes ambivalent representations, fascinating and repulsive at the same time (Geistdoerfer 1998). For a long time, it has been thought of as the domain of monsters (Le Bras-Chopard 2000), of the damned, of the dead, and the fears embodied by necrophagous fish are still alive.

Corbin (2005) explains how relations with the sea have been shaped over time, a world that was at first unknown, then a medium for anxieties and imaginations linked to the abyss and perils of the sea, the scene of the flood for fishermen and sailors exposed to the risks of Hell. The marine world does not belong to the divine but is part of a vision opposed to that of the earthly paradise, of Eden. These representations were progressively transformed, more positive from the 18th century with the increase of the frequentation of the seaside for its beneficial and therapeutic effects. While the image of a "tamed" sea was taking shape and the "desire for the shore" was awakening among land lovers in search of visual and sensory emotions, the coastal fringes became a place of vacation and relaxation where fishing and boating were practiced (Corbin 1990). In the Dutch and Italian paintings of the end of the 18th century, this slow metamorphosis is felt and the figures of the agitated sea made people want to "see the sea" and to "eat fish". It was henceforth a sea which made a spectacle of itself and which marked at the same time, under the influence of the Christian church, the separation between Mankind and the cosmos with the favor of a "natural theology" (Corbin 1997).

The inversion of these representations was also illustrated by the promotion of "sea bathing", especially on the British coasts, whose "cold, salty and choppy" water was reputed to be a remedy for the ills of the soul and the body, or even female sterility; the analogy between the composition of sea water and that of blood serum was also supposed to explain, among the inhabitants of the coast, the fertility of fishermen. At the dawn of the 19th century, the virtues of warm water and sunbathing were proclaimed, inspired by Mediterranean habits, and in less than a century a new relationship with the sea and with bathing was established, as illustrated by the effervescence of seaside resorts during the summer (Geistdoerfer 2003). As Corbin (1997) points out, this is not a question of a clean break, because all these representations continue to exist and are still intermingled in the minds of contemporaries. At the time of the awareness of global warming, of the pollution of the seas and oceans, it is the figures of the deluge and the threats on the fringes of the continents which are reactivated. We are worried about the irreversible stigma left by a sea "coveted" for its living or energy resources.

These representations are particularly intertwined in the minds of the French, who, although they have access to a long coastline, are above all "land lovers", contrary to the countries of northern Europe or in particular, Portugal (Geistdorfer 2007). The French are attracted to the sea especially during vacation periods to see the last wilderness or an "ecologized" shoreline staged by tourism agents. Sincere attraction to the people and jobs of the sea is less common, even where we might expect it, such as in local fish distribution channels (Lazuech and Debucquet 2017).

1.2.3. *The representations associated with marine foods: the historical legacy*

By going back to the origins of the opposition between "flesh (associated with meat)" and "fish", Vialles (1998) shows that this distinction is not based on the objective or anatomical characteristics of the products but on their origin and, therefore, on the systems of representations associated with the sources of these foods. Thus, it is necessary to resituate this opposition in the binary prescriptions of Roman Christianity distinguishing "lean foods" for the days of Lent from "meat" or "fatty foods" for the days of carnage. During Lent, fish is one of the few animal foods tolerated. To abstain from meat during Lent means to deprive ourselves of fat and blood as well as to abstain from all food pleasures (Rousseau 2005). It is because of its lesser symbolic proximity to human and human flesh that fish food is tolerated; unlike the meat of land animals, it is less able to regenerate bodies and "make blood" (Rousseau 2005). In Aristotelian representations, partly taken up in Christian beliefs, fish are indeed "cold beings", "devoid of blood", their vital force being supposed to come from the water itself (Vialles 1998; Rousseau 2005)[1]. It is interesting to note the survival of these representations among contemporary eaters;

1. Another illustration of the system of symbolic thought that relates sea water and blood is seen in the representations associated with sailors: "The 'land lovers' do not understand the particularities of the practices and thoughts of sailors [...] [t]he 'land lovers' represent these 'beings' who have 'sea water in their veins instead of blood'. It cannot be otherwise, in the opinion of several authors, because to be able to live at sea, with the sea [...] one must come from it, be a 'sailor', etymologically 'of the sea'" (Geistdoerfer 1998, p. 32; 2007).

the presence of blood is spontaneously evoked only in fish such as salmon or tuna, whose meat is red[2].

Thus, the appropriation of fish by land lovers and the shaping of its status in the food space were largely determined according to a mechanism of symbolic devaluation in relation to the meat of land animals. Fish is the default food for "lean days" and is thus considered less nourishing than meat. However, the dietetic discourse which has prevailed since the end of the 20th century has made it possible to raise the value given to this food and to charge it with new meanings.

1.3. Eating fish

1.3.1. *Dealing with animality*

According to EUFOMA data (2017 (2015 data)), the average consumption in Europe is 25.1 kg of fish and seafood per capita per year but with great disparities. Portugal leads with 55.9 kg, Spain is second with 45.2 kg and finally France is third with 33.9 kg. The average French consumption actually hides important contrasts observed since time immemorial between the coastal fringe and inland (Salembien 1958; INCA 3 2017), with coastal inhabitants consuming more and more regularly than the latter, for reasons of access to the resource and familiarity with fish. According to the INCA 3 survey, the fish consumption rate is 40.2% among 18–44 year olds, 41.8% among 45–64 year olds and 50% among the over 65 year olds compared to 62.5%, 73.4% and 72.9%, respectively, for meat. Finally, the fish consumption rate is lower among those with a lower rate of education: it is around 40% (compared to 70% for meat) among those with at most a bachelor's degree and 46% (compared to 63% for meat) among those with at least a bachelor's degree or higher. Since the price factor is not sufficient to explain the differences, we observe that the "fish" is more valued among the older and higher classes because it is considered to be low-fat. As Boltanski (1971) has shown, fatty foods are subject to contrasting interpretations between the working classes and the upper classes, the former consuming them for their satiating, fortifying and energetic qualities, while the latter deplore their "heaviness" and their propensity to induce "nausea" and "indigestion". Moreover, the concern for the body among the upper classes, coupled

2. In a research program devoted to fish parasites, particularly those of the family *Anisakidae*, consumers interviewed associated the presence of parasites with the presence of blood, parasites being, for them, an incarnation of *spontaneous generation*. The presence of parasites in white fish came as a great surprise to the subjects and was counterintuitive to most of them since white fish were perceived to be bloodless (Debucquet 2011).

with a greater awareness of the role of food in health, has increased in recent decades, and the rate of fish consumption is a significant marker of this.

In 2020, the French consumer's seafood basket was composed, in volume, of 31.2% of fresh products (of which 56.5% fish and 43.5% seafood), 30.5% of chilled delicatessen products, 19.1% frozen and finally 19.2% canned (FranceAgriMer 2020). Of the volume of fresh fish, the share of whole and cut fish is, respectively, 27.9% and 69.8%. Over the last decades, the share of canned fish has decreased significantly and that of chilled delicatessen products has increased significantly, while the share of fresh fish has increased slightly since 2017 after a downward trend since the 1960s (Guillotreau et al. 2008).

Changes in lifestyles help explain the increase in the share of "ready-to-eat" fish and prepared fish dishes (INCA 3 2017), but the very low share of whole fish (about 5% of total purchases) still reveals a certain number of specific characteristics of fish consumption. Indeed, consumers very frequently mention the difficulties in preparing fish (scaling, skinning, gutting, filleting, etc.), the lack of expertise and knowledge of species (Lazuech and Debucquet 2017). Even today, those who eat fish have difficulty spontaneously naming more than three species, which poses a real challenge when it comes to incorporating unknown species, such as those from the great depths that fishmongers strive to make "good to eat" to offset the scarcity of more common species such as pollock, cod or hake (Geistdoerfer 2007). In addition, apart from the price factor (Guillotreau et al. 2008), obstacles to the purchasing of fresh whole fish are often linked to fears, whether conscious or not, of not knowing how to deal with a form of marine animality, the treatment of which, in the minds of land lovers, is the responsibility of the men of the sea for "white-flesh" fish and above all for red-flesh fish[3]. The success of "fish fillets" can thus be explained, beyond its practical side, by the symbolic distance that most French eaters want to keep with marine foods.

1.3.2. *Fish today, between pleasure and nutrition*

The negative background described above has been partially eclipsed by the health-dietary discourse promoted in the last two decades and highlighted the

3. This animality in the representations of the marine world remains, for the reasons mentioned above, ambiguous. Vialles (1998) explains how tuna fishermen align, through their practices, "the real with the ideal" (p. 5), by eliminating the apparent blood of tunas that are actually "fishes". The treatment of the blood deemed inappropriate and the absence of "real" bleeding contribute to the exclusion of tuna from the "meat" category and make it conform with the representations of fish. As Vialles (1998) puts it, it is a matter of "fishifying" it (p. 5).

nutritional qualities of fish compared to meat. Fish is indeed an important source of long polyunsaturated fatty acids for the human diet (Médale 2009) – representing 58% of EPA and 67% of DHA (INCA 3 2017), and its consumption is recommended for the proper functioning of the nervous system and the prevention of cardiovascular diseases. The French National Agency for Food, Environmental and Occupational Health Safety (*Agence nationale de sécurité sanitaire de l'alimentation, de l'environnement et du travail*, Anses) recommends eating fish twice a week by combining a fatty fish with a high omega-3 content (salmon, sardines, mackerel, herring, trout) and a lean fish, while specifying certain risks related to certain chemical contaminants (methyl-mercury, PCBs in particular) for sensitive populations (pregnant and breastfeeding women, children under three years of age)[4].

The more positive view of fish is reflected in the spontaneous evocations collected from consumers, since fish consumption is associated with the idea of health and freshness, as well as with pleasure, gustatory pleasure and festive consumption (Gouin et al. 2009). The symbolism of the sea, invigorating and purifying, contributes to the construction of representations of fish, a food that is strongly revalued today for its healthy and balanced character, and perceived as "lighter" than the meat of land animals (Paquotte 2004). The image of fish is thus in a way ennobled, even if the more negative images of polluted seas, the effects of global warming or intensive aquaculture somewhat tarnish the representations (Paquotte 2004; Michel-Guillou 2014). Moreover, studies show, still today, the lack of knowledge of the population on the species and their intrinsic qualities. Certain paradoxes can be noted, such as the declared attraction for "wild" and "lean" fish, while salmon, a fatty fish that is mostly farmed, remains the leading fish consumed (26% of the volume of fresh fish), followed by cod (14%) and pollock (8%). It is interesting to note that these three species are largely the most consumed, while all the others (trout, sea bream, monkfish, mackerel, whiting, sardines, hake, sea bass, blue ling, etc.), among which excellent "fatty" fish for their omega 3 content, share the remaining volumes (FranceAgriMer 2020). This paradox is explained by the difficulty of making sense, for most French people, with this lean/fat opposition when it concerns fish, fat being more associated with land animal meat and food eaten outside of Lent. Finally, the lack of knowledge of polyunsaturated fatty acids is not surprising, as a detailed understanding of these nutritional issues is not within

4. According to the latest INCA 3 survey (2017), 36% of respondents report being familiar with the PNNS fish benchmark ("recommended consumption two times a week"), 40% among women versus 33% among men, 41% among 18–44 year olds versus 32% among 45–64 year olds. This benchmark is also better known among those with a higher education.

reach of all eaters despite the multiplication of claims on food packaging over the past two decades (Masson et al. 2016). In doing so, the lesser knowledge of fish species and the *nutritionalization* of fish contribute to confirm the intuition of Geistdoerfer et al. (2003): many contemporary citizens eat fish, i.e. a material for its nutritional qualities and not an animal. Recent developments in consumption and supply patterns seem to be slightly changing this relationship with fish.

1.4. Reconnecting with the fish animal?

1.4.1. *A culinary exoticism: raw fish*

As Lévi-Strauss (1964) has shown, cooking occupies a central place in culinary practices in that it facilitates the passage from nature to culture. Williot (2015) shows that the opposition between raw meat eaters and cooked meat eaters is nowadays expressed in multiple ways, revealing personal (taste preferences), religious, culinary habits or precautions linked to intuitive or hygienic knowledge. While the consumption of raw meat tartar is historically common in France, the very rapid progression of raw fish consumption raises questions.

According to the INCA 3 survey (periods 2014–2015), the consumption of raw foods has increased significantly and more strongly for fish. Indeed, the rate of consumption of raw fish, consumed in the form of sushi or ceviche, has doubled since the last INCA 2 survey (periods 2006 and 2007), rising from 15% to 31% of the population concerned. It is more common among adults aged 18–45 years, men, those with a higher education and those in higher professions and socio-professional categories (SPCs). Historically an uncommon practice in France, it now occupies an important place among French eaters, fond of a certain food exoticism (Régnier 2015). These ways of consuming fish, which originated in Asia, Oceania and South America, are associated with festive and convivial moments and contribute to a reconnection with the sea and its symbols, naturalness, freshness and health. This consumption is a source of hedonism and curiosity for those who wish to reconnect with the foods of the sea. Is eating it raw a way to change the status of fish or to make it become "*flesh*" symbolically? As mentioned earlier, most contemporary eaters still share the image of fish as a bloodless being, which excludes it from the category of meat. The possible presence of parasites is often counter-intuitive for raw fish eaters, because it confronts them with a form of animality somewhat buried in the collective unconscious (Debucquet 2011). Still, in contrast to the success of prepared fish dishes (FranceAgriMer 2020), the growing consumption of raw fish seems to allow the eater to appreciate and rediscover the textures and unique tastes of different species of fish, such as salmon, tuna, sea bass, sea bream, mackerel, etc.

1.4.2. *The local distribution channels of fish*

The SARS-CoV-2 pandemic and subsequent lockdowns have revitalized home-made and local and/or direct sourcing. The purchase of aquatic products increased by 6% in volume between 2019 and 2020, and by 3% for fresh fish in the seafood department in particular (+ 2% for whole fish) (FranceAgriMer 2020). Even if these increases are moderate, they are part, beyond the contingencies of the pandemic, of a context of food anxiety and the desire to reclaim food via raw products. This is illustrated by the return of local or short-distance supply chains. According to the INCA 3 survey (2017), fish purchases are made primarily in supermarkets (58.9% of households), in convenience stores and fishmongers (19.6%) and finally in markets/direct sales (9.6%), the last two channels more particularly among the most educated. However, the last decade has seen the emergence of various short direct supply chains, such as the Poiscaille[5], Ô'Poisson[6] or the AMAP-Poisson de Yeu-Nantes websites, which promote fresh, local, "wild" fish from French coasts, "line-caught" fish and also "noble and forgotten" species[7]. These two modes of supply are very successful, especially in the western regions, especially in the hinterland, where street fish sales almost disappeared at the end of the last century. Recent surveys show that the motivations are plural, economic, social and culinary (Salladarré et al. 2018). The case of the AMAP Yeu-Nantes is interesting in this respect because the members ultimately show little interest in the men and trades of the sea, the techniques and fishing areas. They are more interested in acquiring knowledge about fish in order to be able to better differentiate them, name them and above all gain know-how to prepare and cook them (Lazuech and Debucquet 2017). Among the exclusive buyers of these channels, the main motivation is relational for exchanges and reciprocity with other citizens who wish to support locally regarding fish recipes. These forms of supply help to overcome the obstacles related to the lack of knowledge, expertise and self-confidence to reappropriate the foods of the sea[8]. Communities of "fish eaters" are thus being consolidated for the horizontal transmission of knowledge about the fish animal and the fish food (Debucquet et al. 2020).

5. See: https://poiscaille.fr/.

6. See: https://www.o-poisson.fr/bar-de-ligne.html.

7. See the home page of the Poiscaille website op. cit.

8. More than one in two adults (57%) consider that they do not eat enough fish, because of the high price (for 43% of them), because they do not know how to prepare it (22%), because they do not like fish (20%) and finally because of the odors released during cooking (20%) (INCA 3 2017).

1.5. Conclusion

This chapter has highlighted the duality of the relationship with fish through the lens of the imaginary of the sea and sea creatures, contemporary dietary injunctions and finally social and commercial innovations to restore a certain proximity with marine foods. Eating fish has some peculiarities since the status of fish and its animality remains endowed with a strong symbolic valence; historically associated with a time of penitence throwing out the criteria of exclusion from the category of meat, fish has then become one of the emblems of food hygiene in Western societies. The food pleasure has not been rested for pescatarians as for certain occasional eaters ready to pay an additional cost to obtain fish whose animality is preserved until plate serving (the setting in scene of whole fish, fresh, “line caught”, etc.). For the rest of the contemporary eaters, industrial preparations of fish-based dishes allow them to manage the symbolic distance with the sea and its living resources or to reify the fish animal.

In the light of the dietary transition requiring a reduction in the consumption of animal proteins and the awareness of the pollution of the seas and the consequences of overfishing, we can ask ourselves what will happen to the relationship between Western eaters and fish. Will we eat less “fish” and/or rediscover the diversity of “fishes”, while gaining knowledge and know-how to better appreciate their taste and health qualities? The degree of change in eating habits will tell us.

1.6. References

Barthélémy-Saint Hilaire, J. (1883). *Histoire des animaux d'Aristote*. Hachette, Paris.

Boltanski, L. (1971). Les usages sociaux du corps. In *Annales. Economies, sociétés, civilisations*, 26(1), 205–233.

Byl, S. and Schouls, S. (1990). Quelques préjugés d'Aristote en ichtyologie et leur survivance chez certains de ses successeurs jusqu'au XVIIIe siècle. *Revue belge de philologie et d'histoire*, 68(2), 305–314.

Corbin, A. (1990). *Le territoire du vide : l'Occident et le désir du rivage (1750–1840)*. Flammarion, Paris.

Corbin, A. (1997). La mer et l'émergence du désir du rivage. *Aménagement et nature*, 125, 5–15.

Corbin, A. (2005). *Le ciel et la mer*. Bayard, Paris.

Debucquet, G. (2011). Perception du risque parasitaire par le consommateur. In *Approche intégrée de la sécurité et de la qualité des aliments*, Le Bail, P. (ed.). INRA, Programme de recherche AISQAL, Région Pays de la Loire.

Debucquet, G., Guillotreau, P., Lazuech, G., Salladarré, F., Troiville, J. (2020). Sense of belonging and commitment to a community-supported fishery. The case of Yeu Island, France. *Review of Agricultural, Food and Environmental Studies*, 101(4), 439–459.

Douglas, M. (2005). *De la souillure : essai sur les notions de pollution et de tabou.* La Découverte, Paris.

Elkays, R. (2018). La science populaire et l'imaginaire des profondeurs au XIXème siècle [Online]. Available at: https://hal.archives-ouvertes.fr/hal-01674278.

EUFOMA (2017). Le marché européen du poisson en 2017. Report, European Market Observatory for Fisheries and Aquaculture Products, European Commission.

Ferrières, M. (2002). *Histoire des peurs alimentaires. Du Moyen Age à l'aube du XXe siècle.* Le Seuil, Paris.

Fischler, C. (1990). *L'Homnivore.* Odile Jacob, Paris.

Fischler, C. (1994). Editorial : magie, charmes et aliments. In *Manger magique. Aliments sorciers, croyances comestibles*, Fischler, C. (ed.). Éditions Autrement, Paris.

FranceAgriMer (2020). Consommation des produits de la pêche et de l'aquaculture. Study report, FranceAgriMer.

Geistdoerfer, A. (1998). La mer coule dans leurs veines – Les marins pêcheurs, de la race des insoumis ! *Bulletin de psychologie*, 50(432), 651–668.

Geistdoerfer, A. (2007). L'anthropologie maritime : un domaine en évolution : hors cadre traditionnel de l'anthropologie sociale. Zainak. *Cuadernos de Antropología-Etnografía*, 29, 23–38.

Geistdoerfer, A., Ivanoff, J., Matras-Guin, J. (eds) (2003). *La mer dévorée. Le poisson bon à manger. Le poisson bon à penser.* CETMA-Anthropologie maritime, Paris.

Gouin, S., Fournis, M., Hadouni, L., Mesnildrey, L., Lesueur, M., Letellier, I., Bodilis, C. (2009). Étude des attentes des consommateurs de produits de la mer frais et de leurs comportements selon les circuits de distribution. Research report, Programme Cogépêche, Agrocampus Ouest.

Guillotreau, P., Le Grel, L., Tuncel, M. (2008). La demande de produits de la mer. In *Mare economicum : enjeux et avenir de la France maritime et littorale*, Guillotreau, P. (ed.). Presses universitaires de Rennes.

INCA 2 (2009). *Étude individuelle nationale des consommations alimentaires 2 (2006–2007).* AFSSA, Paris.

INCA 3 (2017). *Étude individuelle nationale des consommations alimentaires 3 (2014–2015).* ANSES, Paris.

Lazuech, G. and Debucquet, G. (2017). Culture alimentaire et accord marchand local : une enquête au sein de l'AMAP "Poisson" Yeu-Continent. *Terrain & Travaux*, 31(2), 131–158.

Le Bras-Chopard, A. (2000). *Le zoo des philosophes : de la bestialisation à l'exclusion*. Plon, Paris.

Leach, E.R. (1980). *L'unité de l'homme et autres essais*. Gallimard, Paris.

Lévi-Strauss, C. (1964). *Le cru et le cuit*, Volume Mythologiques 1. Plon, Paris.

Louis, P. (1971). Animaux anonymes chez Aristote. *Bulletin de l'Association Guillaume Budé*, 2, 211–217.

Masson, E., Debucquet, G., Fischler, C., Merdji, M. (2016). French consumers' perceptions of nutrition and health claims: A psychosocial-anthropological approach. *Appetite*, 105, 618–629.

Médale, F. (2009). Teneur en lipides et composition en acides gras de la chair de poissons issus de la pêche et de l'élevage. *Cahiers de nutrition et de diététique*, 44(4), 173–181.

Michel-Guillou, É. (2014). La représentation sociale du changement climatique : enquête dans le sens commun, auprès de gestionnaires de l'eau. *Les cahiers internationaux de psychologie sociale*, 4, 647–669.

Paquotte, P. (2004). Image et perception des produits d'origine aquatique. In *Entretiens Science et Ethique 2004 : la pêche et les avancées scientifiques*, 3B Conseils, November 5th–6th.

Régnier, F. (2015). *L'exotisme culinaire : essai sur les saveurs de l'Autre*. Presses universitaires de France, Paris.

Rousseau, V. (2005). *Le goût du sang*. Armand Colin, Paris.

Rozin, P., Millman, L., Nemeroff, C. (1986). Operation of the laws of sympathetic magic in disgust and other domains. *Journal of Personality and Social Psychology*, 50(4), 703.

Salembien, E. (1958). La consommation de poisson en France. *Consommation*, 2.

Salladarré, F., Guillotreau, P., Debucquet, G., Lazuech, G. (2018). Some good reasons for buying fish exclusively from community-supported fisheries: The case of Yeu Island in France. *Ecological Economics*, 153, 172–180.

Vialles, N. (1998). Chair ou poisson. *Journal des anthropologues*, 74 [Online]. Available at: http://journals.openedition.org/jda/2672.

Williot, J.P. (2015). Le cru et le cuit. In *L'Alimentation à découvert*, Esnouf, C., Fioramonti, J., Laurioux, B. (eds). CNRS Editions, Paris.

2

Fish Quality and Freshness

Grethe HYLDIG
Technical University of Denmark, Lyngby, Denmark

2.1. Introduction

Freshness and quality is a very complex concept and needs to be defined, as the definition changes with the context where it is applied. The concept is frequently described using terms related to nutritional, microbiological and physiochemical characteristics alone, but none of these terms serve as adequate indices of quality – sensory perception and consumer acceptability must be included (Nielsen et al. 2001). Sensory quality is important in each chain link, all the way from catch to consumer. Analysis of sensory quality attributes describing product characteristics is important, and objective sensory analysis can give a full description including appearance, taste, smell and texture in one analysis. In the food industry, sensory knowledge and training is essential and can be used not only for sensory evaluation, but also when managing raw material. Therefore, it can be used in all of the different steps in the production chain.

This chapter will look into what affects the sensory quality and freshness of fish, and describe some of the sensory methods that can be used for fish as objective markers of quality in the chain, from catch to consumer.

2.2. Factors that affect the sensory quality

Compared with other food commodities, fish are very different in terms of their methods of harvesting, fragility of the product during transport to processing sites, and further down the chain, temperature dependency and variety of species.

Therefore, it is necessary to monitor sensory quality in the whole chain. Sensory assessment of fish and fish products has therefore played a natural part of the seafood chain for years. Factors affecting sensory quality are shown in Figure 2.1.

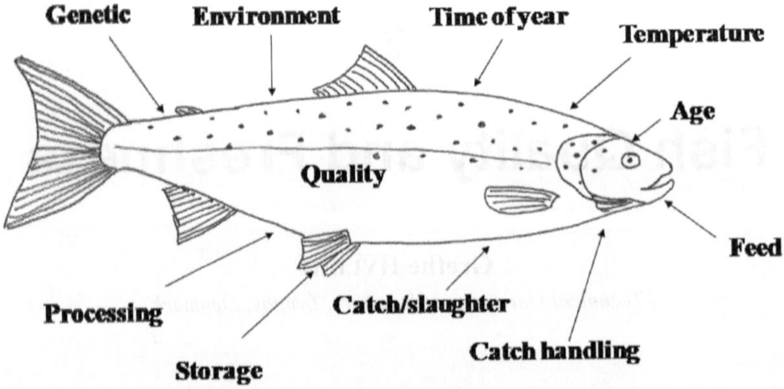

Figure 2.1. *Factors affecting the sensory quality*

2.2.1. *Genetic*

2.2.1.1. *Fish (vertebrates)*

Fish are generally defined as aquatic vertebrates that use gills to obtain oxygen from water, with fins consisting of a variable number of skeletal elements called fin rays. Fish are the most numerous of the vertebrates, with at least 20,000 known species, with more than half of the species found in the marine environment (Thurman and Webber 1984). Each species is identified by a scientific name which has two parts – the genus and the specific epithet (binominal nomenclature). For example, the scientific (species) name of the common cod is *Gadus morhua*.

2.2.1.2. *Fish anatomy*

The backbone of the fish runs from the head to the tail and consists of segments (vertebrae). The vertebrates are extended dorsally to form neural spines, and in the trunk region they have lateral processes that bear ribs. The ribs are cartilaginous or bony structures in the connective tissue between the muscle segments. Many species also have a corresponding number of false ribs or "pin bones" extending more or less horizontally into the muscle tissue. These bones have to be removed if a bone free fillet is the goal. Fish has muscle (see Figure 2.2) cells running in parallel and connected to sheaths of connective tissue (myocommata) anchored to the skeleton and the skin. The bundles of parallel muscle cells are called myotomes.

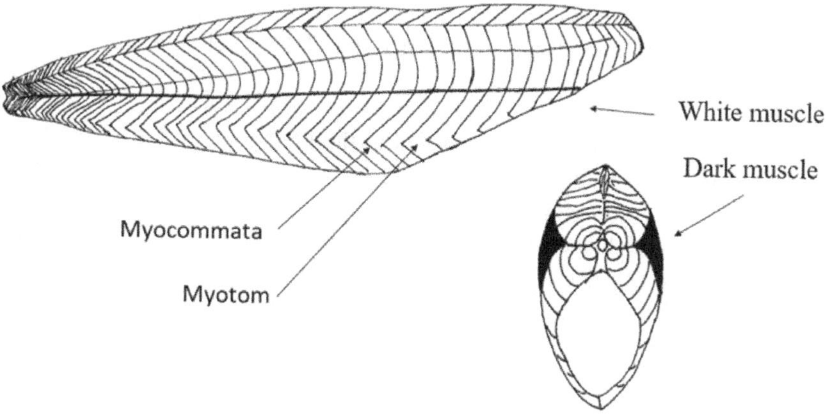

Figure 2.2. *Fish muscle*

The muscle can be white or have a light color (white muscle), but the muscle also has a certain amount of dark tissue that is a brown or reddish color (dark muscle). The dark muscle has a higher level of lipids and myoglobin. The location of the dark muscle is just under the skin along the side of the body. The proportion of dark muscle varies with the activity of the fish. The more active the fish, the larger the dark muscle. In some products, the high lipid content of dark muscle is important and can give high intensities of desirable sensory attributes, but the high lipid content can also give a shorter shelf life due to lipid oxidation, resulting in rancid and sour attributes, among others. A typical pelagic fish such as herring will contain approximately 25% of the dark muscle required for prolonged aerobic muscle activity. The proportion of dark muscle varies with the activity, the more active the fish, the larger the dark muscle. Demersal species such as cod have a white flaky and tender muscle; textural changes such as toughness and chewiness are the most notable problem during storage. Fish can contain natural antioxidants such as tocopherol and astaxanthin; these can also give the fish muscle a red color.

2.2.1.3. *Species*

There can be large variation between the different fish species, but there can also be considerable variation between individuals. Within the same species, there can also be variation due to their living conditions. The lipid in the fish differs from lipid in mammals by a high content of long-chain polyunsaturated fatty acids, including ω-3 fatty acids. The unsaturated fatty acids are easily oxidized and become rancid. Marine fish and some freshwater fish species contain trimethylaminoxid (TMAO).

Bacteria decompose TMAO to trimethylamine (TMA) and by enzymes, in some species, to dimethylamine DMA. Both TMA and DMA smell and taste like rotten fish.

Fish species are often divided according to the lipid content; lean, semi-fat and fat. Some example of lean fish species with 0–2g lipid/100g fish are cod (*Gadus morhua*), hake (*Merluccius merluccius*) and place (*Pleuronectes platessa*). Some example of semi-fat fish species with 2–8g lipid/100g fish are trout (Oncorhynchus mykiss), Nile perch (*Lates niloticus*) and halibut (*Hippoglossus hippoglossus*). Some example of fat fish species with 8–35g lipid/100g fish are herring (*Clupea harengus*), mackerel (*Scomber scombrus*) and salmon (*Salmo salar*). The lipid content will have an influence on the sensory and technological quality, shelf life and commercial value.

In a study with farmed rainbow trout (*Oncorhynchus mykiss*), fed and handled in the same way, groups of fish were sampled three different times during a production day. The rainbow trout were analyzed by the objective sensory profiling of minced fillets that were heat-treated. The results showed significant differences in the sensory profiles of individual fish within all three groups as well as significant differences between the groups. The results showed a significant negative correlation between the lipid content and firm texture, but in general, the chemical and physical measurements could not explain the differences in sensory profiling or mechanical texture measurements (Green-Petersen and Hyldig 2010). For wild caught fish such as herring (*Clupea harengus*), there can also be a wide variation in lipid content. A single catch can contain herring with a lipid content ranging from 1 to 25%. This variation is due to heterogeneity caused by mixing between stocks (Nielsen et al. 2005a).

2.2.2. *Age*

During growth, the sizes of each muscle cell increase, not the number of muscle cells. Also, the proportion of connective tissue increases with age. Fish become sexually mature when they reach the characteristic size for the species and the growth rate decreases after the fish has reached maturity. Mature fish use energy to build up gonads (roe and milk).

2.2.3. *Time of year*

Fish will have a natural period of starvation when they are migrating or spawning, as well as due to a shortage of feed. Such a starvation will give a variation in the sensory quality of the fish. The length of the spawning season varies greatly between species.

In herring, the lipid content will reflect the season and become low during the spawning period; this was shown in studies by Nielsen et al. (2005a) and Hyldig et al. (2012). In the study by Hyldig et al. (2012), they also concluded that both catching location and season had an effect on the development of the sensory quality of herring.

2.2.4. *Environment and aquaculture*

Some off-flavors can come from feeding on different contaminants or organisms. Two of the most known off-flavors in freshwater fish are geosmin (trans-1,10-dimethyl-trans-9-decalol) and 2-methylisoborneol (exo-1,2,7,7-tetramethylF2.2.1]-heptan-2-ol, MIB), which are produced by different microorganisms. The off-flavor from MIB gives a moldy and musty smell and taste, and geosmin gives a muddy smell and taste that it is difficult to remove from the mouth (Petersen et al. 2011).

Theoretically, there are no differences in the content, quality and shelf life of aquaculture fish and wild fish of the same species, given that they have the same feed and physical activity. To a certain extent, the fish farmer is able to design the fish by selecting the farming conditions. It has been reported that factors such as feed composition, environment, fish size and genetic traits all have an impact on the composition and quality (Green-Petersen et al. 2014).

2.2.5. *Feed*

Variation in the composition of the fish is affected by the feed intake, migration and spawn period. It is natural for fish to fast due to spawning and migration, as well as if there is a scarcity of feed. In this period, the fish will use their lipid store; they can also use protein if it is a long migration.

In a diet study (Baron et al. 2009) with rainbow trout (*Oncorhynchus mykiss*) feed, a total of six different diets were studied, containing either fish oil or rapeseed oil, with or without 200 mg/kg of carotenoid (either astaxanthin or canthaxanthin). The farming conditions were the same for all of the six different groups. The fish were slaughtered and stored as butterfly fillets in individual polyethylene bags for 22 months at -20°C. The fish fed fish oil had a grainier, more fibrous and firmer texture compared to the fish fed vegetable oil. In contrast, the "fish fed" vegetable oil was juicier. In general, they also had a less rancid odor and flavor compared to the fish fed fish oil. Fish fed vegetable oil had cooked potato, sweet and mushroom flavors, whereas fish-fed fish oil had more sour flavors. For the fish fed fish oil, the highest rancid odor and flavor were observed when astaxanthin was present in the feed. In contrast, the lower score for rancidity was observed when canthaxanthin was present in the feed. The picture was more complex and difficult to describe for fish-fed vegetable oil, but fish receiving the no-pigment feed had a lower score for both rancid odor and flavor than both samples that were fed carotenoids.

In a feed experiment by Green-Petersen et al. (2014) with organic rainbow trout (*Oncorhynchus mykiss*), they used six different organic diets. The control feed contained mainly fish meal as the protein source and fish oil as the lipid source. In the other five tested diets, 47% of the fish meal protein was replaced by a matrix of organic vegetable protein concentrates containing peas, horse beans and rapeseed (in the ratio 1.07:1.0:0.66). In four of these diets, fish oil was replaced by organic linseed, sunflower, rapeseed or grape seed vegetable oils, while one of the diets had fish oil as the lipid source. All feed types had an analyzed protein content of 45–46% w.w. and a lipid content of 28% w.w. The sensory quality was measured with objective sensory profiling conducted with a tested and trained sensory panel. Their results showed that the dietary lipid sources influenced the sensory characteristics of the trout in different ways during storage in ice. They found differences in texture after three days of ice storage, but after five days of ice storage, the fish showed more similarity in the sensory profile. However, after seven days in ice, some differences appeared. The trout which had rapeseed oil and grape seed oil generally had a more neutral flavor and odor profile with a lower intensity of mushroom odor, boiled potato odor, fresh fish oil and sweet flavor. Furthermore, the texture was less flaky, firm, juicy, stringy and oily compared to the other samples. The ice storage of trout fed with rapeseed or grape seed oil resulted in a more neutral odor and flavor, and a reduced texture quality compared to fish fed with fish oil, linseed oil, or sunflower oil.

2.2.6. *Catch handling and slaughter*

The balance in lipids and pro- and antioxidants changes when a fish dies and the initiation of lipases and proteases begins. This means that the oxidation and breakdown of lipids and protein structure take place, and these processes can results in a low sensory quality due to the formation of off-flavors and a soft texture. It must therefore be pointed out that it is important to have a fast and effective handling procedure resulting in the rapid chilling of the catch. A whole gutted fish has a longer shelf life than a fish fillet because the fish skin protects the muscle. The size and fat content will also affect shelf life.

In some fisheries, a uniform white fillet is desirable and therefore bleeding of the fish is very important. Bleeding is most affected by time on-board prior to bleeding/gutting, and the best bleeding is obtained if the fish is cut before it enters rigor mortis, since it is the muscle contractions that force the blood out of the tissues.

Physical rough handling in the net (long trawling time, very large catches) or on the deck (fishermen stepping on the fish or throwing boxes, containers and other items on top of the fish) may cause bruises, rupture of blood vessels and blood oozing into the muscle tissue (hematoma).

2.2.6.1. *Rigor*

Rigor mortis begins immediately or shortly after death depending on species, size, physical condition, catch method and catch handling. It also depends on whether the fish is starved, which results in depleted glycogen reserves, or if the fish is stressed. The method used for stunning and killing the fish also influences the onset of rigor. The technological significance of rigor mortis and whether the fish has been filleted before or in rigor are also of importance (Cappeln and Jessen 2002). In rigor, the fish body will be completely stiff; the filleting yield will drop significantly, and rough handling can cause gaping in the fillets. If the fillets are removed from the bone pre-rigor, the muscle can contract freely and the fillets will shorten following the onset of rigor.

If the fish is cooked pre-rigor, the texture will be very soft and pasty. In contrast, the texture is tough but not dry when the fish is cooked in rigor. Post-rigor, the flesh will become firm, succulent and elastic.

2.2.6.2. *Stress*

The stress level of the fish affects the sensory quality of the fish product, and therefore the handling of the fish is important for keeping the quality of the fish.

In a study by Sveinsdóttir et al. (2001b, 2010a), they introduced two levels of stress during handling of the farmed fish. Two sample groups of farmed cod (*G. morhua*) were used for the experiment. Cod was farmed in the same cage, and half of the batch was individually harvested and immediately stunned by a blow to the head, followed by a gill cut and bleeding in ice water (group with low stress). The sensory differences between the two groups were not very large, but the main differences were in texture. The low stress group had more a meaty and rubbery texture, but the high stress group was more tender. In addition, marginal significance was found for flaky, juicy and clammy texture, with the low stress group being slightly less flaky and juicy, but clammier.

In a study by Jessen and Hyldig (unpublished results), it was also found that texture was affected by the stress level of trout. The results showed that unstressed trout was juicier and less fibrous than stressed trout and that the unstressed trout remained juicy during the whole ice storage period, whereas the stressed trout became even more dry and fibrous during ice storage.

2.2.7. *Temperature and storage*

To keep the sensory quality, it is important to keep a low temperature during storage. The fish should be kept in ice, meaning the fish should lay on ice and be covered with ice to keep a temperature of approximately 0°C. To keep sensory quality, the fish should be kept cold all the way from the fishmonger to the home of the consumer.

Changes in the sensory quality during storage can be described in four phases:

– Phase 1: the fish is very fresh and has a sweet and delicate taste, like seaweed.

– Phase 2: there is a loss of the characteristic odor and taste. The flesh becomes neutral but has no off-flavors. The texture is still pleasant.

– Phase 3: there are signs of spoilage, and a range of volatile, unpleasant smelling substances is produced depending on the fish species and type of spoilage (aerobic, anaerobic). During the later stages, sickening sweet, cabbage-like, ammoniac, sulfurous and rancid smells develop. The texture becomes either soft and watery, or tough and dry.

– Phase 4: the fish can be characterized as spoiled and putrid.

The development of fish odor can be divided into three groups: for fresh fish, cucumber, champignon, seaweed and green odors come from C6–C9 alcohols and carbonyl connections. Odors developed from microbiological spoilage can be sweet, fruity, fishy, ammoniac and sulfurous, and come from ethanol, ammonia, TMA, acetate and other organic acids, and sulfuric compounds. The oxidized rancid, paint-like and wet dog odors come from lipid oxidation, e.g. ketones, aldehydes and products from protein oxidation.

If the fish is cooled without ice, the appearance will not change so much as if the fish was cooled in ice, but the fish will spoil faster and it is necessary to conduct a sensory test on the heat-treated samples. Therefore, it is important to have knowledge about the time/temperatures the fish have been exposed to until landing. Even though the lipid content is low in lean fish, the lipid will oxidize during frozen storage. When fish such as cod goes rancid, it will often taste somewhat like wet cardboard (cold-storage flavor).

The spoilage rate and shelf life of fish are affected by many parameters, and fish spoil at different rates. In general, it can be stated that larger fish spoil more slowly than small fish, lean fish keep longer than fatty fish under aerobic storage and bony fish are edible longer than cartilaginous fish. Rough handling will also result in a faster spoilage rate due to the physical damage to the fish, resulting in easy access for enzymes and spoilage bacteria.

When developing schemes for the sensory method, the quality index method (QIM) can be used, and the estimated shelf life is also detected. In Table 2.2, the estimated shelf life for some fish species can be seen. Later in this chapter, there will be a description of the QIM.

Species	Estimated shelf life in ice
Haddock, *Melanogrammus aeglefinus*	15 days
Salmon, *Salmo salar*	20 days
Turbot, *Scophthalmus maximus*	15 days
Redfish, *Sebastes mentella/marinus*	18 days
Plaice, *Pleuronectes platessa*	13 days
Pollock, *Pollachius virens*	18 days
Herring, *Clupea harengus*	8 days
Brill, *Rhombus laevis*	14 days
Cod, *Gadus morhua*	15 days
Sole, *Solea vulgaris*	15 days

Table 2.1. *Estimated shelf life for some fish species (Martinsdóttir et al. 2001)*

2.2.8. *Processing and products*

The fish species that can be used for different product types will depend on the fat content, functional properties, bone structure, size and shelf life.

Fish have less connective tissue than meat, and the connective tissue melts at a lower temperature. Fish therefore do not need a long heat treatment to become tender. In addition, fish proteins denature at lower temperatures than proteins in meat. Too long heat treatment will result in water loss and gives a dry texture. The fish should be heat-treated at 80°C to an internal temperature of 65–70°C. The cooking time naturally depends on both species and size, and whether it is whole or filleted.

Sensory testing can be both objective and subjective. The objective tests include discriminative (triangle test, forced choice) and descriptive (profiling, structured scaling) sensory tests. Both groups of test are analytical measurements of the intrinsic quality of the product, whereas affective (subjective test) methods are used for consumer testing, and to measure the attitude and emotional response of the consumer towards the product. All of these methods are used in the fishery chain (see Figure 2.3). Each link in the chain has its own version, as indicated by the numbers of the different methods (discriminative test1, discriminative test2, etc.), and a co-ordination is needed.

In a study by Baron et al. (2009) on the effects of feed on the quality of rainbow trout, they found that the oil type, the carotenoids and the level of antioxidants (tocopherols) in the feed were directly reflected in the fish muscle composition. The fish fed vegetable oil were more stable than those fed fish oil and became less rancid compared to the fish fed fish oil over storage time.

The influence of dietary protein and lipid sources on the quality of organic rainbow trout (*Oncorhynchus mykiss*) was studied by Green-Petersen et al. (2014). The protein and oil sources were fishmeal, fish oil and organic vegetable protein and oils. Sensory profiling was performed during 3–14 days of ice storage, along with lipid analyses of the fillet. Overall, the results showed that the sensory characteristics of the trout were affected in different ways during ice storage. The source of lipid seemed to affect the sensory quality at the beginning of the storage period, while the protein source seemed to have a more pronounced impact at the end of the storage period.

Different vegetable oils in trout diets influenced the sensory profile of trout differently. The impact of the diet was strongly influenced by the storage time in ice. The use of grape seed oil resulted in a different flavor and texture profile at the beginning of an ice storage period (three days) compared to the use of fish oil. However, using linseed oil resulted in a relatively similar sensory profile to that obtained when using fish oil. The sensory differences observed after three days of storage in ice could be explained by the fatty acid profiles of the trout. The sensory characteristics of the trout changed significantly during ice storage. Initially, the sensory profile was dominated by positive sensory descriptors, while after longer storage time, the profile became more neutral, and finally the negative sensory descriptors dominated. However, the results revealed that the development from a positive to a more neutral profile was influenced by the type of vegetable oil used in the diet. Thus, trout that were fed grape seed oil or rapeseed oil had a more neutral sensory profile compared to the other feed types after seven days of storage in ice. The dietary protein type used in the feed positively influenced the shelf life of trout during storage as compared to the fish meal/fish oil diet, but only during the last part of the storage period (i.e. after seven days).

2.3. The use of sensory measurements in the quality control in the fish industry

Sensory analysis plays an important role in quality control and assurance not only in the fish sector, but in all parts of the food sector. Therefore, sensory analysis of fish and fish products has for years been a natural part of the fish chain from catch/slaughter to consumer and plays an important role in quality control and quality assurance in the fish sector.

Sensory analysis is special because there are both objective and subjective sensory methods. The objective methods include discriminative (triangle test, forced choice) and descriptive (profiling, structured scaling) sensory methods. The objective sensory methods use a trained panel that objectively describes attributes of products, whereas consumer studies use untrained human beings that answer subjectively, for example, how much they like/dislike a product. Objective sensory analytical methods do not depend on whether assessors like or dislike a certain item. Instead, they operate by determining the intensities of attributes such as sweetness, amine, sourness, softness, etc. There are standards (e.g. ISO, NMKL, Codex Alimentarius) for conduction objective sensory analysis.

Two of the most often used objective sensory methods for the analysis of freshness and sensory quality in the fish industry are QIM and sensory profiling (see Figure 2.3). The QIM is a significant tool for managing the raw material and can be used all the way from catch to consumer because it contains important key properties that trigger acceptance and can help to bring a good quality fish from step to step in the processing chain towards the consumer. QIM is also suggested as a reference tool for verifying a quality labeling of fish. The objective sensory profile is a method that can characterize the different sensory attributes within the fish samples in detail. It is a valuable tool in quality control programs.

Several conditions are unique for sensory analysis of seafood. When selecting and training judges for sensory analysis, it is very important to be aware of the fact that some people cannot taste rancid flavor, iodine and geosmin, and some have a very low response to cold-storage flavor and rancidity. Also, some people are allergic or hypersensitive to different fish proteins, shellfish or histamine.

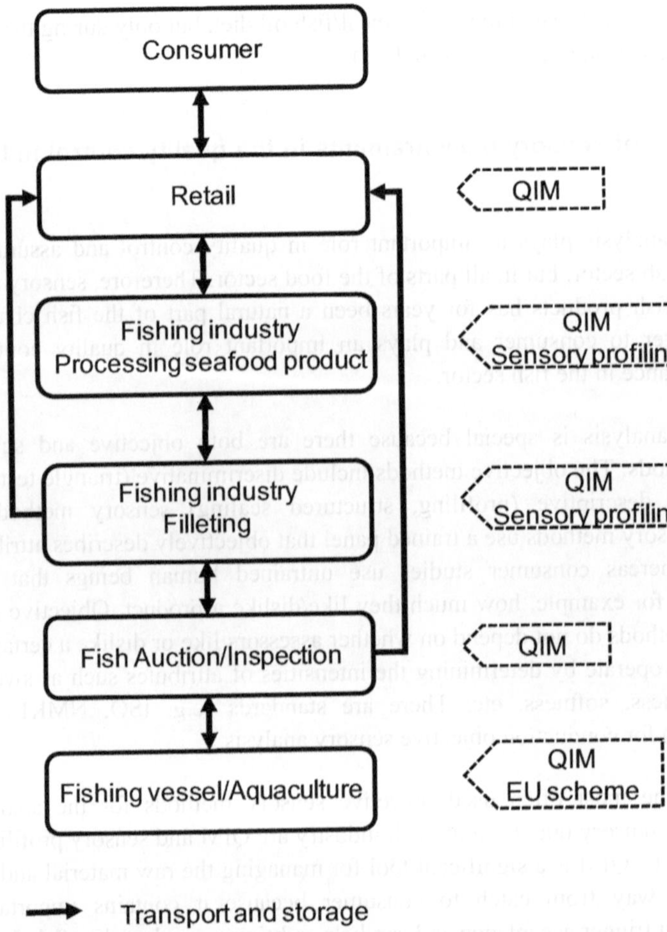

Figure 2.3. *The fish chain from catch to consumer with control points and the sensory methods for analysis of freshness and sensory quality*

2.3.1. *Sensory methods used for measuring freshness and sensory quality*

2.3.1.1. *QIM and the EU scheme*

Structured scaling is most often used. Structured scaling gives the assessors an actual scale showing several degrees of intensity. A few detailed attributes are chosen, often based on work from a fully trained descriptive panel or from experts. Descriptive words must be carefully selected and panelists trained so that they agree with each other about the terms. For raw fish, a large number of schemes for sensory analysis have been developed.

When fish is landed in Europe today, the most commonly used method for quality assessment of raw fish in the inspection service is the EU scheme (Anon 1996). In the EU scheme, according to the Council Regulation (EC) No. 2406/96 November 26, 1996 (Anon 1996), there are three quality levels, E (Extra), A and B, where E is the highest quality and below B is the level where fish is discarded for human consumption. The EU scheme does not take into account the differences between species as it only uses general parameters, and there are also problems with mixing subjective and objective sensory in the scheme.

The QIM is a structured scaling for measuring sensory quality and freshness. In the mid-1980s, the Tasmanian Food Research Unit originally began the development of QIM (Bremner et al. 1987), and since then, many research institutes working with seafood in the Nordic and European countries developed several QIM schemes (Larsen et al. 1992; Sveinsdóttir et al. 2003; Nielsen and Hyldig 2004; Nielsen and Green 2007; Sant'Ana et al. 2011) and schemes for frozen cod (Warm et al. 1998). Many of the QIM schemes can be found at: http://www.qim-eurofish.com.

Several studies have shown that the QIM proved to be more reliable in assessing sensory changes of different fish species as compared to the EU grading scheme (Teiqui and Bouchriti 2003).

QIM is a very fast sensory method, which can be used for the prediction of shelf life. QIM is based on significant sensory parameters for whole fish using many weighted parameters and a score system from 0 to 3 demerit points. The scores for all the characteristics are added to give an overall sensory score, the so-called quality index. QIM gives scores of zero for very fresh fish and an increasingly larger total result as the fish deteriorate. An example of the evaluation scheme for farmed salmon is seen in Table 2.2.

The selection of parameters for QIM is determined as a combination of the best descriptors for the spoiling fish, which also fulfil the aim that it is possible to predict the remaining shelf life. The remaining shelf life (in days on ice) can be calculated on the basis of this line and knowledge about the corresponding quality index at the time of rejection. There is a linear correlation between the sensory quality expressed as the quality score and storage time on ice, which makes it possible to predict the remaining shelf life on ice.

Quality parameters		Description	Score
Skin:	Color/appearance	Pearly-shine all over the skin	0
		The skin has a less pearly-shine	1
		The fish is yellowish, mainly near the abdomen	2
	Mucus	Clear, not clotted	0
		Milky, clotted	1
		Yellow, clotted	2
	Odor	Fresh seaweed, neutral	0
		Cucumber, metal, hay	1
		Sour, dish cloth	2
		Rotten	3
	Texture	In rigor	0
		Finger marks disappear rapidly	1
		Finger leaves mark over three seconds	2
Eyes:	Pupils	Clear and black, metallic and shiny	0
		Dark gray	1
		Matt, gray	2
	Form	Convex	0
		Flat	1
		Sunken	2
Gills:	Color/appearance	Red/dark brown	0
		Light red, pink/hazel	1
		Gray-brown, brown, gray, green	2
	Mucus	Transparent	0
		Milky, clotted	1
		Brown, clotted	2
	Odor	Fresh, seaweed	0
		Metal, cucumber	1
		Sour, moldy	2
		Rotten	3
Abdomen:	Blood in abdomen	Blood red/not present	0
		Blood more brown, yellowish	1
	Odor	Neutral	0
		Cucumber, melon	1
		Sour, reminiscent of fermentation	2
		Rotten/rotten kale	3
Quality index (0–24)			

Table 2.2. *Quality index scheme for whole farmed salmon (*Salmo salar*) (Martinsdottir et al. 2001; Hyldig and Green-Petersen 2005)*

When all parameters in the scheme have been measured and given a score, all scores are summarized to give an overall sensory score, the Quality Index (QI). The scientific development of QIM schemes for various species aims to have the QI increase linearly with storage time in ice (see Figure 2.4). Due to this linear correlation, the remaining shelf life can be estimated.

The theoretical development in QI for salmon and cod is shown in Figure 2.4. The estimation of the remaining shelf life is also shown in Figure 2.4.

When estimating the remaining shelf life, it should be kept in mind that individual fish spoil at different rates even if they come from the same catch or batch. It is recommend that 3–10 fish, collected randomly from different places in fish boxes from a homogenous lot, are included in one measurement. Increasing the amount of fish per sample increases the precision of the evaluation.

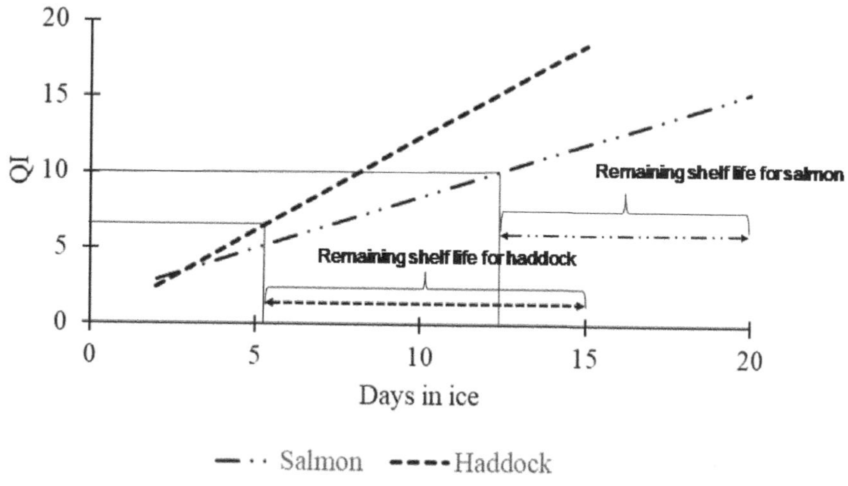

Figure 2.4. *Use of QIM in estimating past and remaining storage time*

Assessors for QIM

QIM assessors must be selected based on a test of their ability to evaluate appearance, color, odor and texture. It is also necessary that they are healthy and possess normally sensitive taste and odor senses (Martinsdóttír et al. 2001; Hyldig et al. 2008; ISO 8589:2012; ISO 11132:2021). The assessors' personal characteristics are also important, such as conscientiousness and accuracy, and they must be able to work in a group without disturbing the other assessors, because the

QIM measurement must often be done in groups. It is recommended to have six tested and trained assessors and to use four to six assessors for the QIM analysis.

For the training of the assessors, the QIM scheme for the actual species must be used. Table 2.2 is the QIM scheme for farmed salmon that should be used if the species was salmon. All quality parameters in the QIM scheme have to be presented to all assessors to make sure that they all have the same definition of the different quality parameters and the use of the category scale described in the QIM scheme. Furthermore, continuous training using reference samples can reduce the risk of drifting. If there is a long time between an assessor being used in the QIM measurements, the assessor must be re-trained.

Trained assessors can make a QIM assessment of 40 fish in 20 minutes, and QIM is non-destructive, meaning that the samples can be used after the assessment.

Preparation of fish samples and QIM analysis

For the QIM evaluations, samples have to be taken from a homogenous lot. A homogeneous lot could be from a catching day or a batch. Samples for QIM assessment have to be kept cold during the analysis. Samples are placed in random order and given a unique three digital number by a person that is not in the sensory panel (a panel leader). The assessors are instructed that they need to handle the samples with care so that the mucus or gills are not destroyed and that they need to measure all parameters for a fish before beginning the measurement of the next fish. All the assessors make the QIM analysis of all fish samples, and it is controlled so that all parameters in the QIM scheme are measured.

In the fish industry, it is not always possible to have a sensory lab, and the QIM analysis is often carried out in the processing halls of fish factories or at auction sites. It would be optimal to have special facilities or rooms for sensory evaluation, but this is not always possible. If this is not possible, the testing area should adhere to the ISO 8589 standard as closely as possible.

2.3.1.2. *Sensory profiling*

Descriptive sensory analysis such as profiling can be used to characterize post-mortem changes of all kinds of seafood. Descriptive testing provides quantitative data and can be both very simple and used for the assessment of a single attribute, or more complex and give a total characterization of sensory quality. Objective descriptive sensory requires a tested and trained panel, together with a test room (ISO 8589:2007; ISO 8586:2012; ISO 13299:2016; ISO 11132:2021). In the sensory method, profiling, a series of attributes are developed to describe the sensory characteristics describing appearance, smell, taste and texture (see the previous chapter). The panel must be trained in identification of the attributes, and furthermore, the panel must be trained in using a scale for describing the intensity of the attributes. In a session, samples will be served in a random order, and the panel will be served a palate cleanser in between samples. Several conditions are unique for the objective sensory profile of seafood and will be described in the following.

The samples are collected and prepared by a person who is not in the sensory panel (a panel leader). Sampling for sensory analysis must be as representative as possible. All samples must be marked individually with three-digit or four-digit codes.

Assessors for sensory profiling and training

When selecting and training assessors for sensory profiling of seafood, it is important to be aware of the fact that some people cannot taste rancid flavor, iodine, geosmin, and that some have a very low response to cold-storage flavor and rancidity.

There can be large variation between individual fish from the same batch and it can be a challenge to have homogeneous samples. Therefore, it is necessary to have 6–10 assessors in the sensory panel for sensory profiling. In all cases, the assessors require intensive training and a detailed briefing before each session.

The training should provide all assessors with a common vocabulary describing the different attributes that are used for the actual fish species. The criteria for the selection of sensory attributes are that they can discriminate between samples, are relevant for the specific seafood, and must be non-redundant and cognitively clear to the assessors. Also, training in using the scale for measuring the intensity of each attribute is important. Must often, it is an unstructured line scale with anchor points that is used, but it could also be a category scale from 0 to 5 or from 0 to 7. Continuous training using reference material can reduce the risk of drifting.

In a study of marinated herring, Nielsen et al. (2005b) used a reference material that was served both as mince and fillet portions. The sensory evaluations took place in five trials over a time period of 19 months. It was therefore important to prevent "drifting" of profiling assessors, and all sessions began with the evaluation of the reference. To exclude variation between fillets, 7–10 fillets were minced together and then served in 10–15 g portions in porcelain trays. The assessors evaluated the mince and "calibrated" their evaluations according to a completed list of assessments of odor and flavor of the herring mince performed by an in-house expert panel. The reference material was also served as a coded (unknown) fillet sample in order to check the performance of the panel. The samples were treated in all ways like the "real" sample material.

By serving the same product as a reference before each session, assessors can recapitulate the descriptors easily and recalibrate their evaluations to the same scale. The additional use of coded reference samples served together with the actual samples makes it possible to monitor the performance of the panel.

Preparation of fish samples and conducting the sensory analysis

For sensory profiling, the samples have to be heat-treated. The samples are placed in individual containers, for instance porcelain bowls with lids, and given heat treatment to an internal core temperature of 70°C. Heating should be done gently because of the delicate muscle structure of fish.

Each assessor evaluated the samples in the same way. All samples must be served to each assessor in a randomized order to minimize possible carry-over effects between samples. The number of samples for each session depends on the number of attributes and the type of sample. If there are many attributes, it is necessary to cut down the number of samples in each session. Depending on the samples, different items can be used as palate cleansers between samples. Tap water can be used if it is not too bitter and served at room temperature, together with unsalted crackers or crisp bread.

There are international and national standards and guidelines for the design and construction of sensory assessment rooms, but if that is not possible, the testing area should adhere to the ISO 8589 standard as closely as possible. It is most important that each assessor has some space to evaluate the samples without disturbance.

2.4. References

Anon (1996). Council Regulation (EC) No. 2406/96 of 26. November 1996 laying down common marketing standards for certain fishery products. *Official Journal of the European Communities*, L334, 1–14.

Baron, C.P., Hyldig, G., Jacobsen, C. (2009). Does feed composition affect oxidation of rainbow trout (*Oncorhynchus mykiss*) during frozen storage? *Journal of Agricultural and Food Chemistry*, 57, 4185–4194.

Bremner, H.A., Olley, J., Vail, A.M.A. (1987). Estimating time-temperature effects by a rapid systematic sensory method. In *Advances in Food Research: Seafood Quality Determination*. Elsevier Press, Amsterdam.

Cappeln, G. and Jessen, F. (2002). ATP, IMP, and glycogen in cod muscle at onset and during development of rigor mortis depend on the sampling location. *Journal of Food Science*, 67(3), 991–995.

Green-Petersen, D.M.B. and Hyldig, G. (2010). Variation in sensory profile between individual rainbow trout (*Oncorhynchus mykiss*) from the same production batch. *Journal of Food Science*, 75(9), S499–S505.

Green-Petersen, D.M.B., Hyldig, G., Jacobsen, C., Baron, C.P., Lund, L., Nielsen, H.H., Jokumsen, A. (2014). Influence of dietary lipid and protein sources on the sensory quality of organic rainbow trout (*Oncorhynchus mykiss*) after ice storage. *Journal of Aquatic Food Product Technology*, 23(4), 333–346.

Hyldig G. and Green-Petersen, D.M.B. (2005). Quality index method – An objective tool for determination of sensory quality. *Journal of Aquatic Food Product Technology*, 13(4), 71–80.

Hyldig, G., Carlehög, M., Martinsdóttir, E., Sveinsdóttir, K,. Lilleberg, L. (2008). Guide for sensory analysis of fish and shellfish. NMKL Procedure No. 21.

Hyldig, G., Jørgensen, B.M., Undeland, I., Olsen, R.E., Jónsson, Á., Nielsen, H.H. (2012). Sensory properties of frozen herring (*Clupea harengus*) from different catch seasons and locations. *Journal of Food Science*, 77(9), S288–S293.

ISO 8589:2007 (2007). Sensory analysis – General guidance for the design of test rooms. Standard, International Organization for Standardization, Switzerland.

ISO 8586:2012 (2012). Sensory analysis – General guidance for the selection, training and monitoring of selected assessors and expert sensory assessors. Standard, International Organization for Standardization, Switzerland.

ISO 13299:2016 (2016). Sensory analysis – Methodology – General guidance for establishing a sensory profile. Standard, International Organization for Standardization, Switzerland.

ISO 11132:2021 (2021). Sensory analysis – Methodology – Guidelines for the measurement of the performance of a quantitative descriptive sensory panel. Standard, International Organization for Standardization, Switzerland.

Martinsdóttir, E., Sveinsdottir, K., Luten, J., Schelvis-Smit R., Hyldig, G. (2001). Sensory evaluation of fish freshness. A reference manual for the fish industry, available in 11 European languages [Online]. Available at: www.qim-eurofish.com.

Nielsen, J., Hyldig, G., Larsen, E. (2001). Eating quality of fish – A review. *Journal of Aquatic Food Product Technology*, 11(3/4), 125–141.

Nielsen, D., Hyldig, G., Nielsen, J., Nielsen, H.H. (2005a). Lipid content in herring (*Clupea harengus* L.) – Influence of biological factors and comparison of different methods of analyses: Solvent extraction, fatmeter, NIR and NMR. *Lebensm-wiss U-Technol.*, 38, 537–548.

Nielsen, D., Hyldig, G., Sørensen, R. (2005b). An effective way to minimize drifting and monitor the performance of a sensory panel during long-term projects – A case study from a project on herring quality. *Journal of Sensory Studies*, 20, 35–47.

Petersen, M.A., Hyldig, G., Strobel, B.W., Henriksen, N.H., Jørgensen, N.O.G. (2011). Chemical and sensory quantification of geosmin and 2-methylisoborneol in rainbow trout (*Oncorhynchus mykiss*) from recirculated aquacultures in relation to concentrations in basin water. *Journal of Agricultural and Food Chemistry*, 59(23), 12561–12568.

Sveinsdóttir, K., Martinsdóttir, E., Thórsdóttir, F., Schlvis, R., Kole, A., Thórsdóttir I. (2010a). Evaluation of farmed cod products by a trained sensory panel and consumers in different test settings. *Journal of Sensory Studies*, 25, 280–293.

Sveinsdóttir, K., Martinsdóttir, E., Hyldig, G., Sigurgísladóttir, S. (2010b). Sensory characteristics of different cod products. *Journal of Sensory Studies*, 25(2), 294–314.

Thurman, H.V. and Webber, H.H. (1984). *Marine Biology*. Charles E. Merrill Publishing, CA, and Bell and Howell Co., Columbus, OH.

Triqui, R. and Bouchriti, N. (2003). Freshness assessment of Moroccan sardine (*Sardina pilchardus*): Comparison of overall sensory changes to instrumentally determined volatiles. *Journal of Agricultural and Food Chemistry*, 51, 7540–7546.

3

Nutritional Value of Finfish

Irineu BATISTA and Carla PIRES
Instituto Português do Mar e da Atmosfera (IPMA), Algés, Portugal

3.1. Introduction

Total world fisheries and aquaculture production accounted for 178.5 million tons in 2018 (FAO 2020). Of this global production, 116.2 million tons were finfish, of which 54.3 million tons were produced in aquaculture and 71.9 million tons resulted from marine catches. Most of the aquaculture finfish is produced in inland aquaculture, which accounts for 47.0 million tons and only 7.3 million tons come from marine and coastal aquaculture. A gradual increase in the per capita apparent consumption (kg) from an average of 13.4 in the period 1986–1995 and 20.5 in 2018 was recorded. In 2017, freshwater and diadromous fish were the main groups of finfish consumed (8.1 kg per capita), followed by pelagic species (3.1 kg per capita) and demersal species (1.8 kg per capita).

3.2. Nutritional value of finfish

Multiple works have highlighted the high nutritional value of fish and fish products namely due to their high-quality proteins and essential amino acids, highly unsaturated fatty acids and micronutrients such as minerals (iron, calcium, zinc and selenium) and vitamins (A, B and D). Tables 3.1–3.3 present the results of proximate composition, fatty acid and elemental profiles, respectively, of finfish published in scientific articles from 2010 onwards. It was intended to show the nutritional value of the greatest number of species. Therefore, the articles were selected in order to avoid repetitions, and preference was given to the most recent

Current Challenges for the Aquatic Products Processing Industry,
coordinated by Véronique VERREZ-BAGNIS.

ones that provided more information. In addition to these articles, other studies on the nutritional value of a large number of finfish species published during this period are also mentioned. The total number of consumable species shown in these tables together with those studied in the other studies accounted for 228 species, captured in different ecosystems, namely fresh, brackish or marine waters. Moreover, the FAO/INFOODS food composition database (FAO 2016) and U.S. Department of Agriculture database (USDA 2021) should be mentioned as essential sources of information on the nutritional value of finfish.

3.2.1. *Proximate composition*

The values presented in this table are in some cases the mean proximate composition of specimens caught in different places or seasons. Table 3.1 includes the carbohydrate contents of some species as mentioned by the authors, and caloric values were calculated based on general Atwater factors, i.e. 4 kcal g^{-1} for protein and 9 kcal g^{-1} for fat.

Chedoloh et al. (2011) determined the fat content of seven freshwater fish (FW) species (shown in Table 3.2) and 20 marine fish (MW) species collected in Southern Thailand. The fat content of MW species varied from 1.89% and 3.36% and only one species (ornate threadfin bream, *Nemipterus hexodon*) was extra lean fish (fat content less than 2%) and the others were lean fish (fat content between 2% and 5%). Li et al. (2011) investigated the fat content of 12 MW species and 10 wild and cultured FW species from the Zhejiang province, China. The fat content variation ranges of MW and FW species were 0.58–7.83% and 1.71–7.43%, respectively. Eight MW species were fatty fish (fat content between 5% and 10%), and an equal number of species was extra lean and lean fish. FW species also included fatty fish (5), lean fish (4) and only one extra lean fish (cultured Asian swamp eel, *Monopterus albus*). Most of the MW had higher fat content than the FW. Nurnadia et al. (2011) analyzed 20 fish species (10 pelagic and 10 demersal) from the West Coast of Peninsular Malaysia. The variation ranges of the main constituents of pelagic fishes were: 59.31–82.12% moisture; 17.46–20.86% protein; 1.05–23.15% fat; and 0.93–1.59% ash. In demersal species, these variation ranges were: 74.61–81.66% moisture; 16.61–22.22% protein; 0.70–6.89% fat; and 0.96–2.08% ash. An equal number of these species were extra lean and lean fish, one was a fatty fish and another was extra fatty fish (fat content $\geq$ 10%). Prato and Biandolino (2012) analyzed the FA profiles of 11 commercially important fish species from the Mediterranean Sea. Of these species, two were extra lean, six were lean and three were fat. The lowest fat content (1.67%) was determined in salema (*Sarpa salpa*) and the highest (8.12%) in red mullet (*Mullus barbatus*). Ahmed et al. (2012)

analyzed the proximate composition of 35 FW species collected from six different water-logged areas of Bangladesh and produced by two culture systems: semi-intensive and waterlogged paddy fields. These species were distributed by the following groups: (i) snakehead fishes (three species); (ii) minor carps (five species); (iii) other small indigenous fishes (SIFs, 10 species); (iv) catfishes (four species); (v) introduced major carps (six species); and (vi) introduced exotic species (seven species). These species presented the following variation ranges of the main constituents: 75.02–83.70% moisture; 14.60–19.50% protein; 1.40–4.52% fat; and 0.26–3.56% ash. Most of these species were lean fish and 11 were extra lean fish. The study by Bogard et al. (2015) was also focused on evaluating the nutrient composition of 41 inland captured and aquaculture species and eight captured MW species collected in different local markets and fish landing sites in Bangladesh. The inland captures were distributed by the following groups: (i) wild small indigenous species (29); (ii) wild large fish species (three); (iii) farmed indigenous major carps (three species); and (iv) farmed introduced fish species (seven). These groups of fish presented the following variation ranges: 60.2–85.4% moisture; 11.9–20.5% protein; 0.3–18.3% fat; and 0.7–5.2% ash. In MW species, these variation ranges were the following: 67.8–82.1% moisture; 16.1–20.6% protein; 0.3–14.3% fat; and 0.7–5.3% ash. Most fish species were lean or extra lean and only nine species were fatty or extra fatty fish. Mohanty et al. (2019) reported the proximate composition of 40 species from different ecosystems and collected from landing stations in India. The variation ranges of the main constituents were: 66.9–87.5% moisture; 8.2–23.9% protein; 0.6–10.5% fat; and 0.9–5.3% ash. Thirteen of these species were extra lean fish and 22 species were lean fish. The others were fatty fish (four) and extra fatty fish (hilsa shad, *Tenualosa ilisha*). Zhang et al. (2020) evaluated the fat content and fatty acid profile of 22 fish species from the Pearl River Estuary, China. The fat content of these species varied from 0.51% in silver sillago (*Sillago sihama*) and 7.35% in Kammal thryssa (*Thryssa kammalensis*). Most of these species were extra lean (four) or lean fish (10) and the remaining species were fatty fish.

In these 228 finfish species, the variation range of the main constituents was: 59.55–87.5% moisture content in longtail shad (Nurnadia et al. 2011) and Bombay-duck (*Harpadon nehereus*) (Mohanty et al. 2019), respectively; 8.2–23.9% protein content in Bombay-duck and yellowfin tuna (*Thunnus albacares*) (Mohanty et al. 2019), respectively; 0.09–19.70% fat content in Walleye pollock and twait shad, respectively; and 0.05–5.3% ash content in kuria labeo and climbing perch (Mohanty et al. 2019), respectively. Most of these species (99) were lean fish, 76 species were extra lean fish, 32 species were fatty fish and 17 species were extra fatty fish.

Common name	Scientific name	Moisture	Protein	Fat	Ash	Carbohydrates	Caloric value	Source
Freshwater fishes								
Wallago	*Wallago attu*	78.80	17.00	0.85	2.90	–	76	Memon et al. (2010)
Orangefin labeo	*Labeo calbasu*	77.00	18.83	0.98	2.30	–	84	
Goonch	*Bagarius bagarius*	73.20	18.05	8.25	0.50	–	146	
Kuria labeo	*Labeo gonius*	79.45	19.00	1.50	0.05	–	90	
Garua bachcha	*Clupisoma garua*	72.30	18.40	3.75	4.95	–	107	
Long-whiskered catfish	*Aorichthys aor*	77.03	19.05	1.78	2.02	–	92	
Batchwa vacha	*Eutropiichthys vacha*	59.95	20.09	18.32	1.45	–	245	
Flatwhiskered catfish (barbado)	*Pinirampus pirinampu*	75.31	16.39	7.30	1.01	–	131	Ramos Filho et al. (2010)
Gilded catfish (jaú)	*Paulicea luetkeni*	72.23	16.92	10.44	0.98	–	162	
Duckbill catfish (jurupensém)	*Sorubim lima*	71.78	16.07	11.19	1.14	–	165	
Porthole shovelnose catfish (jurupoca)	*Hemisorubim platyrhynchos*	74.69	18.10	6.35	0.94	–	130	
Mandi-prateado	*Pimelodus argenteus*	75.97	17.70	5.55	0.90	–	121	
Mandi-amarelo	*Pimelodus maculates*	81.83	16.73	0.35	0.89	–	70	
Palmito	*Ageneiosus brevifilis*	80.74	17.77	0.36	1.07	–	74	
Rainbow trout (w/f)	*Oncorhynchus mykiss*	78.4/74.5	19.2/18.7	1.12/5.11	1.30/1.26	–	87/121	Fallah et al. (2011)
Common carp	*Cyprinus carpio*	78.69	10.55	2.56	1.60	6.61	92	Jabeen and Chaudhry (2011)
Rohu	*Labeo rohita*	79.00	11.32	2.38	1.12	5.96	91	
Mozambique tilapia	*Oreochromis mossambicus*	75.24	11.15	4.01	2.83	6.71	108	

Common name	Scientific name	Moisture	Protein	Fat	Ash	Carbohydrates	Caloric value	Source
Freshwater fishes								
Pike	*Esox lucius*	77.12	18.14	1.23	2.64	0.87	87	Özden and Erkan (2011)
Nile tilapia	*Oreochromis niloticus*	81.2	16.4	2.0	0.5	–	84	Usydus et al. (2011)
Giant river-catfish	*Sperata seenghala*	79.40	20.06	1.40	0.90	–	93	Mohanty et al. (2012a)
Trout	*Salmo macrostigma*	72.44	17.49	9.35	1.40	–	154	Zotos and Vouzanidou (2012)
Striped catfish	*Pangasianodon hypophthalmus*	75.3	17.2	7.8	0.8	–	139	Manthey-Karl et al. (2016)
Brackish water and marine fishes								
European seabass (w/f)	*Dicentrarchus labrax*	77.49/75.63	21.61/18.25	1.04/4.48	1.26/1.21	–	96/113	Fuentes et al. (2010)
Pomfret	*Pampus punctatissimus*	75.35	18.6	4.95	1.36	–	119	Zhao et al. (2010)
Atlantic horse mackerel	*Trachurus trachurus*	72.50	14.78	10.51	1.80	0.48	155	Boran and Karaçam (2011)
Golden mullet	*Chelon auratus*	76.67	16.19	4.78	1.90	0.46	110	
Twaite shad	*Alosa fallax*	61.52	16.43	19.70	1.98	0.37	245	
Indian mackerel	*Rastrelliger kanagurta*	73.23	18.80	5.31	1.20	1.20	129	Nisa and Asadullah (2011)
Herring	*Clupea harengus*	76.7	18.1	3.7	1.4	–	106	Usydus et al. (2011)
Walleye pollock	*Theragra chalcogramma*	86.7	12.2	0.09	1.0	–	50	
Sole	*Limanda aspera*	85.0	13.4	0.5	1.1	–	58	
Silver pomfret (w/f)	*Pampus argenteus*	72.46/71.79	17.18/16.74	8.61/9.61	1.56/1.65	–	146/153	Hossain et al. (2012)
Orange-spotted grouper (w/f)	*Epinephelus coioides*	75.07/75.89	18.72/18.69	4.36/3.80	1.54/1.46	–	114/109	

Common name	Scientific name	Moisture	Protein	Fat	Ash	Carbohydrates	Caloric value	Source
Freshwater fishes								
Sobaity seabream (w/f)	*Sparidentex hasta*	76.12/78.84	17.62/16.25	4.60/3.51	1.21/1.21	–	112/97	
Yellowfin seabream (w/f)	*Acanthopagrus latus*	74.90/74.83	18.83/18.41	4.14/4.43	1.58/1.49	–	113/114	
Salmon (w/f)	*Salmo salar*	71.0/64.3	19.8/18.3	6.3/12.3	–	–	136/184	Jensen et al. (2012)
Anchovy	*Engraulis encrasicolus*	66.95	16.31	13.90	1.03	1.94	202	Kocatepe and Turan (2012)
Garfish	*Belone belone*	73.97	20.26	3.64	1.25	0.71	117	
Black scorpionfish	*Scorpaena porcus*	80.10	16.91	1.26	0.92	0.95	83	
Bluefish	*Pomatomus saltatrix*	70.87	15.24	12.29	0.76	0.87	177	
Red mullet	*Mullus barbatus*	75.25	14.54	8.32	1.14	0.66	134	
Common dentex (w/f)	*Dentex dentex*	75.6/69.8	18.7/17.1	1.32/7.7	4.00/5.20	–	87/138	Rigos et al. (2012)
Baltic sprat	*Sprattus sprattus balticus*	71.53	16.80	10.63	–	–	163	Usydus et al. (2012)
Bogue	*Boops boops*	75.17	18.88	5.68	1.55	–	127	Zotos and Vouzanidou (2012)
European pilchard	*Sardina pilchardus*	71.76	17.62	10.03	1.69	–	161	
European hake	*Merluccius merluccius*	79.99	17.90	1.31	1.36	–	83	
Atlantic chub mackerel	*Scombrus colias*	71.57	20.23	7.74	1.39	–	151	
Cod (w/f)	*Gadus morhua*	80.9/78.0	15.7/18.6	0.9/1.0	1.3/1.3	–	71/83	Jensen et al. (2013)
Yellowfin tuna	*Thunnus albacares*	73.57	23.52	1.93	1.54	–	112	Peng et al. (2013)
Bigeye tuna	*T. obesus*	72.89	23.72	2.06	1.77	–	113	
Allis shad	*Alosa alosa*	66.4	18.8	14.5	1.3	–	211	Pereira et al. (2013)
Commerson anchovy	*Stolephorus commersonii*	78.03	15.79	1.88	4.35	–	80	Sankar et al. (2013)

Common name	Scientific name	Moisture	Protein	Fat	Ash	Carbohydrates	Caloric value	Source
			Freshwater fishes					
Ballyhoo halfbeak	*Hemiramphosus brasiliensis*	79.40	18.36	1.05	1.06	–	83	Fernandes et al. (2014)
Common half beak	*Hyporhamphosus unifasciatus*	78.39	19.34	1.21	1.03	–	88	
Atlantic thread herring	*Opisthonema oglinum*	71.13	18.10	9.03	1.73	–	154	
King mackerel	*Scomberomorus cavalla*	76.37	19.87	1.96	1.14	–	100	
Spiny dogfish	*Squalus suckleyi*	70.6	18.4	9.2	1.8	–	156	Oliveira et al. (2014)
Yellowtail (w/f)	*Seriolla lalandi*	71.09/72.14	23.04/22.80	4.29/3.72	1.31/1.75	–	131/125	O'Neill et al. (2015)
Turbot (w/f)	*Scophthalmus maximus*	80.2/78.9	19.0/19.8	1.0/1.6	1.1/1.2	–	85/94	Manthey-Karl et al. (2016)
Barramundi (w/f)	*Lates calcarifer*	77.4/78.4	19.9/18.2	2.2/2.3	1.0/1.1	–	99/94	
Longnek croaker	*Johnius dussumieri*	81.40	15.40	1.30	1.00	–	73	Mehta and Nayak (2017)

w/f – wild/farmed fish

Table 3.1. *Proximate chemical composition of finfish species (g 100 g^{-1}) and caloric value (kcal 100 g^{-1})*

The following species shown in Table 3.1 had low cholesterol content (between 31 and 60 mg 100 g^{-1}): rainbow trout: 38.8 wild and 40.1 farmed; trout: 46.56; bogue: 56.40; and European pilchard: 60.00. Atlantic chub mackerel (62.81) and spiny dogfish (74) are medium cholesterol content species (between 61 and 90), but European hake with a level of 94.92 is a high cholesterol content species. In general, fish are typically species with low or medium cholesterol content.

3.2.2. *Fatty acids*

Table 3.2 shows the fatty acid (FA) profile of the vast majority of species listed in Table 3.1 plus a few more for a total of 86 species. When the authors did not report the fatty acid contents, expressed in mg 100 g^{-1}, they were estimated using the formula by Weihrauch et al. (1977) and the conversion factor 0.673 for fish with a fat content lower than 0.55 g 100 g^{-1}. Most of the wild FW species shown had total saturated fatty acids (SFAs) as the predominant group in 22 species. Total monounsaturated fatty acids (MUFAs) were the major group in 11 species and total polyunsaturated fatty acids (PUFAs) in seven species. Moreover, total UFAs (PUFAs and MUFAs) were predominant in 46% of total species. Palmitic acid (PAM, 16:0) was predominant in 27 FW species, followed by oleic acid (OLA, 18:1n-9) in 11 and docosahexaenoic acid (DHA, 22:6n-3) in two species. Within SFAs, PAM was the major FA in all species, while OLA was the major MUFA in 39 species and palmitoleic acid (POA, 16:1n-7) was only predominant in creek chub. Within PUFAs, DHA was predominant in 17 species, linoleic acid (LNA, 18:2n-6) in 12 species, arachidonic acid (ARA) in seven species, eicosapentaenoic acid (EPA, 20:5n-3) in creek chub (4.7%) and orangefin labeo (18.53%), α-linolenic acid (ALA, 18:3n-3) in North African catfish (7.21%), eicosatrienoic acid (ETE, 20:3n-3) in garua bachcha (7.39%) and docosapentaenoic acid (DPA, 22:5n-3) in long-whiskered catfish (6.38%).

Common name	Fat	SFAs	MUFAs	PUFAs	EPA	DHA	n-3/n-6	Source
				Freshwater fish				
Perch (*Perca fluviatilis*) (w)	–	27.78	29.20	43.02	7.29	16.71	2.22	Jankowska et al. (2010)
Perch (*P. fluviatilis*) (f)	–	27.08	37.94	34.98	7.24	16.82	4.56	
Wallago	–	40.17 (261.1)	24.55 (159.6)	33.67 (218.9)	2.77 (18.0)	10.95 (71.2)	1.50	Memon et al. (2010)
Orangefin labeo	–	40.19 (310.0)	27.52 (212.3)	31.79 (245.2)	18.53 (142.9)	1.33 (10.3)	3.71	
Goonch	–	25.04 (1891.6)	48.35 (3652.5)	17.06 (1288.8)	2.80 (211.5)	3.11 (234.9)	1.71	
Kuria labeo	–	37.38 (469.7)	26.02 (326.9)	35.34 (444.0)	4.56 (57.3)	6.51 (81.8)	1.72	
Garua bachcha	–	32.72 (1098.0)	33.14 (1112.1)	29.64 (994.6)	1.16 (38.9)	3.50 (117.5)	1.15	
Long-whiskered catfish	–	29.25 (443.9)	36.93 (560.5)	29.30 (444.7)	2.45 (37.2)	5.60 (85.0)	1.91	
Batchwa vacha	–	35.73 (6056.1)	41.81 (7086.6)	15.72 (2664.5)	1.66 (281.4)	2.04 (345.8)	1.19	

Common name	Fat	SFAs	MUFAs	PUFAs	EPA	DHA	n-3/n-6	Source
Freshwater fish								
Flatwhiskered catfish	–	40.74 (2716.5)	37.94 (2529.8)	9.34 (622.8)	0.35 (23.3)	1.05 (70.0)	0.56	Ramos Filho et al. (2010)
Gilded catfish	–	36.48 (3501.2)	41.07 (3941.7)	9.80 (940.6)	0.46 (44.1)	0.79 (75.8)	0.47	
Duckbill catfish	–	35.39 (3644.2)	43.51 (4480.3)	13.94 (1435.4)	0.20 (20.6)	0.75 (77.2)	0.28	
Porthole shovelnose catfish	–	35.46 (2050.1)	42.75 (2471.6)	12.41 (717.5)	0.32 (18.5)	0.70 (40.5)	0.53	
Mandi-prateado	–	38.70 (1948.6)	37.86 (1906.3)	14.30 (720.0)	0.44 (22.2)	1.88 (94.7)	0.77	
Mandi-amarelo	–	36.80 (86.7)	28.93 (68.1)	26.33 (62.0)	1.28 (3.0)	8.19 (19.3)	0.69	
Palmito	–	35.69 (86.5)	26.89 (65.1)	19.52 (47.3)	0.68 (1.6)	8.67 (21.0)	1.12	
Walking catfish (*Clarias macrocephalus*)	1.03	32.65 (267.1)	20.71 (169.4)	45.21 (369.8)	3.16 (25.8)	6.52 (53.3)	0.61	Karapanagiotidis et al. (2010)
Climbing perch (*Anabas testudineus*)	0.99	34.51 (269.4)	22.50 (175.7)	38.95 (304.1)	1.06 (8.3)	8.94 (69.8)	0.58	
Blackline rasbora (*Rasbora borapetensis*)	0.86	38.20 (251.9)	23.01 (151.7)	36.92 (243.4)	2.28 (15.0)	9.72 (64.1)	1.19	
Swamp barb (*Puntius brevis*)	0.90	34.90 (243.1)	25.21 (175.6)	38.72 (269.8)	2.26 (15.7)	5.17 (36.0)	0.67	

Common name	Fat	SFAs	MUFAs	PUFAs	EPA	DHA	n-3/n-6	Source
Freshwater fish								
Malacanthus (*Hoplolatilus* sp.)	2.56	42.2 (947.6)	32.2 (723.0)	25.6 (574.8)	2.4 (53.9)	2.9 (65.1)	0.44	Chedoloh et al. (2011)
Asian redtail catfish (*Cryptarius truncatus*)	2.61	44.5 (1020.0)	22.3 (511.1)	33.3 (763.3)	1.8 (41.3)	3.9 (89.4)	0.59	
Philippine catfish (*Clarias batrachus*)	2.77	33.1 (808.1)	26.8 (654.3)	40.1 (979.0)	1.5 (36.6)	23.3 (568.8)	2.77	
Kissing gourami (*Helostoma temmincki*)	1.08	45.3 (391.7)	22.9 (198.0)	31.7 (274.1)	5.0 (43.2)	12.6 (108.9)	2.16	
Bronze featherback (*Notopterus notopterus*)	1.49	49.4 (616.1)	24.9 (310.5)	25.9 (323.0)	0.5 (6.2)	13.0 (162.1)	2.02	
Striped snakehead (*Channa striata*)	1.47	39.4 (484.0)	19.6 (240.8)	41.2 (506.1)	0.6 (7.4)	13.0 (159.7)	2.82	
Silver barb (*Barbonymus gonionotus*)	1.63	24.1 (332.0)	40.5 (558.0)	35.4 (487.7)	2.6 (35.8)	16.0 (220.4)	1.87	
Rainbow trout (w)	–	25.6 (230.9)	23.2 (209.3)	43.7 (394.2)	8.19 (73.9)	21.80 (196.6)	3.88	Fallah et al. (2011)
Rainbow trout (f)	–	26.5 (1225.5)	34.3 (1586.2)	32.5 (1503.0)	4.54 (210.0)	13.40 (619.7)	1.72	
Common carp	–	55.66 (1249.8)	32.90 (738.8)	11.42 (256.4)	0.34 (7.6)	0.36 (8.1)	0.27	Jabeen and Chaudhry (2011)
Rohu	–	50.54 (1050.0)	27.23 (565.7)	22.21 (461.4)	0.58 (12.0)	1.27 (26.4)	0.23	
Mozambique tilapia	–	63.00 (2266.9)	24.83 (893.5)	12.16 (437.6)	0.42 (15.1)	0.35 (12.6)	0.23	

Common name	Fat	SFAs	MUFAs	PUFAs	EPA	DHA	n-3/n-6	Source
Freshwater fish								
Nile tilapia	–	35.4 (676.8)	33.1 (632.8)	31.5 (608.0)	0.5 (9.6)	3.2 (61.2)	0.5	Usydus et al. (2011)
Giant river-catfish	–	36.08 (419.7)	28.36 (329.9)	35.53 (413.3)	4.37 (50.8)	6.19 (72.0)	1.29	Mohanty et al. (2012a)
North African catfish (*Clarias gariepinus*)	3.81	24.12 (822.9)	51.94 (1772.1)	23.96 (817.5)	1.08 (36.8)	2.53 (86.3)	1.10	Abouel-Yazeed (2013)
Tench (*Tinca tinca*)	–	35.34	29.92	34.26	0.19	17.93	1.19	Citil et al. (2014)
Rudd (*Scardinius erythrophthalmus*)	–	33.52	34.98	30.73	0.34	17.94	2.15	
Crucian carp (*Carassius carassius*)	–	26.41	37.82	35.76	10.24	14.17	2.82	
Creek chub (*Semotilus stromaculatus*)	1.88	52.00 (837.7)	32.00 (515.5)	16.40 (264.2)	4.70 (75.7)	2.40 (38.7)	2.42	Tenyang et al. (2014)
Spotted catfish (*Arius maculates*)	5.41	52.63 (2581.3)	26.70 (1309.5)	20.88 (1024.1)	4.60 (225.6)	6.05 (296.7)	2.40	
Red disc (*Symphysadon discus*)	2.14	47.00 (871.2)	33.10 (613.5)	20.00 (370.7)	3.90 (72.3)	7.20 (133.5)	3.00	
Striped catfish (*Pangasianodon hypophthalmus*)	–	43.4 (3096.3)	33.3 (2375.8)	15.3 (1091.6)	1.2 (85.6)	6.4 (456.6)	1.4	Manthey-Karl et al. (2016)
Brackish water and marine fishes								
European seabass (w)	–	37.38 (309.3)	23.54 (194.8)	39.08 (323.3)	12.17 (100.7)	16.62 (137.5)	3.27	Fuentes et al. (2010)
European seabass (f)	–	30.48 (1230.4)	36.40 (1469.4)	31.73 (1280.9)	8.54 (344.7)	7.85 (316.9)	1.16	

Common name	Fat	SFAs	MUFAs	PUFAs	EPA	DHA	n-3/n-6	Source
				Freshwater fish				
Pomfret	–	48.3 (2161.6)	32.5 (1454.5)	19.3 (863.7)	4.47 (200.0)	12.2 546.0)	8.04	Zhao et al. (2010)
European gilthead seabream (*Sparus aurata*) (w)	2.50	27.36 (599.0)	37.67 (824.8)	23.08 (505.3)	0.28 (6.1)	9.54 (208.9)	2.22	Lenas et al. (2011)
European gilthead seabream (f)	8.85	20.18 (1637.4)	39.47 (3202.6)	32.09 (2603.8)	4.49 (364.3)	9.19 (745.7)	1.64	
Indian mackerel	–	38.07 (1831.6)	23.49 (1130.2)	31.74 (1527.1)	9.10 (437.8)	14.13 (679.8)	–	Nisa and Asadullah (2011)
Herring	–	28.6 (946.4)	30.7 (1015.9)	40.7 (1346.8)	6.2 (205.2)	20.4 (675.1)	5.5	Usydus et al. (2011)
Walleye pollock	–	20.8 (12.6)	8.5 (5.1)	70.7 (42.8)	24.6 (14.9)	40.5 (24.5)	23.4	
Sole	–	24.4 (82.1)	16.4 (55.2)	59.2 (199.2)	25.4 (85.5)	18.0 (60.6)	10.8	

Common name	Fat	SFAs	MUFAs	PUFAs	EPA	DHA	n-3/n-6	Source
Freshwater fish								
Silver pomfret (w)	–	53.05 (4185.7)	31.34 (2472.8)	14.29 (1127.5)	1.45 (114.4)	7.21 (568.9)	14.60	Hossain et al. (2012)
Silver pomfret (f)	–	40.24 (3550.4)	39.11 (3450.7)	20.82 (1837.0)	3.77 (332.6)	5.09 (449.1)	0.99	
Orange-spotted grouper (w)	–	57.35 (2250.9)	21.58 (847.0)	21.61 (848.2)	2.42 (95.0)	7.64 (299.9)	1.09	
Orange-spotted grouper (f)	–	43.70 (1486.8)	30.66 (1043.2)	25.64 (872.4)	3.38 (115.0)	4.70 (159.9)	0.56	
Sobaity seabream (w)	–	57.06 (2367.3)	26.35 (1093.2)	16.59 (688.3)	5.31 (220.30)	6.51 (270.09)	3.10	
Sobaity seabream (f)	–	33.97 (1063.9)	38.60 (1208.9)	27.42 (858.7)	4.88 (152.8)	10.92 (342.0)	1.69	
Yellowfin seabream (w)	–	49.69 (1848.3)	29.35 (1091.7)	19.64 (730.5)	4.61 (171.5)	10.29 (382.8)	3.82	
Yellowfin seabream (f)	–	38.96 (1554.6)	31.98 (1276.1)	29.06 (1159.5)	5.33 (212.7)	12.63 (504.0	1.73	
Salmon (w)	–	23.2 (920)	45.6 (1840)	26.2 (1020)	6.5 (260)	12.5 (520)	12.50	Jensen et al. (2012)
Salmon (f)	–	19.7 (1440)	47.9 (3520)	30.4 (2240)	5.5 (410)	8.4 (620)	2.27	

Common name	Fat	SFAs	MUFAs	PUFAs	EPA	DHA	n-3/n-6	Source
Freshwater fish								
Hilsa shad (*Tenualosa ilisha*) (L<40 cm)	9.19	43.73 (3687.0)	33.48 (2822.8)	22.95 (1935.0)	2.68 (226.0)	8.68 (731.8)	2.40	Mohanty et al. (2012b)
Hilsa shad (*T. ilisha*) (L>40 cm)	14.62	57.99 (7827.2)	27.59 (3723.9)	14.75 (1990.9)	8.22 (1109.5)	2.02 (272.6)	6.41	
Anchovy	–	35.07 (4498.0)	19.50 (2501.0)	32.43 (4159.4)	11.06 (1418.5)	15.99 (2050.8)	9.91	Kocatepe and Turan (2012)
Garfish	–	29.56 (961.6)	30.34 (987.0)	40.02 (1301.9)	4.68 (152.3)	28.52 (927.8)	10.01	
Black scorpionfish	–	35.91 (370.8)	25.79 (266.3)	26.38 (272.4)	6.08 (62.8)	8.91 (92.0)	7.87	
Bluefish	–	32.86 (3720.9)	30.5 (3453.7)	20.15 (2281.7)	4.70 (532.2)	10.75 (1217.3)	7.27	
Red mullet	–	30.46 (2320.9)	33.80 (2575.4)	15.25 (1162.0)	5.52 (420.7)	5.41 (412.2)	7.79	
Common dentex (w)	–	29.0 (315.7)	32.6 (354.9)	38.5 (419.1)	8.81 (95.9)	15.40 (167.6)	4.03	Rigos et al. (2012)
Common dentex (f)	–	32.4 (2281.3)	25.7 (1809.6)	41.9 (2950.2)	9.34 (657.6)	21.50 (1513.8)	6.36	
Baltic sprat	–	28.70 (2805.4)	33.32 (3257.0)	37.76 (3691.0)	7.14 (697.9)	17.92 (1751.6)	6.90	Usydus et al. (2012)

Common name	Fat	SFAs	MUFAs	PUFAs	EPA	DHA	n-3/n-6	Source
				Freshwater fish				
Bogue	–	(1421.7)	(1261.7)	(1568.3)	(551.7)	(846.7)	13.45	Zotos and Vouzanidou (2012)
European pilchard	–	(2535.0)	(2570.0)	(2400.0)	(868.3)	(1148.3)	9.32	
European hake	–	(281.7)	(326.7)	(363.3)	(713.3)	(183.3)	11.47	
Atlantic chub mackerel	–	(1928.3)	(2328.3)	(1528.3)	(138.3)	(668.3)	10.8	
Goldenblotch grouper (*Epinephelus costae*)	2.24	26.28 (511.7)	44.97 (875.5)	28.78 (365.6)	0.32 (6.2)	15.74 (306.5)	1.95	Abouel-Yazeed (2013)
Cod (w)	–	30.4 (211.8)	13.7 (95.4)	49.4 (344.2)	11.9 (82.9)	32.3 (225.0)	43.91	Jensen et al. (2013)
Cod (f)	–	30.5 (241.0)	14.1 (111.4)	49.6 (391.8)	16.0 (126.4)	25.7 (203.0)	11.40	
Yellowfin tuna	–	44.99 (745.8)	26.66 (441.9)	28.35 (470.0)	2.39 (39.6)	16.91 (280.3)	3.29	Peng et al. (2013)
Bigeye tuna	–	37.00 (658.2)	32.40 (576.4)	30.26 (538.3)	3.27 (58.2)	20.22 (359.7)	4.56	
Allis shad	–	29.61 (3962.8)	39.73 (5318.0)	29.49 (3947.4)	8.40 (1124.4)	13.11 (1754.4)	7.98	Pereira et al. (2013)
Commerson anchovy	–	40.12 (646.3)	25.90 (417.3)	33.45 (538.9)	3.64 (58.6)	15.40 (248.1)	3.04	Sankar et al. (2013)

Common name	Fat	SFAs	MUFAs	PUFAs	EPA	DHA	n-3/n-6	Source
Freshwater fish								
Ballyhoo halfbeak	–	18.28 (152.9)	48.31 (404.2)	20.26 (169.5)	6.41 (53.6)	9.12 (76.3)	7.40	Fernandes et al. (2014)
Common halfbeak	–	19.59 (193.1)	48.17 (474.9)	21.32 (210.2)	6.80 (67.0)	9.91 (97.7)	7.08	
Atlantic threadfin herring	–	33.26 (2754.6)	14.10 (1167.8)	48.81 (4042.4)	10.66 (882.9)	30.20 (2501.2)	12.95	
King mackerel	–	30.29 (510.6)	13.96 (235.3)	35.66 (769.7)	7.51 (126.6)	27.54 (464.2)	7.88	
Green jack (*Caranx caballus*)	3.74	37.77 (1263.9)	28.66 (959.1)	33.66 (1126.4)	7.09 (237.3)	19.65 (657.6)	8.12	Murillo et al. (2014)
Cachema weakfish (*Cynoscion phoxocephalus*)	2.75	42.50 (1029.7)	33.47 (810.9)	23.59 (571.5)	6.24 (151.2)	10.82 (262.1)	4.86	
Spotted rose snapper (*Lutjanus guttatus*)	1.81	39.79 (615.0)	32.48 (502.1)	27.90 (431.3)	4.84 (74.8)	16.11 (249.0)	5.90	
Pacific sierra (*Scomberomorus sierra*)	3.09	39.11 (1071.6)	32.64 (894.3)	28.13 (770.8)	6.80 (186.3)	14.64 (401.1)	5.62	
Spiny dogfish	–	26.45 (1508.8)	41.94 (2392.0)	30.81 (1757.2)	8.90 (507.8)	14.97 (853.8)	5.91	Oliveira et al. (2014)
Yellowtail (w)	–	37.71 (1618)	29.13 (1250)	33.16 (1423)	6.18 (265)	22.05 (946)	13.23	O´Neill et al. (2015)
Yellowtail (f)	–	32.88 (1222)	27.37 (1017)	39.75 (1477)	12.67 (471)	13.11 (487)	2.68	

Common name	Fat	SFAs	MUFAs	PUFAs	EPA	DHA	n-3/n-6	Source
				Freshwater fish				
Turbot (w)	–	25.2 (199.1)	21.3 (168.3)	39.1 (308.9)	6.3 (49.8)	22.6 (178.5)	7.1	
Turbot (f)	–	25.5 (344.2)	22.1 (298.3)	41.1 (554.8)	11.5 (155.2)	15.8 (213.3)	4.9	Manthey-Karl et al. (2016)
Barramundi (w)	–	42.3 (807.8)	25.2 (481.2)	21.8 (416.3)	2.9 (55.4)	8.7 (166.1)	2.5	
Barramundi (f)	–	32.5 (650.9)	34.1 (683.0)	23.6 (472.7)	3.0 (60.1)	6.2 (124.2)	1.3	
Mediterranean moray (*Muraena helena*)	–	34.26	36.15	29.60	3.92	14.85	3.14	
European conger (*Conger conger*)	–	32.94	30.67	36.41	3.60	21.54	3.11	Đikić et al. (2017)
European eel (*Anguilla anguilla*)	–	28.02	53.88	18.12	2.21	1.96	0.80	
Longneck croaker	–	70.07 (749.7)	10.23 (109.5)	19.71 (210.9)	3.55 (38.0)	11.36 (121.5)	3.11	Mehta and Nayak (2017)

Table 3.2. *Total SFAs, MUFAs, PUFAs, EPA and DHA as the % of total fatty acids and mg 100 g^{-1} (in brackets) in finfish species*

Concerning wild brackish water and MW species, total SFAs were dominant in 20 species, MUFAs in 12 species and PUFAs in 14 species. About 56% of these species had total UFAs as the major group. PAM was the major FA in 26 species, OLA predominated in 11 species, DHA in seven species and EPA was the major FA in sole. In the case of hilsa shad, myristic acid (14:0) was dominant in fish smaller than approximately 40 cm (approximately 38%), whereas PAM was the major FA in longer fish. Within SFAs, PAM was predominant in all species with the exception of shad hilsa with the smallest sizes. OLA was the predominant MUFA in 43 species and POA in sole and Atlantic threadfin herring accounting for 7.9% and 7.45%, respectively. Regarding PUFAs, DHA predominated in 41 species, EPA in red mullet (5.52%), hilsa shad longer than 40 cm (8.22%) and sole (25.4%), LNA in orange-spotted grouper (10.32%) and ARA in European eel (3.88%). The variation range of n-3/n-6 ratio in wild FW fish was between 0.23 and 4.05. However, in wild brackish and MW, this range was between 1.09 and 43.9.

Apart from the species mentioned in Table 3.1, Chedoloh et al. (2011) analyzed the FA profile of 10 MW and six of these species had PUFAs as the main FA group. SFA group was dominant in the other species. PAM predominated in eight species and DHA in bigeye scad (*Selar crumenophthalmus*) and chacunda gizzard shad (*Anodontostoma chacunda*). Within SFAs and MUFAs, the major fatty acids in all species were PAM and OLA, respectively. DHA was the predominant PUFA in nine species and LNA in spotted scat (*Scatophagus argus*). The n-3/n-6 ratio was in the range 0.61–3.72.

SFAs were the major fatty acid group of seven MW analyzed by Li et al. (2011), followed by MUFAs in three species and PUFAs in two species. PAM was predominant in eight species, OLA in three and DHA in one species (Pacific rudderfish, *Psenopsis anomala*). Within SFAs and PUFAs, PAM and DHA were the major fatty acids, respectively. OLA was the major MUFA in all species with the exception of silver croaker (*Pennahia argentata*) where POA was dominant. All MW species had a higher percentage of n-3 PUFA than that of n-6 PUFA, with a variation range of n-3/n-6 ratio being 5.3–17.1. MUFAs were dominant in three wild freshwater species and PUFAs in one, while equal number of farmed species (5) had MUFAs and PUFAs as major groups. OLA was predominant in all wild species but in farmed species PAM predominated in three species and OLA in seven. These two fatty acids predominated in SFAs and MUFAs, respectively, but within PUFAs, LNA was predominant in four species, DGA in five and EPA in one. Farmed FW species had higher percentages of EPA+DHA than the wild FW fish. The n-3/n-6 ratios were 0.6–2.0 and 0.4–2.9 in wild and farmed FW fish, respectively. SFAs were the major FA group in all species analyzed by Prato and Biandolino (2012) as well as PAM. OLA was the major MUFA in all species with the exception of sand smelt (*Atherina boyeri*) where POA predominated. Within PUFAs, DHA was the main FA in all species and the n-3/n-6 ratio range of 2.46–2.58.

Bogard et al. (2015) analyzed the FA profile of species from Bangladesh with a total fat content > 6%. In those species, SFAs were predominant in four species, MUFAs in three and PUFAs in one species. PAM and OLA predominated in equal number of species and both were the major fatty acids in the SFAs and MUFAs groups, respectively. Within PUFAs, LNA was the main FA in six species and EPA in two. The n-3/n-6 ratio varied between 0.11 and 4.19.

Mohanty et al. (2016a) analyzed the FA profile of 34 fish species from India. These species were included in the following fish groups: marine (12), brackish water (3), freshwater (8), small indigenous (6) and cold water (5). SFAs were the major group in 13 MW and brackish water species and PUFAs in two species. Within SFAs, PAM was predominant in all species and OLA was the main MUFA in 14 species and POA in flathead gray mullet (*Mugil cephalus*). DHA was the major PUFA in 10 species and EPA in Indian oil sardine (*Sardinella longiceps*) (12.3%) and skipjack (*Katsuwonus pelamis*) (5.8%). In the species of the other groups, SFAs were the main FA in 12 species and within this FA group, PAM was the major FA in 10 species. However, other SFAs such as myristic acid (four species), stearic acid (STA, 18:0) (two species), capric acid (10:0) and the unusual 19:0 and 21:0 in one species, respectively, were also the major SFAs. OLA predominated within the MUFA of 14 species but POA and 20:1 were the major MUFAs in three and two species, respectively. Within PUFAs, LNA was the major FA in eight species, ALA in four and DHA in hilsa shad (8.9%) and rita (*Rita rita*) (5.0%). The n-3/n-6 ratio varied from 0.2 in mrigal (*Cirrhinus mrigala*) and 7.9 in climbing perch.

In 17 fish species from Pearl River Estuary analyzed by Zhang et al. (2020), SFAs were dominant and their content ranged from 33.4% and 47.7%, whereas MUFAs predominated in five species ranging from 19.3% and 46.0%. PAM was the major FA in all species. Within MUFAs, OLA was predominant in 13 species and POA in nine species. DHA was the major PUFA in 13 species and EPA in nine species. The variation range of the n-3/n-6 ratio was between 2.04 and 11.11.

The n-3/n-6 ratio in the different FW species here reported ranged from 0.2 to 7.9 and in brackish and MW it was 0.61–43.91. Zhang et al. (2020) reported n-6/n-3 ratio value below 4.0 in a diet as recommended for cardiovascular disease prevention. Thus, only mrigal had n-6/n-3 ratio below 4.0 but all MW had lower n-6/n-3 values. The PUFA/SFA ratio has also been used as a fat quality index. As also mentioned by Zhang et al. (2020), foods with a PUFA/SFA ratio below 0.45 have been considered undesirable for human health because of their potential to increase blood cholesterol levels. FW species with PUFA/SFA<0.45 accounted for 27% of these species, while in MW species this percentage was 18%. The recognized importance of highly unsaturated FAs in preventing various chronic diseases has led to various levels of EPA and DHA intake being recommended. Taking into account the daily intake of 500 mg of EPA+DHA recommended by ISSFAL (International Society for the Study of Fatty Acids and Lipids), the percentage contribution from a serving dose (85 g, approximately 3 ounces) of each species was calculated. In the case of freshwater species, it was estimated that 46% of these species had a contribution higher than 25%. However, the same contribution of brackish and marine species would be provided by about 90% of these species.

Farmed fish show significant differences in the proximate and biochemical composition compared to their wild counterparts. In general, farmed fish are fattier but in turn have lower protein content than wild ones. In wild fish, PAM is the predominant in most species, but in farmed fish OLA becomes the major FA in an increasing number of species. In farmed fish, n-6 fatty acids gain greater significance due particularly to the higher contents of LNA. This increase in n-6 PUFA levels led to the decrease of the n-3/n-6 ratio in farmed fish. These changes in chemical composition are fundamentally due to the lipid profile of feeds used.

The external appearance and sensory characteristics of farmed and wild fish have been evaluated in some species. Farmed gilthead seabream and large yellow croaker showed a fatty body shape and differences in the skin color from their wild counterparts. Wild fish of these two species generally have more pleasant taste and firmer texture than farmed fish. However, these differences were not so accentuated between wild and farmed European seabass (Grigorakis 2007; Ma et al. 2020).

3.2.3. *Amino acids*

Özden and Erkan (2011) studied the amino acid profile of several marine fish species consumed in Turkey. Total essential amino acids (EAAs) varied between 8.69 g 100 g^{-1} in angler fish and 15.24 g 100 g^{-1} in European hake and non-essential amino acids (NEAAs) range was 7.1 g 100 g^{-1} in tub gurnard and 21.14 g 100 g^{-1} in

John Dory. The lowest EAA/NEAA ratio (0.63) was recorded in angler fish and the highest (1.13) in turbot fish. The amino acid profiles of wild and farmed ussuri catfish (*Pseudobagrus ussuriensis*) were compared by Wang et al. (2014). These authors found that the EAA/NEAA ratio in wild and farmed catfish was 0.83 and 0.98, respectively, and concluded that this could be due to the type of diet and environmental conditions. Sankar et al. (2013) evaluated the amino acid profile of Commerson's anchovy (*Stolephorus commersonii*) in three different size groups (3–5 g, 6–10 g, 25–30 g) from Kerala coast, India. This profile followed the same pattern in all size groups with major contributions of glutamic acid, glycine, aspartic acid, alanine and leucine and the mean EAA/NEAA ratio was 0.67. The amino acid composition of Indian fish species analyzed by Mohanty et al. (2019) varied largely among them. Major carps were rich in glutamic acid, aspartic acid and glycine. In general, coldwater species were rich in lysine and aspartic acid, marine fishes in leucine, SIFs in histidine, and carps and catfishes in glutamic acid and glycine. The amount of EAAs and conditionally essential amino acids (Arg, Cys, Gly, Gln, Pro, and Tyr) was higher than NEAAs in different groups of fishes studied.

Fish proteins have a well-balanced amino acid composition and are a good source of essential amino acids, particularly lysine, methionine, cystine, threonine and tryptophan. Moreover, aspartic and glutamic acids are also reported as the major amino acids in fish. The differences in the amino acid profile among species have been related to the fish species, type of diets and environmental conditions.

3.2.4. *Vitamins*

Vitamins A, D, E and K contents of several fish species from India had shown that the SIFs were much richer in vitamins than in the other species (Mohanty et al. 2019). Mora carplet (*Amblypharyngodon mola*) and *Epinephelus* spp. were rich in vitamins A, D, E and K. Indian oil sardine was rich in vitamins A, E and K, pool barb (*Puntius sophore*) in vitamins D, E and K, and hilsa shad in vitamin D. Bogard et al. (2015) evaluated the content of vitamins A, B12, D3, E and folate in several fishes. The ranges of these vitamins were (µg 100 g^{-1}): 0–2053 vitamin A; 0.05–14 vitamin B12; 0–34 vitamin D3; and 0–18 folate. For vitamin E, the range content was 0–1.9 mg 100 g^{-1}.

Fish is considered to be a reasonable source of vitamins, especially vitamins A, D, E and the B vitamins. Fattier fish contain modest amounts of vitamin A and fish oils contain high amounts of fat-soluble vitamins (A, D and E). Some fishes are considered as the best natural source of vitamin D, which is not found in many foods.

3.2.5. *Minerals*

Table 3.3 shows the elemental profile of a few species from Tables 3.1 and 3.2. The "good" or "excellent" contributions of each element according to the recommended daily allowance (RDA) are displayed in italics or bold based on the Dietary Guidelines Advisory Committee (2020). The "good" sources have a contribution of 10–19% of the RDA, and the "excellent" sources provide 20% or more of a particular element. Nurnadia et al. (2013) evaluated the mineral content of 20 marine species of Malaysia. The variation ranges (mg 100 g^{-1}) of the elemental contents of these species were: 22.42–76.76 Na; 6.58–20.42 K; 618.38–1314.50 Mg; 12.89–127.59 Ca; 0.176–1.358 Fe; 0.04–0.197 Cu; 0.148–0.464 Zn; and 0.008–0.060 Mn. The variation ranges (mg 100 g^{-1}) of elemental composition of species included in the four fish groups reported by Bogard et al. (2015) were: 26–110 Na; 58–350 K; 110–910 P; 160–270 S; 21–57 Mg; 8.6–1,900 Ca; 0.34–19 Fe; 0.015–0.12 Cu; 0.60–4.7 Zn; 0.010–2.3 Mn; 0.006–0.081 I; and 0.005–0.110 Se. Mohanty et al. (2016b) analyzed the elemental profile of 35 species and the variation ranges (mg 100 g^{-1}) of the macro- and microminerals were: 43.4–364 Na; 120.0–2,397 K; 27–220 Mg; 64.1–5310 Ca; 113.2–3,900 P; 0.5–36.5 Fe; 0.1–2.0 Cu; 0.4–26.0 Zn; 0.04–4.61 Mn; and 0.1–1.7 Se. Finfish are important sources of minerals, and their good or excellent dietary contribution of Se, Fe and Zn is noteworthy.

	Na	K	P	Mg	Ca	Fe	Cu	Zn	Mn	I	Se	Source
Freshwater species												
Pike	58.2	358.8	**580.9**	*42.2*	**615.1**	**10.96**	-	-	-	0.14	**0.02**	Özden and Erkan (2011)
Rainbow trout (w/f)	34/36	438/427	**249/245**	31/32	66/36	0.70/0.33	0.09/**0.11**	0.82/0.77	0.08/0.10	-	**0.015/0.013**	Fallah et al. (2011)
Giant river-catfish	198.3	**1378.0**	-	**120.0**	**458.1**	**4.51**	**0.53**	**2.94**	0.24	-	-	Mohanty et al. (2012a)
Trout	-	-	-	37.39	-	1.00	*0.13*	**2.59**	0.06	-	-	Zotos and Vouzanidou (2012)
Striped catfish	40.5	300.6	-	24.4	16.2	-	-	0.59	-	-	**0.02**	Manthey-Karl et al. (2016)
Brackish water and marine fish												
Bogue	-	-	-	41.2	-	*2.82*	0.06	**2.61**	0.086	-	-	Zotos and Vouzanidou (2012)
European pilchard	-	-	-	*42.7*	-	*2.49*	**0.20**	**2.53**	0.059	-	-	
European hake	-	-	-	*60.0*	-	*2.46*	*0.09*	**2.38**	0.098	-	-	
Atlantic chub mackerel	-	-	-	37.7	-	*2.68*	**0.15**	**2.71**	0.080	-	-	
Turbot (w/f)	108.4/127.5	283.6/294.2	-	24.0/21.3	9.0/7.0	-	-	0.61/0.62	-	-	**0.04/0.03**	Manthey-Karl et al. (2016)
Barramundi (w/f)	105.3/94.4	303.5/313.4	-	24.1/26.6	18.3/27.7	-	-	0.45/0.48	-	-	**0.03/0.02**	
Recommended daily allowance	-	-	700	420	1,000	8	0.9	11	-	-	0.055	
Adequate intake	-	4,700	-	-	-	-	-	-	2.3	-	-	

Values in italics are "good sources", and values in bold are "excellent sources" based on RDA.

Table 3.3. *Elemental profile of finfish species (mg 100 g^{-1})*

3.3. Future trends

Much information is already available on the nutritional value of fish in response to the increasing importance of this food item in the diets of many countries. However, there are still many gaps in this information, particularly with regard to micronutrient content (mineral elements and vitamins), both in unprocessed fish and in processed or simply cooked fish products. In addition, there are, with some regularity, new species appearing on the market which are imported or not usually consumed and whose nutritional value is insufficiently known or even unknown. The growing importance of aquaculture production is also a challenge for producers, fish nutritionists and food researchers to meet consumer expectations better. Therefore, knowledge of nutritional value continues to be of great significance to fish processing or marketing companies, especially for nutrition labeling, as well as to food technologists, nutritionists and consumers. Moreover, the study of the beneficial health effects of fish consumption is another area that continues to receive great attention.

3.4. References

Abouel-Yazeed, A.M. (2013). Fatty acids profile of some marine water and freshwater fish. *Journal of the Arabian Aquaculture Society*, 8(2), 283–292.

Ahmed, S., Rahman, A.F.M.A., Mustafa, M.G., Hossain, M.B., Nahar, N. (2012). Nutrient composition of indigenous and exotic fishes of rainfed waterlogged paddy fields in Lakshmipur, Bangladesh. *World Journal of Zoology*, 7(2), 135–140.

Bogard, J.R., Thilsted, S.H., Marks, G.C., Wahab, M.A., Hossain, M.A.R., Jakobsen, J., Stangoulis, J. (2015). Nutrient composition of important fish species in Bangladesh and potential contribution to recommended nutrient intakes. *Journal of Food Composition and Analysis*, 42, 120–133.

Boran, G. and Karaçam, H. (2011). Seasonal changes in proximate composition of some fish species from the Black Sea. *Turkish Journal of Fisheries and Aquatic Sciences*, 11, 1–5.

Chedoloh, R., Karrila, T.T., Pakdeechanuan, P. (2011). Fatty acid composition of important aquatic animals in Southern Thailand. *International Food Research Journal*, 18, 783–790.

Citil, O.B., Kalyoncu, L., Kahraman, O. (2014). Fatty acid composition of the muscle lipids of five fish species in Işikli and Karacaören Dam Lake, Turkey. *Veterinary Medicine International*, 936091, 5.

Dietary Guidelines Advisory Committee (2020). Scientific report of the 2020 dietary guidelines advisory committee [Online]. Available at: https://www.dietaryguidelines.gov/2020-advisorycommittee-report.

Đikić, D., Landeka, I., Fuchs, R., Skaramuca, D., Matić-Skoko, S., Tutman, P., Franić, Z., Cvetković, I., Skaramuca, B. (2017). Lipid profiles of Mediterranean moray, *Muraena helena*, European conger, *Conger conger*, and European eel, *Anguilla anguilla* (Actinopterygii: anguilliformes). *Acta Ichthyologica et Piscatoria*, 47(1), 1–11.

Fallah, A.A., Saei-Dehkordi, S.S., Nematollahi, A. (2011). Comparative assessment of proximate composition, physicochemical parameters, fatty acid profile and mineral content in farmed and wild rainbow trout (*Oncorhynchus mykiss*). *International Journal of Food Science and Technology*, 46, 767–773.

FAO (2020). The state of the world fisheries and aquaculture 2020. Sustainability in action. Report, FAO, Rome. doi:10.4060/ca9229en.

FAO of the United Nations (2016). FAO/INFOODS Global Food Composition Database for Fish and Shellfish Version 1.0 – uFiSh1.0. FAO, Rome, Italy.

Fernandes, C.E., Vasconcelos, M.A.S., Ribeiro, M.A., Sarubbo, L.A., Andrade, S.A.C., Filho, A.B.M. (2014). Nutritional and lipid profiles in marine fish species from Brazil. *Food Chemistry*, 160, 67–71.

Fuentes, A., Fernández-Segovia, I., Serra, J.A., Barat, J.M. (2010). Comparison of wild and cultured sea bass (*Dicentrarchus labrax*) quality. *Food Chemistry*, 119, 1514–1518.

Grigorakis, K. (2007). Compositional and organoleptic quality of farmed and wild gilthead sea bream (*Sparus aurata*) and sea bass (*Dicentrarchus labrax*) and factors affecting it: A review. *Aquaculture*, 272, 55–75.

Hossain, M.A., Almatar, S.M., Al-Abdul-Elah, K.M., Yaseen, S.B. (2012). Comparison of proximate composition and fatty acid profiles in cultured and wild marine fishes in Kuwait. *Journal of Applied Aquaculture*, 24(3), 199–209.

Jabeen, F. and Chaudhry, A.S. (2011). Chemical compositions and fatty acid profiles of three freshwater fish species. *Food Chemistry*, 125, 991–996.

Jankowska, B., Zakęś, Z., Żmijewski, T., Szczepkowski, M. (2010). Fatty acid profile of muscles, liver and mesenteric fat in wild and reared perch (*Perca fluviatilis* L.). *Food Chemistry*, 118, 764–768.

Jensen, I.J., Mæhre, H.K., Tømmerås, S., Eilertsen, K.E., Olsen R.L., Elvevoll, E.O. (2012). Farmed Atlantic salmon (*Salmo salar* L.) is a good source of long chain omega-3 fatty acids. *Nutrition Bulletin*, 37, 25–29.

Jensen, I.-J., Larsen, R., Rustad, T., Eilertsen, K.-E. (2013). Nutritional content and bioactive properties of wild and farmed cod (*Gadus morhua* L.) subjected to food preparation. *Journal of Food Composition and Analysis*, 31, 212–216.

Karapanagiotidis, I.T., Yakupitiyage, A., Little, D.C., Bell, M.V., Mente, E. (2010). The nutritional value of lipids in various tropical aquatic animals from rice–fish farming systems in northeast Thailand. *Journal of Food Composition and Analysis*, 23, 1–8.

Kocatepe, D. and Turan, H. (2012). Proximate and fatty acid composition of some commercially important fish species from the Sinop region of the Black Sea. *Lipids*, 47, 635–641.

Lenas, D.S., Triantafillou, D.J., Chatziantoniou, S., Nathanailides, C. (2011). Fatty acid profile of wild and farmed gilthead sea bream (*Sparus aurata*). *Journal of Consumer Protection and Food Safety*, 6, 435–440.

Li, G., Sinclair, A.J., Li, D. (2011). Comparison of lipid content and fatty acid composition in the edible meat of wild and cultured freshwater and marine fish and shrimps from China. *Journal of Agricultural and Food Chemistry*, 59, 1871–1881.

Ma, R., Meng, Y., Zhang, W., Mai, K. (2020). Comparative study on the organoleptic quality of wild and farmed large yellow croaker *Larimichthys crocea*. *Journal of Oceanology and Limnology*, 38(1), 260–274.

Manthey-Karl, M., Lehmann, I., Ostermeyer, U., Schröder, U. (2016). Natural chemical composition of commercial fish species: Characterisation of pangasius, wild and farmed turbot and barramundi. *Foods*, 5, 58. doi:10.3390/foods5030058.

Mehta, N.K. and Nayak, B.B. (2017). Bio-chemical composition, functional, and rheological properties of fresh meat from fish, squid, and shrimp: A comparative study. *International Journal of Food Properties*, 20(S1), S707–S721.

Memon, N.N., Talpur, F.N., Bhanger, M.I. (2010). A comparison of proximate composition and fatty acid profile of Indus River fish species. *International Journal of Food Properties*, 13(2), 328–337.

Mohanty, B.P., Paria, P., Das, D., Ganguly, S., Mitra, P., Verma, A., Sahoo, S., Mahanty, A., Aftabuddin, M., Behera, B.K. et al. (2012a). Nutrient profile of giant river-catfish *Sperata seenghala* (Sykes). *National Academy Science Letters*, 35(3), 155–161.

Mohanty, B.P., Paria, P., Mahanty, A., Behera, B.K., Mathew, S., Sankar, T.V., Sharma, A.P. (2012b). Fatty acid profile of Indian shad *Tenualosa ilisha* oil and its dietary significance. *National Academy Science Letters*, 35(4), 263–269.

Mohanty, B.P., Ganguly, S., Mahanty, A., Sankar, T.V., Anandan, R., Chakraborty, K., Paul, B.N., Sarma, D., Dayal, J.S., Venkateshwarlu, G. et al. (2016a). DHA and EPA content and fatty acid profile of 39 food fishes from India. *Hindawi Publishing Corporation BioMed Research International*, 4027437, 14.

Mohanty, B.P., Sankar, T.V., Ganguly, S., Mahanty, A., Anandan, R., Chakraborty, K., Paul, B.N., Sarma, D., Dayal, J.S., Mathew, S. et al. (2016b). Micronutrient composition of 35 food fishes from India and their significance in human nutrition. *Biological Trace Element Research*. doi: 10.1007/s12011-016-0714-3.

Mohanty, B.P., Mahanty, A., Ganguly, S., Mitra, T., Karunakaran, D., Anandan, R. (2019). Nutritional composition of food fishes and their importance in providing food and nutritional security. *Food Chemistry*, 293, 561–570.

Murillo, E., Rao, K.S., Durant, A.A. (2014). The lipid content and fatty acid composition of four eastern central Pacific native fish species. *Journal of Food Composition and Analysis*, 33, 1–5.

Nisa, K. and Asadullah, K. (2011). Seasonal variation in chemical composition of the Indian mackerel (*Rastrelliger kanagurta*) from Karachi Coast. *Iranian Journal of Fisheries Sciences*, 10(1), 67–74.

Nowak, V., Rittenschober, D., Exler, J., Charrondiere, U.R. (2014). Proposal on the usage of conversion factors for fatty acids in fish and shellfish. *Food Chemistry*, 153, 457–463.

Nurnadia, A.A., Azrina, A., Amin, I. (2011). Proximate composition and energetic value of selected marine fish and shellfish from the West coast of Peninsular Malaysia. *International Food Research Journal*, 18, 137–148.

Nurnadia, A.A., Azrina, A., Amin, I., Yunus, A.S.M., Effendi, H.M.I. (2013). Mineral contents of selected marine fish and shellfish from the west coast of Peninsular Malaysia. *International Food Research Journal*, 20(1), 1337–1343.

Oliveira, A.C.M., Bechtel, P.J., Morey, A., Brenner, K., Lapis, T.J. (2014). Chemical composition of spiny dogfish (*Squalus suckleyi*) harvested in Alaska. *Journal of Food Processing and Preservation*, 38(1), 600–606.

O'Neill, B., Le Roux, A., Hoffman, L.C. (2015). Comparative study of the nutritional composition of wild versus farmed yellowtail (*Seriola lalandi*). *Aquaculture*, 448, 169–175.

Özden Ö. and Erkan, N. (2011). A preliminary study of amino acid and mineral profiles of important and estimable 21 seafood species. *British Food Journal*, 113(4), 457–469.

Peng, S., Chen, C., Shi, Z., Wang, L. (2013). Amino acid and fatty acid composition of the muscle tissue of yellowfin tuna (*Thunnus albacares*) and bigeye tuna (*Thunnus obesus*). *Journal of Food and Nutrition Research*, 1(4), 42–45.

Pereira, T.G., Batista, I., Bandarra, N.M., Ferreira, J., Fradinho, N., Afonso, F. (2013). Chemical composition and nutritional value of raw and fried allis shad (*Alosa alosa*). *International Journal of Food Science and Technology*, 48, 1303–1308.

Prato, E. and Biandolino F. (2012). Total lipid content and fatty acid composition of commercially important fish species from the Mediterranean, Mar Grande Sea. *Food Chemistry*, 131, 1233–1239.

Ramos Filho, M.M., Ramos, M.I.L., Hiane, P.A., de Souza, E.M.T. (2010). Nutritional value of seven freshwater fish species from the Brazilian Pantanal. *Journal of the American Oil Chemists' Society*, 87, 1461–1467.

Rigos, G., Grigorakis, K., Koutsodimou, M., Gialamas, I., Fountoulaki, E., Nengas, I. (2012). Comparison of muscle fatty acid and vitamin composition between wild and farmed common dentex (*Dentex dentex*). *Journal of Biological Research-Thessaloniki*, 17, 26–32.

Sankar, T.V., Anandan, R., Mathew, S., Asha, K.K., Lakshmanan, P.T., Varkey, J., Aneesh, P.A., Mohanty, B.P. (2013). Chemical composition and nutritional value of anchovy (*Stolephorus commersonii*) caught from Kerala coast, India. *European Journal of Experimental Biology*, 3(1), 85–89.

Tenyang, N., Womeni, H.M., Linder, M., Tiencheu, B., Villeneuve, P., Mbiapo, F.T. (2014). The chemical composition, fatty acid, amino acid profiles and mineral content of six fish species commercialized on the Wouri river coast in Cameroon. *La rivista italiana delle sostanze grasse*, XCI(April/June), 129–138.

USDA (2021). FoodDataCentral [Online]. Available at: https://fdc.nal.usda.gov [Accessed 5 January 2021].

Usydus, Z., Szlinder-Richert, J., Adamczyk, M., Szatkowska, U. (2011). Marine and farmed fish in the Polish market: Comparison of the nutritional value. *Food Chemistry*, 126, 78–84.

Usydus, Z., Szlifder-Richert, J., Adamczyk, M. (2012). Variations in proximate composition and fatty acid profiles of Baltic sprat (*Sprattus sprattus balticus*). *Food Chemistry*, 130, 97–103.

Wang, Y., Yu, S., Ma, G., Chen, S., Shi, Y., Yang, Y. (2014). Comparative study of proximate composition and amino acid in farmed and wild *Pseudobagrus ussuriensis* muscles. *International Journal of Food Science and Technology*, 49, 983–989.

Weihrauch, J.L., Posati, L.P., Anderson, B.A., Exler, J. (1977). Lipid conversion factors for calculating fatty acid contents of foods. *Journal of the American Oil Chemists' Society*, 54(1), 36–40.

Zhang, X., Ning, X., He, X., Sun, X., Yu, X., Cheng, Y., Yu, R.-Q., Wu, Y. (2020). Fatty acid composition analyses of commercially important fish species from the Pearl River Estuary, China. *PLoS ONE*, 15(1), e0228276.

Zhao, F., Zhuang, P., Song, C., Shi, Z.-H., Zhang, L.-Z. (2010). Amino acid and fatty acid compositions and nutritional quality of muscle in the pomfret, *Pampus punctatissimus*. *Food Chemistry*, 118, 224–227.

Zotos, A. and Vouzanidou, M. (2012). Seasonal changes in composition, fatty acid, cholesterol and mineral content of six highly commercial fish species of Greece. *Food Science and Technology International*, 18(2), 1–11.

4

Fish Traceability and Authenticity

Carmen GONZÁLEZ SOTELO[1], Valur GUNNLAUGSSON[2], Ute SCHRÖDER[3], Rogério MENDES[4], Kristina KAPPEL[3], Helena SILVA[4] and Véronique VERREZ-BAGNIS[5]

[1] *Instituto de Investigaciones Marinas–CSIC, Vigo, Spain*

[2] *Matis, Reykjavík, Iceland*

[3] *Max Rubner-Institut, Kiel, Germany*

[4] *Instituto Português do Mar e da Atmosfera (IPMA), Algés, Portugal*

[5] *Ifremer, Nantes, France*

4.1. Introduction

Fishery product consumption has increased drastically worldwide according to FAO's 2020 report, from an average of 9.9 kg per capita in the 1960s to 20.5 kg in 2018. In 2018, total capture fishery production reached the highest level ever recorded at 96.4 million tons, while aquaculture production was 82.1 million tons.

Regional or local sourcing is no longer limiting for retailers and fish industries due to technological improvements and the globalization of trade. Market demand has profoundly changed production, trade and distribution, allowing the industry to source from all over the world, transforming it into an interconnected system with

For a color version of all figures in this chapter, see http://www.iste.co.uk/verrez-bagnis/aquaticproducts.zip.

Current Challenges for the Aquatic Products Processing Industry,
coordinated by Véronique VERREZ-BAGNIS.

complex interactions and trade flows. Due to this complexity, the seafood supply chain is vulnerable and prone to the development of "dark paths" that involve illegal, unreported and unregulated (IUU) fishery activities, incidents with the presence of harmful substances (e.g. heavy metals, histamine), seafood fraud (e.g. mislabeling), unsustainable fishing (e.g. overfishing and pollution) and vulneration of human rights (e.g. unfair payment) (Borit and Olsen 2012; He 2018).

Furthermore, as global demand for fish and fishery products has increased significantly, consumer awareness of sustainability, safety, quality, environmental and social issues has increased in parallel. Consumers are increasingly demanding verifiable product information, so there is an increasing need for transparent information along the whole fish chain, supported by modern traceability methods, improved methods for fish authenticity control, and certification schemes for the growing range of fish and fish products.

To overcome these problems associated with the seafood value chain, most occidental countries or regions, and in particular, the European Union (EU), which is one of the main importers of seafood in the world, have established a regulatory and control framework that has contributed to mitigating some of the main problems mentioned above. The most important policies and regulations concerning seafood in the EU are: the Common fisheries policy (CFP) that establishes guidance and rules to preserve the sustainability of fisheries and aquaculture, and the Common Organization of the Markets in Fishery and Aquaculture products regulation (COM), currently the EU1379/2013, which aims to help stakeholders to implement CFP objectives (D'Amico et al. 2016).

According to the definition put forward by Aung and Chang (2014), traceability could answer the questions of "who (i.e. actor/product), what (i.e. actor/product information), when (i.e. time), where (i.e. location) and why (i.e. cause/reasons)" with regard to food safety, quality and visibility. Authentication can be defined as the act of establishing or confirming something as authentic. The main objectives for authenticity assessment include: i) protection against fraud, to safeguard fair trade; ii) ensuring correct labeling of products (e.g. following the last EU regulations (EC 1379/2013)); iii) protecting endangered species and fighting IUU fisheries; and iv) supporting customs examinations. In addition, authenticity may involve confirming the identity of a product (e.g. which species), its geographical origin (e.g. from which farm), discrimination between production methods (e.g. farmed or wild; or organically aquaculture fish production), and detection of water addition, frozen–thawing and discoloration treatments. Food authenticity control does not only serve consumers, but also the stakeholders who are searching for possibilities to assure their food product labeling compliance and branding (Danezis et al. 2016).

This chapter first reviews the traceability issues and technologies to be used all along the fish industries. Also, both the current analytical methods for the control of seafood traceability and authenticity and the emerging methods are detailed. Current species authentication methods include protein-based methods, such as IEF or ELISA, and DNA-based methods that were used by most control laboratories, such as DNA sequencing and real-time PCR. Emerging technologies include different approaches such as the development of quick DNA-based methods, like microarrays, isothermal amplification or mass spectrometry.

Finally, a review of the development of methods for identifying fishing areas, distinguishing between wild and farmed fish, or frozen–thawed fish products is addressed. The needs and developments of methodologies (e.g. on-site methods to improve traceability and authenticity for fish industries) are also tackled.

4.2. Seafood traceability

Keeping track of product information through the globalized seafood markets can be a challenging task. Information sharing and coordination between the buyer and vendor in the supply chain have been considered as useful strategies to remedy the so-called bullwhip effect and improve supply chain performance, especially because of the often short shelf life of fresh products. The debate is not about whether production information should be shared in the supply chain, but about how to share the right information, at the right time, in the right format, by the right people, under the right environment. This should maximize the mutual benefits of the whole supply chain, as well as the individual business players, and hopefully for the end consumer as well.

Food traceability has become increasingly important in recent years for a number of reasons. Around the turn of the century, the main driver for improved food product traceability was the number of tragic and costly food scandals, which received wide media attention around the world at that time (Olsen and Borit 2013). Among those food scandals where the BSE, or mad cow disease in the early and mid-nineties, the dioxin contamination of chicken feed in Belgium in 1999 and Sudan 1 food color contamination of a number of ready meals in Europe in 2005, producing a massive food recall, to mention some. Product withdrawal and recalls from retail shops with extensive press coverage have increased demands for further documentation and improved traceability and transparency of food products throughout the whole supply chain.

Seafood, in particular, often follows a long and winding path from the ocean, lakes, rivers or land-based aquaculture to the end consumer. Unprocessed seafood is frequently shipped great distances, through multiple ports, changes hands among various brokers and/or fish markets, and is processed by primary and secondary processors before it ends up at retailers or is prepared and served by the HoReCa (Hotels, Restaurants & Catering) sector. On that journey, seafood supply chains are regularly complicated with a number of processing steps, where many food business operators can be involved, sometimes in different countries that can have different regulatory frameworks regarding traceability. Raw materials and ingredients from different sources can be mixed, joined or split up, causing the systematic loss of information and hampering the identification of product provenance, which in turn can prevent the consumer from being able to make an informed and sustainable choice from the labeling provided, but is in accordance with basic legal requirements.

Some consumer studies have demonstrated that there is an increased need for consumers to have access to additional product information, like catch and environmental information (Viðarsson et al. 2015). There is also a need for increased transparency, which can be related to the high dependence on seafood imports into the EU, in order to avoid products from unsustainable or Illegal, Unreported and Unregulated fishing (IUU), where someone is fishing without a license or quota authorized by relevant authorities. To put things into perspective, the EU imports fisheries and aquaculture products from nearly 150 countries around the world, and the internal demand of seafood products in the EU is mostly met through imports, as they cover approximately 60% of the total supply, according to Eumofa.

Currently, the seafood industry is facing a number of challenging local and global issues. These include a sometimes unstable seafood supply, shaky consumer confidence in seafood labeling, product safety and authenticity. There is also uncertainty about sustainability and the high number of different certification schemes, and also how to display sustainable information to the consumer in a standardized way. There are also increasing regulatory demands by local and global markets, environmental concerns about the aquaculture sector and trawler and IUU fisheries, and increasing concern about the general health of the oceans, which is a prerequisite for a healthy planet and healthy communities around the globe. The recent Covid-19 pandemic also had an effect on the market, with changes in supply and demand, especially with the change from consumption in HoReCa facilities to home consumption and increased sales over the Internet.

Today, companies need to be able to meet customers' demands for constant good quality, safe, nutritious and environmentally conscious products. Furthermore, they need to meet demands for supply chain transparency, which can be very challenging in this fragmented sector. The aim of this chapter is to give an overview of the basics in seafood traceability.

4.2.1. *Food safety and traceability legal requirements*

The safety of seafood, as with any foods, is of critical importance to all. Consumers must have confidence and assurance that the food they are going to consume will not harm them or have an adverse effect, and furthermore, that the seafood in the package or on the plate is definitely the same seafood as described on the product label or menu.

Since the application of the European Union Common Food Law regulation 178 from 2002, each food business operator must be able to identify all those who delivered food, feed or ingredients that were used in their products. Furthermore, they have to be able to identify where products were sent and rapidly provide this information to competent authorities, in case of a food scare. Similar requirements for traceability systems are present in the United States and Japan (Bechini et al. 2008). In the US, the Bioterrorism Act requires that when demanded by regulatory bodies, traceability information must be available within 24 hours (US Food and Drug Administration 2002). Moreover, the EU introduced a regulation to prevent, deter and eliminate the import of IUU fishery products into the community, which came into force in 2010. Regulations 1005/2008 and 1010/2009 create new requirements on fish and fishery products entering the EU market from non-EU countries. All importers now need to take steps to ensure the goods they import have been legally caught. As a minimum, the importer needs to ascertain that their export partner is able to provide a validated catch certificate for every consignment. In addition, regulation 1224/2009 includes requirements about the labeling of fish products, pointing out each batch identification number, each vessel's external marketing or name of production site, an FAO-code for each species and the date of catchment or production, as the minimum mandatory information to accomplish the traceability of fish products.

4.2.2. *Definitions of traceability*

Traceability has been defined in international standards of traceability, as well as by standalone definitions in scientific articles. Probably, the most used definition of traceability that comes from the scientific paper by Moe (1998). It is stated that

"Traceability is the ability to track a product batch and its history through the whole, or part, of a production chain from harvest through transport, storage, processing, distribution and sales." Moe introduces the "product batch" as the group of production items that are being traced.

4.2.3. *Traceability main features*

In the world of traceability, there are often two main features of traceability systems, i.e. internal and external traceability. Internal traceability is commonly within one company or link in the supply chain; it allows companies to trace what happens to a product within a business' operation, usually limited to "in house" traceability, while external traceability requires functional interoperability between different stakeholders to facilitate information sharing, and allows us to trace what happens to a product through all parts of the supply chain, or part of the supply chain outside of one business entity.

Internal traceability is the core of functional traceability, everything else depends on each company in the supply chain having traceability systems and standardized methods in recording all of the relevant internal information. Internal traceability usually deals with "batches" or "lots" of raw material or products, but the relationship between those batches must be recorded. Data confidentiality and access is usually not problematic inside a single company since there are several good systems, solutions, standardized practices and standards available to support this work and facilitate functionality.

There are some definitions that are necessary to recognize in the field of food traceability. Tracing upstream means looking back along the supply chain towards the fishing or harvester steps, while tracing downstream looks forward towards distribution, retail and the end consumer. As seafood supply chains become longer and more complex, there is rising demand for data regarding the traceability and sustainability of products, both by regulators and consumers.

EC regulation No. 2406/96 lays down common marketing standards for certain fishery products. There, a "lot" is defined as a quantity of fishery products of a given species, which has been subjected to the same treatment and may have come from the same fishing grounds and the same vessel. "Fishing grounds" will be interpreted as the customary name given by the fishing industry to the place in which catches have been taken, and "presentation" will mean the form in which the fish is marketed, such as whole, gutted or headless.

4.2.4. *Traceability systems*

Traceability is often documented in a different way from simple paper-based traceability records, using electronic traceability systems that are becoming increasingly more common. The paper-based systematic documenting of traceability is on the way out, although it can sometimes find its application in countries trying to get some systematic documentation of traceability going. The paper-based systematic reporting of traceability is occasionally developed into electronic traceability systems by companies.

Electronic traceability can be achieved using several different tools. Most are either Web-based and accessible over the Internet or module-based, installed in computer networks at various points along the supply chain. Added hardware is often also required to get connection and functionality right.

4.2.5. *Why is full seafood chain traceability not more common?*

There are several reasons why seafood chain traceability is not more common, with functional interoperability between different systems and seamless data flow between different stakeholders in the respective chains. Technical issues, like a lack of a common identifier and interoperability, are often the reason for the loss of systematic information in the supply chains of seafood. Interoperability is probably the biggest single technical factor preventing full chain traceability of seafood. Even if data is recorded and transmitted in the chains, the recipient is often in a situation where the data is meaningless, because there is a lack of standards in place, or where the data formats used by the different links in the chain are incompatible, resulting in manual work to get a coherent picture of a chain with multiple stakeholders.

There are other issues like a lack of motivation for companies, and sometimes unclear regulatory frameworks, or even different regulatory frameworks between different countries or areas, which is hampering full chain traceability of seafood. To limit these issues, it is of high importance to highlight the benefits of data sharing and interoperability, the ability of one traceability system to communicate with another system seamlessly, and exchange and interpret key data elements across all critical tracking events in the supply chains. This is facilitated with the usage of common standards both for batch/lot/unit identification (GS1), data structure (EDI standards, like GS1 EPCIS) and for data content (e.g. ISO 12875/12877). These standards for seafood traceability developed and published by organizations such as GS1, the Marine Stewardship Council (MSC) and the International Standards Organization (ISO) have helped us to provide the necessary foundations for full chain traceability and interoperability of traceability systems.

Electronic traceability systems based on automatic data capture and software applications are considered the most effective solution for providing relevant food safety information to the food industry and consumers in a fast and effective way. Electronic data capture can use optical or radio-wave systems, for example, barcodes and RFID (Radio Frequency Identification) technology. Barcodes are used to transfer basic information, since they are simple and there is low implementation cost as they are based on open standards such as EAN-UCC, administered by GS1. RFID systems are an emerging technology that make use of wireless communication to identify products (trade items), pallets (logistic units) or returnable assets throughout the supply chain. The standard for using RFID is based on the EPCglobal standard. A technical concept that has been established and evaluated for chain traceability in the food industry is the electronic product code information services (EPCIS) standard, which originates from EPCglobal. EPCIS is designed to enable EPC-related data sharing within and across enterprises (EPCIS Standard 2007).

However, most of the work in this field presents traceability solutions where only the product packaging is used as a traceable resource unit and tracked through the supply chains, and fails to address the internal traceability issues linked to the production events within a food facility. This includes food product transformations which are so common in the seafood sector, including mixing, joining and splitting up lots and batches. Regattieri et al. (2007) presented the application of an RFID system integrated with alphanumeric code to trace Parmigiano Reggiano cheese through the complete supply chain. Shanahan et al. (2009) proposed the use of RFID for the identification of individual cattle and biometric identifiers for the verification of cattle identity. They also proposed a data structure for RFID tags and a middleware to convert animal identification data to the EPC data structure.

The rapidly increasing application of RFID and other types of EPCs give opportunities to increase the efficiency and accuracy to monitor and control the flow of goods and materials within and between actors in the value chain, where the availability of real-time information is regarded as the main benefit (Bottani and Rizzi 2008). EPC provides a method for unique identification of all items in a supply chain and makes it possible to register internal and external events electronically that are related to the movement of tagged items.

Another standard is the "Global Dialogue on Seafood Traceability", which is an international voluntary industry standard for interoperable seafood traceability. It consists of a business-to-business platform, established to advance a unified framework for interoperable seafood traceability practices. The dialogue brings together a broad spectrum of seafood industry stakeholders from across different parts of the supply chain, as well as relevant civil society experts from diverse regions. The voluntary industry standard should help companies to ensure that the products they buy and sell come from sources that are legal, sustainable and ethical. This effort is intended to make global seafood traceability more reliable and more affordable for companies throughout the supply chain.

To tackle the problem of over-exploited fish stocks, a new communication tool for automatically collecting and disseminating fishery catch data for sustainable fishery management has been developed, called FLUX. FLUX is a free, open standard for communication between countries and organizations, and offers modules for tracking vessels, catches, vessel registry, regulating catch, electronic reporting (ERS), inspection and surveillance, which are all elements that are needed to build a fisheries information system (FIS). It facilitates messages between authorities regarding catches up to the first sale, but it does not cover trade after the first sale. The EU bases the development of its systems on FLUX; this includes their system for IUU catch certification. Although still only partly in place and in limited areas, it has been observed to be quite effective.

Currently, there is much hype around digital ledger technologies (DLTs), in particular blockchain technology and crypto-anchors, which do not prevent fraud from occurring, but reveal who recorded the original data, and make it so that the data cannot subsequently be changed. This may improve transparency, and therefore improve consumer trust in the data provided. Some companies have been trying out this technology with promising results, but usually in small scale trials, without fully implementing it into their traceability procedures. However, there is a promising potential here to be explored further.

4.2.6. *Implementing a traceability system*

Traceability facilitates the determination of the identity, history and source of a product. Traceability is a way of responding to potential risks that can arise in food and feed, to ensure that all food products are safe to consume. It is vital that when national authorities or food businesses identify a risk, they can trace it back to its source in order to swiftly isolate the problem and prevent contaminated products from reaching consumers; this functionality must be included in a traceability system. When implementing a traceability system, the added value of such a system to a company also needs to be considered, as well as being able to fulfil regulatory demands. Such a system and documentation are often a prerequisite to gaining market access. Traceability systems can serve many purposes and have several additional benefits for the company involved. As previously mentioned, it is of upmost importance to be able to perform a fast and targeted product withdrawal, and promptly identify and recall potentially unsafe products, thus protecting consumers and limiting financial impact. With traceability, companies are in a better position to provide internal logistical and quality-related information, thus improving efficiency, supporting supply-chain optimization and reducing food loss. Increasing supply chain transparency to other stakeholders in the chain and consumers can open the door to dissemination of product sustainability and authenticity.

As the implementation of a traceability system often requires considerable investments, companies should take all benefits into consideration before implementing a full-scale traceability system. Sometimes, a large investment is not needed to implement a traceability system, since the procedures are often already in place. It is often just a matter of getting the connection right between the physical processes involved, systematic documentation of those processes and sharing of relevant information to other stakeholders in the chain.

Nowadays, companies usually have effective internal traceability systems up and running, but chain traceability between stakeholders in the chains is often lacking. The main requirements for handling withdrawals and recalls across the supply chains are having reliable data, procedures to exchange data and properly mapped business processes. A good internal traceability system is therefore a prerequisite to a functional chain traceability system. The investments in an internal traceability system will not be wasted in moving towards chain traceability. Chain traceability software should be able to integrate seamlessly into any current internal system in place in companies.

4.2.7. *Example of data gathering and information exchange in the seafood sector*

In Europe, the electronic recording and reporting system (ERS) is used to record, report, process, store and send fisheries data. The key element is the electronic logbook, where the captain of a fishing vessel keeps the records of the fishing operations. The records are then sent to the national authorities, which store the information in a secure database and use the data to update quota for the boat or company. These e-logbooks are basically an electronic edition of the paper-based logbooks that have been used for decades. Usually, it is the captain of the vessel that enters information regarding the catch, by haul or fishing days, depending on the type of fisheries. Catch reports are created with information on the catch volumes by species, catch location, date and sometimes weather conditions and other factors. Today, suppliers of seafood into the EU must also provide a catch certificate with each lot, to prove that their supply is not coming from illegal, unreported or unregulated (IUU) fisheries. The electronic logbooks create large amounts of data concerning the catch that can be of importance for all stakeholders in the value chain, to ascertain how they can get value from the use of electronic logbook data.

When fish is processed on board in freezing trawlers, information gathering is simpler because there is only one stakeholder who is fishing, processing and labeling the product before freezing. So, most of the relevant information is stored in one system and the relevant information is put directly on the boxes. This, however, only covers information components from the vessel and does not contain information regarding where and when the fish was landed, how it was transported to the marketplace, nor information regarding secondary processing. Therefore, information from other stakeholders needs to be included to tell this product story from catch to customer. When fish is landed fresh for processing, the fishermen are sometimes able to forward information that exceeds the data available within the basic version of the electronic logbooks. This is particularly practiced in integrated value chains, where the processor, for example, owns the fishing vessels. Here, data interoperability is of upmost importance for processors to be able to receive relevant information and process it along with the physical processing of the fish (Thakur and Gunnlaugsson 2018).

When companies are vertically integrated, i.e. when the same company does the fishing and primary processing, the people at the processing facilities have access to all the live catch data and it can be used in their planning decisions regarding the fisheries, processing of the catch and sales. This includes catch, trip, haul and gear information, as well as the processing information and some environmental parameters like sea temperature, which can be of interest to the processors (Thakur and Gunnlaugsson 2018).

Usually, fresh fish is processed on land where there are many different enterprise resource planning (ERP) systems used in the fishing industry around the world. In Iceland, the most common systems are the Wise system from Wisedynamics and the Innova system from Marel and SAP systems (Thakur and Gunnlaugsson 2018). These systems vary greatly, but they all manage data from processing, inventory control, lot or batch tracking, logistics and sometimes marketing. These systems are used to record provenance and usage of raw material, work in progress, and sometimes collect all associated production costs and maintain full product records including price, sales and production history. These systems are usually the backbone of each company, connecting data from different sources and therefore making products and information traceable through the different processes involved. After processing, the fish products are usually labeled in a similar manner as the ones frozen-at-sea. Products, however, are not usually labeled with the catching date, just the processing date in the Julian Day Calendar format, for example, 99 days of the year is the same as April 5. For fresh fish transported by plane, the labels include the flight number and destination, so there is more traceability information on these labels beyond the frozen-at sea ones. Still, with only this limited information on the boxes, information from aforementioned systems has to be accessed to complete the story of the product through the value chain (Thakur and Gunnlaugsson 2018).

4.2.8. *The way forward*

Seafood supply chains are often long and complicated, as has been documented here before, and seafood is sometimes tampered with. This can often result in changes regarding species name, provenance of the product, and the product can even go from being farmed to wild caught. Important information, like if the product has been previously frozen or not, gets lost or does not get displayed to the consumer on product labels or by other means.

Technology-enabled end-to-end traceability in food value chains, coupled with multi-stakeholder collaboration, has the potential to fundamentally improve food systems, according to a World Economic Forum report (WEF 2019). To achieve its full potential, stakeholders will need to come together to enable emerging technologies and install a broad system and standards where they recognize the mutual benefits.

It has been covered that there are several reasons why seafood chain traceability is not more common, with functional interoperability between different systems and seamless data flow between different stakeholders in seafood chains. It was mentioned that there are usually technical issues, lack of motivation for companies and sometimes unclear regulatory frameworks. To limit these issues, it is of high importance to highlight the mutual benefits of data sharing and interoperability, and the ability of one traceability system to communicate with another system, seamlessly exchanging and interpreting key data elements across all critical tracking events in the supply chains.

This is of high importance for the seafood sector to promote multi-stakeholder collaboration to increase supply chain transparency and create product stories with validated facts and figures, in order to gain consumer trust and promote that the products originated from sustainable fisheries or aquaculture. If that is achieved, then the sector will grow and prosper.

4.3. Analytical methods for the control of seafood traceability and authenticity

4.3.1. *Species identification tools*

4.3.1.1. *Current methods*

4.3.1.1.1. Isoelectric focusing

While isoelectric focusing (IEF) has paved the way for species identification in protein-based products, such as meat and seafood in control laboratories in the 1980s, its use today is very limited (Kappel and Schröder 2015). Although this application is not very time-intensive and costly – water-soluble proteins are electrophoretically separated in a gel and stained – other protein-based techniques like MALDI-ToF-MS (matrix-assisted laser desorption/ionization-time of flight mass spectrometry) have replaced it.

4.3.1.1.2. Immunological methods

Immunological diagnostic tools are based on the specificity of the antigen–antibody (Ag–Ab) reaction with muscle proteins generally used as antigens (e.g. parvalbumin, tropomyosin complex). Due to the high number of fish species globally consumed (over 1,000), studies and commercially available tests based on immunological methods are rare. It is notable that generally, tests have been developed to detect the presence of fish (cod species mainly) in food products to protect fish-allergic children and adults who represent up to 3% in the general population. Commercially available ELISA (enzyme-linked immunosorbent assay) kits developed for this purpose demonstrated, in general, a limited capacity to detect both raw and heat-processed bony fish, and they cannot detect the presence of cartilaginous fish species.

Even though immunoassays, previously developed, are easy to use, and have high sensitivity and throughput, their use is limited because they require analysis in parallel to reference samples to check their cross-reactivity with other or similar species. Moreover, they are not suitable to analyze processed food products, as processing steps may alter the three-dimensional structure of proteins, losing and/or modifying the targeted protein recognition site(s).

4.3.1.1.3. DNA sequencing (Sanger) and FINS method

Discovering the structure of the DNA molecule and the method for retrieving the information contained in this molecule has been essential and of paramount importance for many different applications, including finding the cure for illnesses, the criminal of a homicide or the existence of fraud in our foods. DNA sequencing consists of finding the order of nucleotide bases in the DNA molecule. Sanger et al. (1977) proposed the first method to sequence DNA, based on the use of DNA polymerases and chemically modified deoxynucleotides (for further information, see Heather and Chain (2016)).

DNA sequencing has been used for fish species identification since 1991 (Bartlett and Davidson 1991); there are two main approaches: BLAST (Basic Local Alignment Search Tool) and FINS (Forensically Informative Nucleotide Sequencing). Both methods require a DNA database, a collection of DNA sequences, which contain information about millions of samples of thousands of living organisms. Both methods involve the comparison of the DNA sequence with the sequences present in the databases in order to identify the species. The DNA sequence that is analyzed presents the following characteristics: it is a specific part of the genome, for example, a fragment of the cytochrome b gene, and the size varies depending on the DNA fragment, from 100 bp to 600 bp. The most used DNA markers are cytochrome b (cytb), cytochrome c oxidase subunit I (COI) and 16S rRNA (Rasmussen and Morrissey 2009; Griffith et al. 2014).

In the case of the BLAST, the problem nucleotide sequence has been uploaded in the query section of BLASTn suite (see: https://blast.ncbi.nlm.nih.gov/Blast.cgi), where an alignment of the query with sequences in Genbank databases is performed. Results show the degree of homology between the query and the percentage of identity; it also presents other useful parameters such as score and e-value (Wheeler and Bhagwat 2007). To identify fish species, the dedicated system BOLD (The Barcode of Life Data System; Ratnasingham and Hebert 2007) provides specialized services based on reference libraries on a 648 bp region of the cytochrome *c* oxidase I (COI).

Forensically Informative Nucleotide Sequencing (FINS) estimates genetic distance between the "problem" DNA sequence and DNA sequences from authentic seafood species previously sequenced. Based on these genetic distances, a phylogenetic tree is constructed to graphically present the topology of the relationships among the sequences analyzed, showing in the same cluster the most similar sequences (see Figure 4.1). Different genetic distance methods can be used, such as Kimura-2, Tamura-Nei and Jukes-Cantor, and different nucleotide substitution types (transitions and/or transversions), nucleotide frequencies or similarity in the rate of nucleotide substitution can be considered (Hellberg et al. 2016).

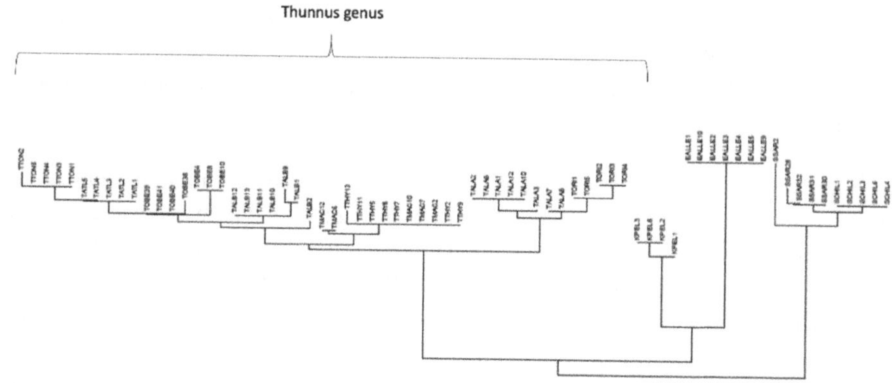

Figure 4.1. *Phylogenetic tree showing the Thunnus species and other genera, like Katusuwomus, Euthynnus and Sarda. For authentication of tuna species, DNA markers such as cytochrome b or the control region can be used to test the identity of the tuna species present in a product*

4.3.1.1.4. Next-generation sequencing: metabarcoding

The simultaneous sequencing of tens of thousands to billions of DNA fragments from a single sample in one reaction solution is commonly referred to as massively parallel sequencing, high-throughput sequencing or next-generation sequencing (NGS). It is achieved by a combination of complicated molecular mechanisms, sophisticated sequencing instruments and complex bioinformatic pipelines. In contrast to conventional (Sanger) DNA sequencing, NGS methods provide the ability to identify species in processed seafood products that may contain multiple species, such as, for example, canned tuna, fish marinades, surimi or fish cakes. They can also be used to analyze all seafood individuals in a package or container within a single analytical experiment by either homogenizing the entire contents or using small pieces of each individual (e.g. fin clips).

The molecular mechanisms underlying NGS analytics vary depending on the platform used. In general, the DNA must first be amplified to generate a separate DNA clone for each DNA template. This may either be accomplished by bridge amplification on a solid support (Illumina) or by generating small droplets (emulsion PCR) (e.g. Ion Torrent). Afterward, the sequences of the DNA clones are determined in most applications with a technique called sequencing by synthesis by either optic measurement of the incorporation of fluorescence-labeled nucleotides during the synthesis of a new DNA strand (e.g. Illumina) or by measuring specific by-products of nucleotide incorporation such as hydrogen ions. The release of hydrogen ions can be detected with semiconductor technology (e.g. Ion Torrent). In addition, there are also third-generation sequencing techniques for single-molecule sequencing that do not require prior DNA amplification, but rely on zero-mode waveguides (Pacific Bioscience) or nanopore grids (Oxford Nanopore Technologies).

Currently, the most common method for species analysis in food samples is metabarcoding, the targeted analysis of specific genetic marker sequences with species information. Similar to conventional DNA barcoding, the DNA marker(s) of choice is/are first amplified by conventional PCR. By using NGS techniques for sequencing the resulting PCR products, different species occurring in one DNA sample can be detected after implementation of specific bioinformatic data processing pipelines with integrated database queries. Despite relatively few publications about the application of metabarcoding for seafood products such as, for example, surimi-based products (Giusti et al. 2017; Noh et al. 2020), tuna samples (Kappel et al. 2017) or salmon products (Wang et al. 2021), metabarcoding-based analysis of seafood samples is already offered by some service laboratories. However, due to possible amplification bias during the first round of PCR and the use of mitochondrial genetic markers, an exact quantification of species proportions in one sample is not feasible. Furthermore, some species may not be amplified at all with the used PCR primers. Thus, a careful validation of the NGS setup with the species in question has to be done beforehand.

4.3.1.1.5. Quantitative PCR (qPCR) and digital PCR

Amplification of DNA using polymerase chain reaction is the corner stone of modern molecular biology and therefore also for most developed authentication techniques (Rasmussen and Morrissey 2009). Real-time or quantitative PCR (qPCR) was developed by Higuchi et al (1993). This technique records the amount of amplification product by real-time monitoring of the fluorescence increase as amplification product accumulates during the reaction. The quantification cycle, the cycle where fluorescence is significantly different from background noise, inversely correlates with the concentration of target DNA, thus allowing the specific quantitation of one particular DNA, provided standard solutions are used (Bustin et al. 2009). Development of different platforms and chemistries for performing this technique and, consequently, applications have soared since then. Many seafood authentication methods based on qPCR have been developed; these include most important species, such as gadoids, hakes, tunas, crustaceans and mollusks (Lopez and Pardo 2005; Sanchez et al. 2009; Herrero et al. 2010; Velasco et al. 2020; Brenn et al. 2021; Klapper and Schröder 2021). Although qPCR has been proposed as a quantitative method for different applications, the quantification of food products containing muscular tissues, such as meat and seafood products, using qPCR may present some issues, especially when mitochondrial genes are used as species markers (Lopez and Pardo 2005). While the validation of qPCR methods for meat products has been largely established in public control laboratories, applications for seafood products have not really taken off. So far, qPCR methods for the detection of certain seafood products seem to be a viable alternative only for private laboratories. However, qPCR enables high sample throughput and lower staff requirements that could be important for performing frequent monitoring. Several recently published studies have described high substitution rates for certain seafood species that require more control and subsequently the use of public control laboratories (Naaum et al. 2016). High-resolution melting (HRM) analysis is an interesting alternative to conventional qPCR with potential in food analysis. Difficult targets with only one SNP (single nucleotide polymorphism) difference or small insertions or deletions can be differentiated, provided that saturation dyes and high-resolution qPCR devices with a specific software are used (Druml and Cichna-Markl 2014). Promising applications were developed among others for *Mytilus* species and their hybrids, for shrimps, hake and puffer fish species (Fernandes et al. 2017, 2018; Jilberto et al. 2017; Chen et al. 2021).

Digital PCR (dPCR) was developed in the 1990s, with its first reported application in 1999 (Vogelstein and Kinzler 1999). dPCR overcomes some of the problems mentioned above, since amplification of individual template DNA molecules can be performed and individually evaluated. In dPCR, target DNA molecules are randomly partitioned in individual PCR reactions, containing none, one or several target DNA molecules. These individual PCR reactions will be performed until endpoint, positive reactions will be recorded versus negative ones, the concentration of target DNA molecules will be calculated as the ratio of positive reactions and the number of reactions analyzed (Pinheiro et al. 2012). Compared to qPCR, dPCR has considerable benefits which could be crucial for its application. The quantification of the target DNA is performed by determining the absolute copy numbers without using standard curves; more sensitivity is reached in low copy number ranges and greater tolerance to some inhibitors.

Different platforms have been developed and are commercialized nowadays; they can be divided grosso modo in those using microfluidic chip technology (Sanders et al. 2011) (Biomark HD dPCR system from Fluidigm) and others using droplet generators coupled with chips (QuantStudioR family from Life Technologies, RainDropTM instrument from RainDance Technologies) or microfluidics approaches (Cai et al. 2017; Shehata et al. 2017) (BioRad QX200TM Droplet DigitalTM PCR system). Some applications of dPCR for seafood identification and quantification have been recently published (Cao et al. 2019; Deconink et al. 2021). The technique has a higher sensitivity than qPCR, but similar specificity; as a drawback, the authors point out that the time required to perform the analysis is longer in the case of dPCR (Cao et al. 2019).

4.3.1.2. *Emergent tools*

4.3.1.2.1. Next-generation sequencing: shotgun sequencing

An alternative to NGS-metabarcoding-based species identification (see above) is the high-throughput shotgun DNA sequencing of whole genomes of the ingredients in complex food samples without prior amplification of specific DNA targets (see, for example, Ripp et al. (2014); Haiminen et al. (2019); Varunjikar et al. (2021)). With this approach, a possible bias introduced by PCR or the use of mitochondrial DNA markers is avoided, which results in a better quantification capability. Using shotgun sequencing, all species present, whether they are fish, mollusks, crustaceans, mammals, plants or even spoilage bacteria can be detected, provided that appropriate reference genome sequence information is available for the species in question (Kobus et al. 2020).

4.3.1.2.2. DNA microarrays

Instead of sequencing the obtained PCR products with the aim of performing a database query or producing a phylogenetic tree, the amplicons can also be analyzed by hybridization to species-specific oligonucleotide probes that are arranged on a DNA microarray/DNA chip. Each probe is present in replicates on specific positions of the DNA microarray, and the species can be determined by analyzing the positions to where the labeled PCR products are bound and show signals. DNA microarray analyses can be performed rapidly with little equipment (a microarray reader is needed) and make time-consuming database queries obsolete. They are already used by some food authenticity authorities for the detection of animal species in meat products (Iwobi et al. 2011) and have also been developed for the identification of seafood species (e.g. Kochzius et al. 2008; Teletchea et al. 2008; Chitipothu et al. 2014; Handy et al. 2014). However, due to the wide variety of seafood species regularly used in seafood products or that may be used as substitutes, a one probe per species-approach is too unspecific. Thus, there have been some first attempts of a pattern recognition-based differentiation of fish or crustacean species that relies on multiple probes per species and the comparison of the probe signal patterns with digital probe signal pattern information of reference specimens (Kappel et al. 2020).

4.3.1.2.3. Isothermal amplification

Isothermal amplification of targeted DNA is used for loop-mediated isothermal amplification (LAMP) assays. LAMP is a simple, low-cost and short-time method requiring only the Bst DNA polymerase, and a set of four specially designed inner primers (FIP/BIP) and outer primers (F3/B3) that recognize a total of six distinct sequences on the target DNA. Compared to PCR methods, LAMP is less affected by PCR inhibitors, and due to the shorter operating time, lower costs and easy-to-handle equipment, LAMP is well-suited for routine fish species authentication. In Germany, there is already a LAMP in use by one official control lab for eel differentiation, with the aim of discovering the importation of prohibited eel, and detecting the critically endangered species *A. anguilla* (Spielmann et al. 2019).

The first report of an application for LAMP in fish authentication was published by Saull et al. (2016). They used a LAMP assay based on the cytochrome b (*cytb*) gene as targeted DNA to discriminate Atlantic cod (*Gadus morhua*) from two closely related species (Pacific cod (*Gadus macrocephalus*) and Greenland cod (*Gadus ogac*)) using a heating block at 63 °C for 60 min and a final step at 80 °C for 1 min to denature the Bst polymerase. Their method can detect by fluorescence 0.1% w/w *Gadus morhua* in a homogenized raw fish mix. In the same way, Li et al. (2021) developed a LAMP assay containing a colorimetric pH indicator to identify cod and oil fish commonly found in the Chinese market. Indeed, a positive amplification leads to the formation of protons, which induces a drop in pH that can be monitored with a colored indicator.

It is also possible to monitor the amplification in real-time by a pre-addition of fluorescent dyes in the reaction mixture; this method is called real-time fluorescent LAMP. This is the method used by Xiong et al. (2021a, 2021b), who developed a real-time fluorescence LAMP assay targeting the *cytb* gene for the quick detection of Atlantic salmon (*Salmo salar*) and skipjack tuna (*Katsuwonus pelamis*) in processed fish products. Wang et al. (2019) also developed two LAMP assays to distinguish three cod species in single or mixed samples by targeting the *cytb* gene or the *Gadidae* family by targeting the 12SrDNA gene.

On-site detection LAMP assays were also developed to identify 12 CITES-listed shark species in shark fin products to monitor illegal trade (But et al. 2020), or to identify selected invasive alien species in an ecosystem (Vythalingam et al. 2021).

In conclusion, this methodology, which allows the identification of one or more fish species quickly in raw or processed food products, with a detection limit of up to 0.1% in some cases, should be developed in the future, not only in the context of authenticity verification, but also to help preserve endangered species.

4.3.1.2.4. PCR-ELISA

Single-stranded DNA can be used as a probe to verify the presence of a specific sequence in a sample. One example of this approach is the immunodetection of previously amplified DNA by using an enzyme-linked immunosorbent assay (ELISA) (Coutlée et al. 1989). The method requires the labeling of the DNA to be analyzed, a hybridization step in which the probe and labeled and amplified DNA are mixed, and a detection step that commonly requires a conjugated antibody, as can be seen in Figure 4.1.

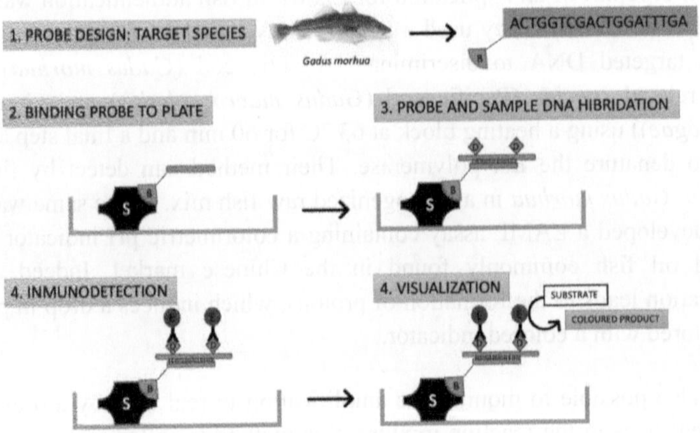

Figure 4.2. *PCR-ELISA steps. The designed probe (1) is linked to a biotin molecule (B) that is used to immobilize the probe by bonding with streptavidin attached to the plate (2). Sample DNA labeled with digoxigenin (D) is added to the plate and hybridization is allowed (3). If the probe and sample DNA hybridize, it is immunodetected using an antibody that is coupled with an enzyme (4). The immunodetected hybridization is visualized with the enzymatic reaction that produced a colored product (4)*

PCR-ELISA has been used for different fish species identification (Asensio et al. 2004; Santaclara et al. 2015; Taboada et al. 2017). PCR-ELISA offers a simplified authentication analysis providing a quick rapid and visual inspection of positives and negatives, and allowing the detection of mixed species present in a food product (Taboada et al. 2017).

4.3.1.2.5. Lateral flow dipstick assay

Lateral flow dipstick assay (LFDA) is a method of visualization of different analytical tests without the need for any equipment; this technique can be applied to DNA analysis methods such as specific PCR or isothermal amplification. The assay consists of one step to obtain a biotin-labeled mixture of DNA fragments, followed by another step to load the mixture on a paper stick (dipstick) and then allow the sample to migrate through the paper using a buffer solution (Taboada et al. 2017).

The use of recombinase for isothermal amplification of sample DNA, combined with a DNA probe and LFDA, was recently used for the specific authentication of *Octopus vulgaris* (Velasco et al. 2021), the common octopus. This technique is also interesting in the case of seafood products; its simplicity permits the rapid check of seafood species by for instance border inspection, markets or industry processing plants; amplification of DNA can be performed isothermally without the requirement of a PCR machine, with just a hot water bath, and positive and negative results are easily and visually seen after few minutes after loading the labeled amplified DNA (see Figure 4.2).

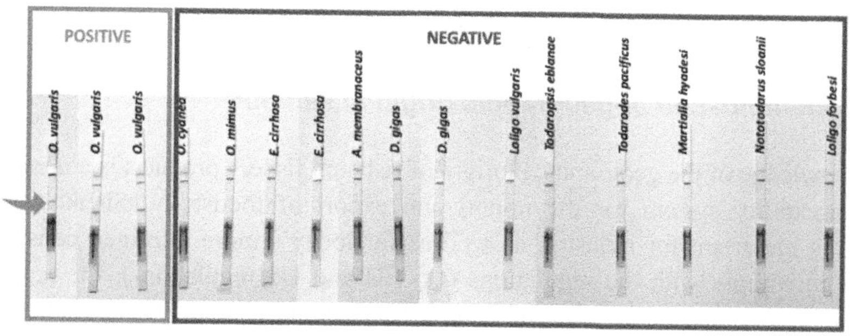

Figure 4.3. *RPA-LFDA detection of the common octopus (Octopus vulgaris). Positive signal band (a green arrow is only seen with DNA samples obtained from Octopus vulgaris; other cephalopod species did not show this band. All samples showed the control band*

4.3.1.2.6. Mass spectrometry methods (MALDI/TOF)

Current mass spectrometry techniques (mainly matrix-assisted laser desorption/ionization-time-of-flight (MALDI-TOF) and electrospray-ion trap (ESI-IT) MS) have been tested with both global (whole protein) and specific (peptide-based) approaches for fish and shellfish identification. Using this latter approach, it has been proven useful for Penaeid shrimp identification (peptide mass finger printing spectra of arginine kinase or tropomyosin, mainly).

MALDI-TOF mass spectrometry has shown its reliability to identify Gadidae and Pleuronectiformes fish based on the signals generated by parvalbumins on muscular protein extracts. In Germany, competent authorities and private companies are currently establishing reference databases for MALDI-TOF profiles for fish species identification (Spielman et al. 2017; Stahl and Schröder 2017; Rau et al. 2021; Valletta et al. 2021). This approach also shows limited application in the identification of the time of storage, which could be useful for freshness evaluation of animal species.

In contrast with these techniques which require sample preparation, ambient mass spectrometry (AMS) needs minimal or no sample preparation, and solid samples can be analyzed. Rapid evaporative ionization mass spectrometry (REIMS), one of the newest forms of AMS, may provide results nearly instantaneously (2–3s). Recently, REIMS has demonstrated its potential to determine the sample species in real time, and additionally, its possibility to distinguish between different catch methods within a species (ability to distinguish line and trawl caught haddock fish), which is also an aspect of fish fraud (Black et al. 2017). Other traceability aspects such as geographic origin and wild/farmed could also be differentiated by this last innovative mass spectrometry technique.

4.3.2. *Identification of geographic origin in seafood*

Knowledge of the geographical origin of fish and fishery products is a demand of the traceability system for the import and export of foodstuffs (Sheikha 2015), which is important for industry, as well as for today's more informed consumers. Either to comply with the regulations (i.e. EU and US regulations), for economic reasons, because of the sustainability and conservation issues or fishery management concerns to protect overexploited and endangered species (Teletchea et al. 2005; Ensing et al. 2013), several methods for determining geographic origin of species have been used in the past decade. These methods can be categorized into physicochemical and biological techniques depending on the type of products used (Ercolini 2004), or can be different according to the state of the product, processed or non-processed (Pimentel et al. 2017). There are several limitations that have to be considered as methods cannot be applied to every processed products or to geographically close populations. Many authors agree that the success of any methodology used for the determination of the geographic origin of fish and other fishery products depends on solid environmental background information to compare sound baseline databases, quality reference materials of known origin, multivariate analysis and specific genetic software to support the results. Frequently, a combination of more than one method is used. Physicochemical techniques are based on chemical compounds that characterize the particular origin of fish (i.e. trace elements, fatty acids, stable isotopes) (Arechavala et al. 2011; Tanner et al. 2011; Stowasser et al. 2012). Biological techniques are based on the DNA and isoenzymes. Each method has a number of advantages or disadvantages depending on the state of the product, as summarized in Leal et al. (2015), and some may change with diet or modifications in the environment (Stowasser et al. 2012). Fatty acid profiles have been used for the identification of origin, but lipids are susceptible to oxidation and there is a temporal variability associated with changes in food availability and environmental conditions (Ezgeta-Balić et al. 2012), and so there is a potential obstacle for using this approach for traceability purposes.

4.3.2.1. *Trace element fingerprints*

Trace element fingerprints (TEF) in the hard structures of fishery products (shells, otoliths and bones) have been successfully used to determine their geographic origin (Standish et al. 2008). TEF availability varies in each ecosystem and reflects the geographic surroundings of the specimen throughout its life. Usually, the method uses ICP-MS (inductively coupled plasma mass spectrometry (LA-ICP-MS) to measure a wide range of element ratios like Ba/Ba, Cd/Ca, Cu/Ca, Mg/Ca and Mn/Ca. Ricardo et al. (2015) separated geographically close populations of cockles based on Ba and Mn, and Yamashita et al. (2008) canned tuna species, littleneck clams and chinook salmon from different locations. Some authors favor the use of scales, which are easily obtained (Wells et al. 2003). The limitations of using TEF are related to the need for reference ranges of the trace elements in foods and its relation with precise knowledge on their geographic origin (Danezis et al. 2016).

4.3.2.2. *Stable isotope analysis (SIA)*

Stable isotope analysis is a method that requires time-consuming sample preparation (Zhou et al. 2017), but it is used more and more to discriminate the geographic origin of seafood (Gopi et al. 2019). Stable isotopes are incorporated into animal tissues through food and water and vary according to the environmental conditions (Li et al. 2016). The more used ratios are H/H, C/C, O7O, N/N, S, S and the results are comparable between laboratories (Camin et al. 2017). They are generally analyzed by isotope ratio mass spectrometry (IRMS) and the ratio obtained comparing the value of the isotopes in its natural form. Scales and otoliths are used because they reflect the events that occurred during fish growth (Ashford and Jones 2007; Matta et al. 2013; Dixon et al. 2015). Tissues can also be tested (Moreno Rojas et al. 2007), but isotope ratios can vary within the same species because of changes in diet (Thomas and Cahoon 1993), with water discharges or even anthropogenic contamination (Kolarova and Napiórkowski 2021). Several authors use stable isotopes to authenticate fish species origin (Morrison et al. 2007;

Wolff et al. 2012), and recently Tulli et al. (2020), with a CF-E-A-IRMS (continuous flow elemental analyzer), and δ18O and δ2H, were able to obtain the origin of sea bass caught from different sites in Italy. Very rarely, a single element provides the full information, and in some cases, several stable isotopes need to be combined (Li et al. 2016). Some isotopes like δC and δN do not offer consensus in their use (Chaguri et al. 2017). For increased accuracy, stable isotopes are used together with TEF (Moreno-Rojas et al. 2008; Curtis et al. 2014). Multi-element and stable isotope analysis (δ13C and δ15N) have been widely used together when verifying the geographical origin of many food products. Different studies (Li et al. 2019) revealed that elemental profiling could identify the country of origin of shrimp from Thailand, India and Vietnam. Also, Zhang et al. (2017) used δ13C and δ15N together with fatty acid analysis to discriminate between different origins of the sea cucumber *A. japonicus*. However, Artetxe-Arrak et al. (2021) used the chemical composition of the otoliths of less than six months *K. pelamis* (skipjack tuna) from different areas and found that genetic analysis or others would be needed to distinguish their origin.

4.3.2.3. *DNA-based techniques*

Conventional PCR amplification followed by Sanger sequencing are widely used, as they are rather easy and cheap methodologies with several applications in fisheries and aquaculture research, with population genetics, stocks and linkage maps (Chistiakov et al. 2006) as examples. Markers like microsatellites, allozymes or mtDNA can be used to separate the effect of migration, genetic drift and local adaptation of populations, or in combination with more than one marker, and stay stable along the different seasons in a specific location. Also, Park et al. (2018) used a multiplex PCR using the 16S rRNA gene and two SNP differentiate *Corbicula japonica* from Korea, China and Japan. Cardeñosa et al. (2019) proposed species-specific primers to distinguish *A. anguilla* from other origin Anguilla species at early life stages and even in their transportation water. Western and eastern Pacific populations of the shark *Alopias pelagicus* were differentiated with mitochondrial DNA COI sequences and seven microsatellite loci (Cardeñosa et al. 2014).

Universally scientific methods for the determination of the geographic origin of a foodstuff do not exist (Boudinar et al. 2016). There are only indirect methods, which often have to be used together in order to increase their accuracy (Boudinar et al. 2016). Several of these methodologies seem promising and may increase their accuracy, as they are based on the existence of credible environmental information as it becomes available, though the high cost and availability of some equipment may prevent its use from being universal.

4.3.3. *Wild and farmed seafood discrimination*

Assessment of seafood origin is crucial in the perception of product quality by final consumers (Güney 2019). Fish supply can originate from wild fisheries or aquaculture. Thus, information of the origin of the seafood products is required by the European Regulation 1379/2013 (European Parliament and Council of the European Union 2013), so that consumers are better informed of the food purchases they make and the effect of production methods on nutrition, food quality and safety. As such, the mislabeling and sale of farmed fish as wild fish is a fraudulent practice. In addition, consumers may wish to have more informed choices that can involve knowledge of wild fish stock sustainability, the biotechnology used in production of farmed fish, the risks of chemical or heavy metal contamination, or the methods of postharvest handling. Despite the relevance of this control, there are no known official standardized methods at present. Techniques developed to discriminate the wild or farmed origin of fish have mainly used the external appearance and morphometry differences (Grigorakis 2007; Arechavala-Lopez et al. 2012), organoleptic characteristics (Šimat et al. 2012; Bodur 2018), quantification of differences between genetics (Louvotikis et al. 2011; Bylemans et al. 2016), chemical characteristics and fatty acid compositions (Attouchi and Sadok 2012; Šimat et al. 2012), proteomic profiles (Piovesana et al. 2016; Chiozzi et al. 2018; Guglielmetti et al. 2021), trace elements and pollutants (Ferreira et al. 2010; Arechavala-Lopez et al. 2016) and stable isotopes (Ottavian et al. 2012). The analysis of the fatty acids and/or lipids has been the most frequently used method, whereas stable isotope fingerprinting is the less used. Both methods share the highest accuracy in the results and also have the highest potential for differentiating wild and farmed fish.

Regarding the use of fatty acids and/or lipids, several studies have explored the differences in lipid content between farmed and wild fish (Attouchi and Sadok 2012; Šimat et al. 2012), which result from the high fat content in the farmed fish diet, intensive feeding regime and lower metabolic turnover. In addition, since the fish fatty acid (FA) profile is a function of the incorporation and accumulation of the diet FAs (Strobel et al. 2012), differences in the lipid composition and dietary FA profiles of the feed have been used to distinguish between wild and farmed fish. In fact, the rising shortage of marine resources rich in key dietary nutrients, like fish meal and fish oil, is leading to an increased use of vegetable oils and alternative plant-based ingredients in the fish feeds, like leaf meals, oilseed cakes and legume seeds (Berge et al. 2009). High levels of FAs of terrestrial origin such as oleic acid (OA, 18:1n-9), α-linolenic acid (LNA, 18:3n-3) or linoleic acid (LA, 18:2n-6), mostly present in marine fish at low levels, and the generally increased n-6/n-3 ratio in the flesh of farmed fish, have been regarded as evidence of the increased level of plant oils as alternative raw materials for fish feeds and, therefore, used to differentiate between wild and farmed fish (Fasolato et al. 2010). Also, significant differences between wild and farmed fish were shown in terms of arachidonic acid (ARA, 20:4n-6), eicosapentaenoic acid (EPA, 20:5n-3), docosahexaenoic acid (DHA, 22:6n-3) and DHA/ARA ratio (Ferreira et al. 2010; Sharma et al. 2010; Lenas et al. 2011; Šimat et al. 2012; Sciuto et al. 2019). Methods most frequently used in the determination of the FA profiles involve gas chromatography (Megdal et al. 2009), although other fat-related analytical methods such as near-infrared spectroscopy (Ottavian et al. 2012; Ghidini et al. 2019), nuclear magnetic resonance (NMR) (Mannina et al. 2008; Del Coco et al. 2009), comprehensive two-dimensional gas chromatography-mass spectrometry (Costa et al. 2017) and direct sampling analysis coupled with a time-of-flight mass spectrometer (Sciuto et al. 2019) have also been used to explore differences in the lipid profile.

The analysis of stable isotope ratio by isotope ratio mass spectrometry is a key method in trophic studies and farmed/wild fish differentiation. Major components of organic matter, like carbon (13C), nitrogen (15N) and oxygen (18O) isotopes, transfer through the food cycle from one organism to another and increase the isotope fingerprint via assimilation and growth. Thus, as a result of this process and according to the selected element, the isotope ratios may reflect, in the case of animals, the isotopic composition of the diet or feed which has been used (Bell et al. 2007; Chaguri et al. 2015). Since farmed and wild fish differ in terms of food availability and origin, and also because farmed fish feed has a lower quantity of marine raw materials and an increased component of vegetable meals and oils with a higher terrestrial carbon input and a lighter isotope content, these differences have been used to determine seafood authenticity. Using particularly the 13C/12C and 15N/14N ratios, different researchers have shown significant differences among wild and farmed fish isotope ratios (Fasolato et al. 2010; Gamboa-Delgado et al. 2014; Molkentin et al. 2015; Olsen et al. 2015) and likewise between wild, organically and conventionally farmed fish (Molkentin et al. 2007). Higher 15N isotopic contents were reported in wild fish muscle than in farmed seabream (Morrison et al. 2007) and seabass (Fasolato et al. 2010), allegedly due to the higher trophic level of wild fish marine feed, and also to the various inputs of terrestrial N sources in the farmed fish feed. Although there is adequate accuracy, effective cost and time advantages in relation to other methods, isotope analysis needs standardization and the development of databases of known isotope values or ranges of wild fish species for comparison. Likewise, as isotope ratios are influenced by different factors (e.g. food source, availability), these need to be further comprehensively and frequently updated.

Though characteristic FA signatures and diverse stable isotope ratios may characterize wild and farmed fish species, the use of multivariate analysis and chemometrics have been recommended as required data processing methods for analyzing the effects of all the variables analyzed in a more robust way, as it was shown in the case of the effective farmed/wild differentiation of saithe and cod (Fernandez-Jover et al. 2011), bogue (Arechavala-Lopez et al. 2010), sea bass (Fasolato et al. 2010; Ghidini et al. 2019) and salmon (Aursand et al. 2009). Despite the different methods used in the wild and farmed fish differentiation, the variability of the laboratory results has deterred a wider use in seafood authentication. Thus, in order to convert these methods into regulations, the consistency and accuracy of the methods need to be assured by a continuous monitoring of fatty acid/stable isotopic composition of both farmed and wild fish, and also by collaborative work targeted to the standardization of the methodologies.

4.3.4. *Detection of water addition, frozen–thawing and discoloration*

Threats to seafood authenticity involve not only the replacement of more valuable species, but also the undue addition of water, thawed seafood sold as unfrozen or the application of additives to prevent seafood color change, among others. The addition of water is done through treatment with phosphates or non-phosphate based additives before freezing or by injection, and though it is not an illegal practice, because it may have technological purposes, it becomes a problem if incorrectly labeled on the final package. Due to physiological reasons, strong correlations exist between protein and moisture levels in meat (Ang et al. 2006) and seafood (Breck 2014), and these were used in some meat standards and regulations. Supported by knowledge of baseline levels of proximate composition of unprocessed products, the moisture/protein ratio (M/P) has also been considered as a good indicator of water addition to seafood products (Seafish 2012; Mendes et al. 2017). However, non-destructive methods that can be executed in seconds are advantageous and have been researched for an expeditious determination of abusive practices. Bioelectrical impedance analysis (BIA), which measures the impedance (resistance and reactance) of a current passed through an organism or tissue (Cox and Hartman 2005), and spectroscopy techniques like near-infrared (NIR), which uses wavelength measurements in the electromagnetic spectrum, between 780 nm and 2500 nm (Huang et al. 2008), and the measurement of the product's dielectric properties in the microwave region as a function of frequency (100 MHz–10 GHz) (Kent and Daschner 2009), are among the most promising rapid methods under development.

BIA is a technique characterized by its sensitivity to changes in the volume of extracellular and cellular materials, and has been used in fish biology and fishery studies mainly focused on determining the body composition in fish. NIR spectroscopy has become extensively applied in the assessment of food quality and food integrity, including the determination of water content in seafood products (Erikson et al. 2012). In what concerns the dielectric properties of the muscle, changes in the amount of water in seafood were reported to affect the dielectric spectrum and its use suggested for control purposes (Mendes et al. 2018; Lee et al. 2019). Centered on this technology, Sequid GmbH (Germany) developed a patented system based on the principle of dielectric spectroscopy, namely the radio frequency quality scan (RFQ-Scan®) (Sequid 2021). The handheld RFQ-Scan® equipment detects the addition of added water, independently of used additives, and detection modules exist for various fish species (pangasius, Alaska pollock, redfish), with preliminary modules offered for others (plaice, Atlantic wolf fish, shrimp).

Differentiation between fresh and frozen–thawed seafood is critical for the control of (i) the illegal supply of thawed seafood as fresh unfrozen products and (ii) the non-compliance with regulations determining that fish to be consumed raw, such as sushi and sashimi will be frozen at -20°C for at least 24 hours (e.g. Regulations (EC) Nos. 178/2002, 853/2004 and 2074/2005) (European Parliament and Council of the European Union 2002, 2004, 2005) for parasite control. The proposed analytical methods can be categorized as enzymatic, physiological and physical. The enzymatic methods were the first to be developed and determined increases in the enzymatic activity of specific enzymes mainly in mitochondria and red blood cells, which leak into tissue exudate following cellular disruption occurring during freezing/thawing. Several disadvantages limited further developments, like the existence of parallel mitochondrial isoenzymes in the cytoplasm, the effect that other stresses imposed on the product (e.g. ice storage) had on the activity of highly sensitive differentiating enzymes (Erikson 2012) and the dependence on the fish species of the most suitable type of enzyme (mitochondrial, lysosomal or extracted from blood cells) used for authentication purposes (Rehbein and Cakli 2000). In terms of physiological methods, these have involved, namely the evaluation of the transparency of the fish eye lens, which becomes turbid or opaque after freezing (Okazaki et al. 2006), the determination of the hematocrit value as an index of blood cell destruction or the level of the integrity of red blood cells by microscopy (Rehbein 1992) and front-face fluorescence spectroscopy analysis of nicotinamide adenine dinucleotide (NADH) coupled with chemometrics (Karoui et al. 2017). The robustness of some of these methods must, however, be checked, as differences were not evident in species like plaice and eel (Duflos et al. 2002).

Physical methods can depend on changes in the electric resistance of tissues and the loss of impedance in fish induced by the freezing/thawing cycle, as used by the Intellectron Fischtester, the RT Fish Freshness Tester or the Torrymeter, which examine only optimum frequencies (Oehlenschläger 2003; Kent and Oehlenschläger 2009), or other bioelectrical impedance spectroscopy methods that analyze the electrical response over an entire frequency region from 1 Hz to 1 MHz (Fuentes et al. 2013). Other physical methods rely on the measurement of dielectric properties in the microwave region as a function of frequency (100 MHz to 10 GHz) and multivariate analysis of the spectra, obtained using principal component analysis (Kent et al. 2004; Mendes et al. 2017). On account of the NMR signal, usual sensitivity to water mobility and its interaction with other molecules, this spectroscopic method has also been considered appropriate for determining freezing and thawing changes in fish flesh in a noninvasive and nondestructive way (Shumilina et al. 2020). Nevertheless, high investment costs, the significant size of the equipment and associated infrastructures needed for operation strongly limit its use as a standard analytical tool. Other spectroscopic techniques have been successfully applied to determine freshness and detect thawed fish or other seafood, rapidly and non-destructively. The use of impedance spectroscopy (Sun et al. 2020), fluorescence (Karoui et al. 2017), visible/near-infrared, near-infrared, mid-infrared (Kimiya et al. 2013; Alamprese et al. 2015), hyperspectral imaging (HSI) and multispectral imaging (Feng et al. 2018), and Raman spectroscopy in combination with multivariate analysis were shown to be promising methods in fresh and frozen–thawed seafood differentiation. HSI and NMR are the most used techniques, followed by NIR and fluorescence spectroscopy (Hassoum et al. 2020). Overall, fresh and frozen–thawed fish can be differentiated by a wide range of technologies, though the development of fast, nondestructive and relatively inexpensive equipment and methods needs to continue, along with research on chemometrics, in order to fully validate the best-performing methodologies, and lastly allow the needed standardization among laboratories.

Other challenges to the control of seafood authenticity are related to the detection of additives unduly used in seafood coloring. Practices, like the injection of carbon monoxide (CO) or vegetable additives, have been used in order to keep the "fresh-looking" aspect of seafood and disguise malpractices, leading to oxidation and/or high levels of histamine in the product, as in the cases of tilapia, tuna and mahi-mahi (Djenane and Roncáles 2018). Methods available for the analysis of CO in food materials include spectrophotometric methods, infrared analysis and gas chromatography. Quantitative determination of carbon monoxide in seafood using GC/MS, following chemical liberation of CO into headspace, has been referred to as one of the most robust and dependable methods (Anderson and Wu 2005; Bartolucci et al. 2010; Droghetti et al. 2011). Still challenging and in need of research are methods for the rapid detection of injected plant extracts used, for example, to give tuna a red color so it looks fresher. Analysis of compounds of vegetable origin, like anthocyanins or betanin using RP-HPLC, HPLC-MS or MS/MS, may be explored (Silva et al. 2019; Demirbas 2020).

4.4. Needs and developments to improved traceability and authenticity for fish industries

4.4.1. *Standardization*

A free trade of seafood products needs rules and also harmonization to overcome trade barriers and simultaneously warranty consumer protection and interests. One essential aspect is to ensure the traceability and authenticity of traded seafood products, so that many prerequisites have to be in place before an appropriate system is established and working. Important frameworks have been established by the European law, such as Regulation (EC) No. 178/2002 laying down general traceability conceptions to prevent fraudulent and deceptive practices, Regulation (EU) No. 1379/2013 on the market organization of fishery and aquaculture products requiring essential labeling aspects, and Regulation (EU) 2017/625 requesting competent authorities to increase their activities on food fraud. Finally, methods and tools need to be developed in order to fulfil the requirements formulated in the European food law and food standards for the traceability and verification of goods. Fish and fishery products are at the top of the list of food where fraudulent practices have occurred, which has led to the publication of a considerable number of scientific papers on newly developed food authentication methods since 2005 (Ulberth 2020). However, before new analytical methods can be used by control laboratories, it is necessary that they undergo a validation and a standardization process to demonstrate that all required performance properties are met. If this is the case, results obtained can be considered as reliable and comparable to those generated by other laboratories using the same validated method. International standards developing organizations (SDOs) such as FAO/WHO Codex Alimentarius, ISO (International Organisation for Standards), AOAC International (Association of Official Agricultural Chemists) and CEN (European Committee for Standardization) make efforts to standardize promising methods for detecting food adulteration. The international trade of seafood products will benefit from these standardized methods, but today there is a significant lack in harmonization

(Ulberth 2020). To overcome this problem, general standard or guidance documents need to be adopted that demonstrate how methods can be validated and verified to find acceptance in control laboratories. As published authentication studies show that in most cases DNA-based methods are used for species identification or traceability aspects, a selection of standardization documents and publications on this topic have already been developed. For example, the ISO 20813 Standard for nucleic acid-based analysis on detection and identification of animal species in foods and food products contains general requirements and definitions, and should be considered during the validation process. Investigations regarding specificity, robustness, precision, LOD (limit of detection) and LOQ (limit of quantification), if applicable, are described, as well as explanations on interlaboratory studies. Attention should also be paid to the Codex Alimentarius Guidance Document CXG 74-2010, which can be downloaded free of charge. This also outlines performance criteria and validation of methods for the detection, identification and quantification of specific DNA sequences in food and, in addition, specific proteins.

Due to the fact that numerous real-time PCR methods were developed over the last two decades for the identification and quantification of GMOs, as well as for species identification in meat and fishery products, several publications and guidelines for single and multiplex in-house validation, or for full validation in a collaborative study, have been written (Bustin et al. 2009; BVL 2016, 2017; Broeder et al. 2014; Kang 2019). Finally, CEN/TS 17329 – Part 1, for validation of a binary real-time PCR method and Part 2 for performing a collaborative study, is the official response to the harmonization of this approach.

Also recently published was the "Guidance document on multiplex real-time PCR" from the European Network of GMO Laboratories, in which practical settings for in-house validation are comprehensibly described (Grohmann et al. 2021). Furthermore, harmonization efforts for DNA sequencing methods including NGS (next-generation sequencing) have been undertaken by ISO, by developing a general standard ISO/WD 22949-1.

An important initiation was the validation of an SOP (Standard Operating Procedure) for DNA-based differentiation of fish species using the gene marker COI by an international ring trial within the European Labelfish project (2012–2015, Interreg Atlantic Area; see: http://labelfish.eu). With the aim of standardization, the SOP was initially transferred into the Technical Specification CEN/TC13703 by combining it with the official German method for species identification of fish and fish products using *cytb* fragments As it is not possible to achieve standardized barcoding methods for all different animal taxa, a German initiative compiled a comprehensive collection of PCR protocols with the specification of primer sequences, tested species and original references, and published it as the Barcoding Table of Animal Species: BaTAnS (Matthes et al. 2020).

4.4.2. *Online iTool FISH-FIT*

To support the European control laboratories in detecting food fraud in seafood and aquaculture products and in validating activities of newly developed DNA-based methods, an online-based Fish Fraud Identification iTool (www.fish-fit.org) was developed within the Interreg Atlantic Area project SEATRACES (2018–2022; see: http://seatraces.eu). FISH-FIT offers a range of authentic material of different fish species, for example, of the families Gadidae, Merlucciidae, Scombridae, Pleuronectiformes and other taxonomic groups, i.e. cephalopods, bivalves and crustaceans, which can be requested by control laboratories for authenticity purposes. Each authentic material is associated with at least one DNA sequence of a gene marker, and the respective information about the material (e.g. origin) used the PCR protocol and primers to generate sequences. In addition, FISH-FIT contains DNA sequences of various gene markers such as *cytb*, COI, 16S rRNA, 18S rRNA and a control region that allows users to verify their own sequence using an alignment tool. For closely related species like tuna species, such a simple alignment is not sufficient for species identification. Therefore, collections of *cytb* sequences from tuna and other scombroid species with different base pair (bp) lengths (126 bp, 130 bp and 413 bp) are available for public control laboratories that can be used for species assignment with the help of an implemented neighbor-joining (NJ) tree construction tool. The use of short sequence fragments is particularly interesting for species identification in tuna cans, where the DNA sequences break down into short fragments < 200 bp during processing (Kappel et al. 2017). In addition to the *cytb* marker, a collection of sequences from the first internal transcribed spacer (ITS1) would also be desirable to consider the presence of introgression, which has been reported for three tuna species (Vinas and Tudela 2009). However, this cannot be done at the moment.

Another interesting option for users is the archive for DNA-based species identification methods such as the SOP for fish and fish products (Labelfish), and the SOP for crustaceans (SEATRACES) that have already been validated beside and for other methods that have proven reliable in practice. Links to useful websites on fish, crustaceans or mollusks such as FishBase (see: www.fishbase.de), SeaLifeBase (see: www.sealifebase.ca) or WoRMS (World Register of Marine Species) (see: www.marinespecies.org), as well as links to the barcoding table BaTAnS, CEN or ISO standards give the user the possibility to obtain comprehensive information.

4.5. References

Alamprese, C. and Casiraghi, E. (2015). Application of FT-NIR and FT-IR spectroscopy to fish fillet authentication. *LWT Food Sci. Technol.*, 63, 720–725.

Anderson, C.R. and Wu, W.H. (2005). Analysis of carbon monoxide in commercially treated tuna (*Thunnus spp.*) and mahi–mahi (*Coryphaena hippurus*) by gas chromatography/mass spectrometry. *J. Agric. Food Chem.*, 53, 7019–7023.

Ang, C.Y.W., Young, L., Wilson, R. (2006). Interrelationships of protein, fat and moisture content of broiler meat. *J. Food Sci.*, 49(2), 359–362.

Arechavala-Lopez, P., Sanchez-Jerez, P., Bayle-Sempere, J.T., Fernandez-Jover, D., Martinez-Rubio, L., Lopez-Jimenez, J.A., Martinez-Lopez, F.J. (2011). Direct interaction between wild fish aggregations at fish farms and fisheries activity at fishing grounds: A case study with *Boops boops*. *Aquac. Res.*, 42, 996–1010.

Arechavala-Lopez, P., Sanchez-Jerez, P., Bayle-Sempere, J.T., Sfakianakis, D.G., Somarakis, S. (2012). Discriminating farmed fish from wild Mediterranean stocks through scales and otoliths. *J. Fish Biol.*, 8, 2159–2175.

Arechavala-Lopez, P., Milošević-González, M., Sanchez-Jerez, P. (2016). Using trace elements in otoliths to discriminate between wild and farmed European sea bass (*Dicentrarchus labrax* L.) and gilthead sea bream (*Sparus aurata* L.). *Int. Aquat. Res.*, 8, 263–273.

Artetxe-Arrate, I., Fraile, I., Farley, J., Darnaude, A.M., Clear, N., Rodríguez-Ezpeleta, N., Dettman, D.L., Pécheyran, C., Krug, I., Médieu, A. et al. (2021). Otolith chemical fingerprints of skipjack tuna (*Katsuwonus pelamis*) in the Indian Ocean: First insights into stock structure delineation. *PLoS ONE*, 16(3), e0249327.

Asensio, L., González, I., Rodríguez, M.A., Hernández, P.E., García, T., Martín, R. (2004). PCR-ELISA for the semiquantitative detection of Nile perch (*Lates niloticus*) in sterilized fish muscle mixtures. *Journal of Agricultural and Food Chemistry*, 52(14), 4419–4422.

Ashford, J. and Jones, C. (2007). Oxygen and carbon stable isotopes in otoliths record spatial isolation of Patagonian toothfish (*Dissostichus eleginoides*). *Geochim. Cosmochim. Acta*, 71, 87–94.

Attouchi, M. and Sadok, S. (2012). The effects of essential oils addition on the quality of wild and farmed sea bream (*Sparus aurata*) stored in ice. *Food Bioprocess. Tech.*, 5, 1803–1816.

Aung, M.M. and Chang, Y.S. (2014). Traceability in a food supply chain: Safety and quality perspectives. *Food Control*, 39, 172–184.

Aursand, I.G., Veliyulin, E., Böcker, U., Ofstad, R., Rustad, T., Erikson, U. (2009). Water and salt distribution in Atlantic salmon (*Salmo salar*) studied by low-field ^{1}H NMR, ^{1}H and ^{23}Na MRI and light microscopy: Effects of raw material quality and brine salting. *Agric. Food Chem.*, 57(1), 46–54.

Bartlett, S.E. and Davidson, W.S. (1991). Identification of *Thunnus* tuna species by the polymerase chain-reaction and direct sequence-analysis of their mitochondrial cytochrome-b genes. *Can. J. Fish Aquat. Sci.*, 48(2), 309–317.

Bartolucci, G., Droghetti, E., Focardi, C., Bambagiotti-Alberti, M., Nocentini, M., Smulevich, G. (2010). High throughput headspace GC-MS quantitative method to measure the amount of carbon monoxide in treated tuna fish. *J. Mass Spectrom.*, 45, 1041–1045.

Bechini, A., Cimino, M.G.C.A., Marcelloni, F., Tomasi, A. (2008). Patterns and technologies for enabling supply chain traceability through collaborative e-business. *Inf. Sofw. Technol.*, 50, 342–359.

Bell, J.G., Preston, T., Henderson, R.J., Strachan, F., Bron, J.E., Cooper, K., Douglas, J.M. (2007). Discrimination of wild and cultured European sea bass (*Dicentrarchus labrax*) using chemical and isotopic analyses. *J. Agric. Food Chem.*, 55, 5934–5941.

Berge, G.M., Witten, P.E., Baeverfjord, G., Vegusdal, A., Wadsworth, S., Ruyter, B. (2009). Diets with different n-6/n-3 fatty acid ratio in diets for juvenile Atlantic salmon, effects on growth, body composition, bone development and eicosanoid production. *Aquaculture*, 296, 299–308.

Black, C., Chevallier, O.P., Haughey, S.A., Balog, J., Stead, S., Pringle, S.D., Riina, M.V., Martucci, F., Acutis, P.L., Morris, M. et al. (2017). A real time metabolomic profiling approach to detecting fish fraud using rapid evaporative ionisation mass spectrometry. *Metabolomics*, 13(12), 1–13.

Bodur, T. (2018). Morphological discrimination of farmed and wild gilthead sea bream (*Sparus aurata* L. 1753) populations of North-East Mediterranean. *Pak. J. Agri. Sci.*, 55(4), 941–947.

Borit, M. and Olsen, P. (2012). Evaluation framework for regulatory requirements related to data recording and traceability designed to prevent illegal, unreported and unregulated fishing. *Mar. Policy*, 36(1), 96–102.

Bottani, E. and Rizzi, A. (2008). Economical assessment of the impact of RFID technology and EPC system on the fast-moving consumer goods supply chain. *Int. J. Prod. Econ.*, 112, 548–569.

Boudinar, A.S., Chaoui, L., Quignard, J.P., Aurelle, D., Kara, M.H. (2016). Otolith shape analysis and mitochondrial DNA markers distinguish three sand smelt species in the *Atherina boyeri* species complex in western Mediterranean. *Estuar. Coast. Shelf. Sci.*, 182, 202–210.

Brenn, C., Schröder, U., Hanel, R., Martínez Arbizu, P. (2021). A multiplex real-time PCR screening assay for routine species identification of four commercially relevant crustaceans. *Food Control*, 125, 107986.

Broeder, S., Huber, I., Grohmann, L., Berben, G., Taverniers, I., Mazzara, M., Roosens, N., Morisset, D. (2014). Guidelines for validation of qualitative real-time PCR methods. *Trends Food Sci. Technol.*, 37(2), 115–126.

Bustin, S., Benes, V., Garson, J., Hellemans, J., Huggett, J., Kubista, M., Mueller, R., Nolan, T., Pfaffl, M., Shipley, G. et al. (2009). The MIQE guidelines: Minimum information for publication of quantitative real-time PCR experiments. *Clin. Chem.*, 55(4), 611–622.

But, G.W.C., Wu, H.Y., Shao, K.T., Shaw, P.C. (2020). Rapid detection of CITES-listed shark fin species by loop-mediated isothermal amplification assay with potential for field use. *Sci. Rep.*, 10(1), 1–14.

BVL (Federal Office of Consumer Protection and Food Safety, Germany) (2016). Guidelines for the single-laboratory validation of qualitative real-time PCR methods [Online]. Available at: http://www.bvl.bund.de.

BVL (Federal Office of Consumer Protection and Food Safety, Germany) (2017). Guidelines for the validation of qualitative real-time PCR methods by means of a collaborative study [Online]. Available at: http://www.bvl.bund.de.

Bylemans, J., Maes, G.E., Diopere, E., Cariani, A., Senn, H., Taylor, M.I., Helyar, S., Bargelloni, L., Bonaldo, A., Carvalho, G. et al. (2016). Evaluating genetic traceability methods for captive-bred marine fish and their applications in fisheries management and wildlife forensics. *Aquacult. Environ. Interact.*, 8, 131–145.

Cai, Y., He, Y., Lu, R., Chen, H., Wang, Q., Pan, L. (2017). Detection and quantification of beef and pork materials in meat products by duplex droplet digital PCR. *PlosOne* 12(8), e0181949.

Cao, W., Li, Y., Chen, X., Chang, Y., Li, L., Shi, L., Bai, W., Ye, L. (2019). Species identification and quantification of silver pomfret using the droplet digital PCR assay. *Food Chemistry*, 302, 125331.

Cardeñosa, D., Hyde, J., Caballero, S. (2014). Genetic diversity and population structure of the pelagic thresher shark (*Alopias pelagicus*) in the Pacific Ocean: Evidence for two evolutionarily significant units. *PLoS ONE*, 9, e110193.

Cardeñosa, D., Gollock, M., Chapman, D. (2019). Development and application of a novel real-time polymerase chain reaction assay to detect illegal trade of the European eel (*Anguilla anguilla*). *Conserv. Sci. Pract.*, 1(5), e39.

CEN/TS 17329-1:2019 (2019a). Foodstuffs – General guidelines for the validation of qualitative real-time PCR methods. Part 1: Single-laboratory validation [Online]. Available at: https://global.ihs.com.

CEN/TS 17329-2:2019 (2019b). Foodstuffs – General guidelines for the validation of qualitative real-time PCR methods. Part 2: Collaborative study [Online]. Available at: https://global.ihs.com.

Chaguri, M.P., Maulvault, A.L., Nunes, M.L., Santiago, D.A., Denadai, J.C., Fogaca, F.H., Sant'Ana, L.S., Ducatti, C., Bandarra, N., Carvalho, M.L. et al. (2015). Different tools to trace geographic origin and seasonality of croaker (*Micropogonias furnieri*). *LWT – Food Sci. Technol.*, 61, 194–200.

Chaguri, M.P., Maulvault, A.L., Costa, S., Gonçalves, A., Nunes, M.L., Carvalho, M.L., Sant'ana, L.S., Bandarra, B., Marques, A. (2017). Chemometrics tools to distinguish wild and farmed meagre (*Argyrosomus regius*). *J. Food Process. Preserv.*, 41(6), e13312.

Chen, C., Ding, Y., Wang, Y., Jiang, Q., Wang, F., Lu, C., Zhang, L., Zhu, C. (2021). High-resolution melting analysis of COI sequences distinguishes pufferfish species (*Takifugu* spp.) in China. *J. Agric. Food Chem.*, 69(2), 794–804.

Chiozzi, R.Z., Capriotti, A.L., Cavaliere, C., La Barbera, G., Montone, C.M., Piovesana, S., Laganà, A. (2018). Label free shotgun proteomics approach to characterize muscle tissue from farmed and wild European sea bass (*Dicentrarchus labrax*). *Food Anal. Method.*, 11(1), 292–301.

Chistiakov, D.A., Hellemans, B., Volckaert, F.A.M. (2006). Microsatellites and their genomic distribution, evolution, function and applications: A review with special reference to fish genetics. *Aquaculture*, 255, 1–29.

Chitipothu, S., Cariani, A., Bertasi, F., Stagioni, M., Kochzius, M., Blohm, D., Tinti, T., Landi, M. (2014). Invertebrate DNA chip: Opportunities and challenges in the development and application of microarrays for marine biodiversity studies. In *Microarrays: Principles, Applications and Technologies*, Rogers, J.V. (ed.). Nova Science Publishers, Hauppauge.

Commission of the European Communities (2005). Commission Regulation (EC) No 2074/2005 of 5 December 2005 laying down implementing measures for certain products under Regulation (EC) No 853/2004 of the European Parliament and of the Council and for the organisation of official controls under Regulation (EC) No 854/2004 of the European Parliament and of the Council and Regulation (EC) No 882/2004 of the European Parliament and of the Council, derogating from Regulation (EC) No 852/2004 of the European Parliament and of the Council and amending Regulations (EC) No 853/2004 and (EC) No 854/2004 (Text with EEA relevance) [Online]. Available at: http://data.europa.eu/eli/reg/2005/2074/oj.

Costa, R., Albergamo, A., Piparo, M., Zaccone, G., Capillo G., Manganaro, A., Dugo, P., Mondello, L. (2017). Multidimensional gas chromatographic techniques applied to the analysis of lipids from wild-caught and farmed marine species. *Eur. J. Lipid Sci. Technol.*, 119, 1600043.

Coutlée, F., Bobo, L., Mayur, K., Yolken, R.H., Viscidi, R.P. (1989). Immunodetection of DNA with biotinylated RNA probes: A study of reactivity of a monoclonal antibody to DNA-RNA hybrids. *Anal. Biochem.*, 181(1), 96–105.

Cox, M.K. and Hartman, K.J. (2005). Non-lethal estimation of proximate composition in fish. *Can. J. Fish. Aquat. Sci.*, 62(2), 269–275.

Curtis, J.M., Stunz, G.W., Overath, R.D., Vega, R.R. (2014). Otolith chemistry can discriminate signatures of hatchery-reared and wild spotted seatrout. *Fish. Res.*, 153, 31–40.

CXG-74-2010 (2010). Guidelines on performance criteria and validation of methods for detection, identification and quantification of specific DNA sequences and specific proteins in food. *Codex Alimentarius* [Online]. Available at: http://www.fao.org/fao-who-codexalimentarius/codex-texts/guidelines/en/.

D'Amico, P., Armani, A., Gianfaldoni, D., Guidi, A. (2016). New provisions for the labelling of fishery and aquaculture products: Difficulties in the implementation of Regulation (EU) no. 1379/2013. *Mar. Policy*, 71, 147–156.

Danezis, G.P., Tsagkaris, A.S., Camin, F., Brusic, V., Georgiou, C.A. (2016). Food authentication: Techniques, trends & emerging approaches. *Trends Anal. Chem.*, 85, 123–132.

Deconinck, D., Hostens, K., Taverniers, I., Volckaert, F.A.M., Robbens, J., Derycke, S. (2021). Identification and semi-quantification of Atlantic salmon in processed and mixed seafood products using droplet digital PCR (DdPCR). *Food Chem. Toxicol.*, 154, 112329.

Del Coco, L., Papadia, P., De Pascali, S.A., Bressani, G., Storelli, C., Zonno, V., Fanizzi, F.P. (2009). Comparison among different gilthead sea bream (*Sparus aurata*) farming systems: Activity of intestinal and hepatic enzymes and 13C-NMR analysis of lipids. *Forum Nutr.*, 1, 291–301.

Demirbas, A. (2020). Red cabbage extracts as inhibitors of lipid oxidation in fresh minced tilapia (*Nile perch*) during refrigerated storage. *Turk. J. Agric. Food Sci. Technol.*, 8(1), 81–88.

Dixon, H.J., Dempson, J.B., Power, M. (2015). Assessing the use of different marine growth zones of adult Atlantic salmon scales for studying marine trophic ecology with stable isotope analysis. *Fish. Res.*, 164, 112–119.

Djenane, D. and Roncalés, P. (2018). Carbon monoxide in meat and fish packaging: Advantages and limits. *Foods*, 7(2), 2.

Droghetti, E., Bartolucci, G.L., Focardi, C., Bambagiotti-Alberti, M., Nocentini, M., Smulevich, G. (2011). Development and validation of a quantitative spectrophotometric method to detect the amount of carbon monoxide in treated tuna fish. *Food Chem.*, 128, 1143–1151.

Druml, B. and Cichna-Markl, M. (2014). High resolution melting (HRM) analysis of DNA – Its role and potential in food analysis. *Food Chem.*, 158, 245–254.

Duflos, G., Le Fur, B., Mulak, V., Becel, P., Malle, P. (2002). Comparison of methods of differentiating between fresh and frozen–thawed fish or fillets. *J. Sci. Food Agric.*, 82, 1341–1345.

El Sheikha, A.F. (2015). New strategies for tracing foodstuffs: Biological barcodes utilizing PCR-DGGE. *Adv. Food Technol. Nutr. Sci. Open J.*, 1, S1–S7.

El Sheikha, A.F. and Montet, D. (2016). How to determine the geographical origin of seafood? *Crit. Rev. Food Sci. Nutr.*, 56(2), 306–317.

Ensing, D., Crozier, W.W., Boylan, P., O'Maoiléidigh, N., Mcginnity, P. (2013). An analysis of genetic stock identification on a small geographical scale using microsatellite markers, and its application in the management of a mixed-stock fishery for Atlantic salmon *Salmo salar* in Ireland. *J. Fish Biol.*, 82(6), 2080–2094.

EPCIS Standard (2007). EPC Information Services Version 1.0.1 Specification [Online]. Available at: http://www.gs1.org/gsmp/kc/epcglobal/epcis/epcis_1_0_1-standard-20070921.pdf.

Ercolini, D. (2004). PCR-DGGE fingerprinting: Novel strategies for detection of microbes in food. *J. Microbiol. Methods*, 56(3), 297–314.

Erickson, M.C. (2012). Chemical measurements. In *Handbook of Frozen Food Processing and Packaging*, Sun, D.W. (ed). CRC Press/Taylor & Francis Group, Boca Raton.

Erikson, U., Standal, I.B., Aursand, I.G., Veliyulina, E., Aursanda, M. (2012). Use of NMR in fish processing optimization: A review of recent progress. *Magn. Reson. Chem.*, 50, 471–480.

EUMOFA (2022). Country profile: European Union. Report, EUMOFA [Online]. Available at: https://www.eumofa.eu/documents/20178/61322/European+Union.pdf.

European Parliament and Council of the European Union (2002). Regulation (EC) No 178/2002 of the European Parliament and of the Council of 28 January 2002 laying down the general principles and requirements of food law, establishing the European Food Safety Authority and laying down procedures in matters of food safety [Online]. Available at: http://data.europa.eu/eli/reg/2002/178/oj.

European Parliament and Council of the European Union (2004). Regulation (EC) No 853/2004 of the European Parliament and of the Council of 29 April 2004 laying down specific hygiene rules for food of animal origin [Online]. Available at: http://data.europa.eu/eli/reg/2004/853/oj.

European Parliament and Council of the European Union (2013). Regulation (EU) No 1379/2013 of the European Parliament and of the Council of 11 December 2013 on the common organisation of the markets in fishery and aquaculture products [Online]. Available at: http://data.europa.eu/eli/reg/2013/1379/oj.

European Parliament and Council of the European Union (2017). Regulation (EU) 2017/625 of the European Parliament and of the Council of 15 March 2017 on official controls and other official activities performed to ensure the application of food and feed law, rules on animal health and welfare, plant health and plant protection products [Online]. Available at: http://data.europa.eu/eli/reg/2017/625/oj.

Ezgeta-Balić, D., Najdek, M., Peharda, M., Blažina, M. (2012). Seasonal fatty acid profile analysis to trace origin of food sources of four commercially important bivalves. *Aquaculture*, 334, 89–100.

FAO (2020). *The State of World Fisheries and Aquaculture*. FAO, Rome.

Fasolato, L., Novelli, E., Salmaso, L., Corain, L., Camin, F., Perini, M., Antonetti, P., Balzan, S. (2010). Application of nonparametric multivariate analyses to the authentication of wild and farmed European sea bass (*Dicentrarchus labrax*). Results of a survey on fish sampled in the retail trade. *J. Agric. Food Chem.*, 58, 10979–10988.

Feng, C.-H., Makino, Y., Oshita, S., García Martín, J.F. (2018). Hyperspectral imaging and multispectral imaging as the novel techniques for detecting defects in raw and processed meat products: Current state-of-the-art research advances. *Food Control*, 84, 165–176.

Fernandes, T.J.R., Silva, C.R., Costa, J., Oliveira, M.B.P.P., Mafra, I. (2017). High resolution melting analysis of a COI mini-barcode as a new approach for Penaeidae shrimp species discrimination. *Food Control*, 82, 8–17.

Fernandes, T.J.R., Costa, J., Oliveira, M.B.P.P., Mafra, I. (2018). COI barcode-HRM as a novel approach for the discrimination of hake species. *Fish. Res.*, 197, 50–59.

Fernandez-Jover, D., Martinez-Rubio, L., Sanchez-Jerez, P., Bayle-Sempere, J.T., Lopez-Jimenez, J.A., Martínez-Lopez, F.J., Pål-Arne, B., Uglem, I., Dempster, T. (2011). Waste feed from coastal fish farms: A trophic subsidy with compositional side-effects for wild gadoids. *Estuar. Coast. Shelf S.*, 91, 568–559.

Ferreira, M., Caetano, M., Antunes, P., Costa, J., Gil, O., Bandarra, N., Pousão-Ferreira, P., Vale, C., Reis-Henrique, M. (2010). Assessment of contaminants and biomarkers of exposure in wild and farmed sea bass. *Ecotox. Environ. Safe.*, 73, 579–588.

Fuentes, A., Masot, R., Fernández-Segovia, I., Ruiz-Rico, M., Alcañiz, M., Barat, J.M. (2013). Differentiation between fresh and frozen–thawed sea bream (*Sparus aurata*) using impedance spectroscopy techniques. *Innov. Food Sci. Emerg.*, 19, 210–217.

Gamboa-Delgado, J., Molina-Poveda, C., Godinez-Siordia, D.E., Villareal-Cavazos, D., Ricque-Marie, D., Cruz-Suarez, L.E. (2014). Application of stable isotope analysis to differentiate shrimp extracted by industrial fishing or produced through aquaculture practices. *Can. J. Fish. Aquat. Sci.*, 71, 1520–1528.

Ghidini, S., Varràa, M.O., Dall'Astaa, C., Badianib, A., Ianieria, A., Zanardia, E. (2019). Rapid authentication of European sea bass (*Dicentrarchus labrax* L.) according to production method, farming system, and geographical origin by near infrared spectroscopy coupled with chemometrics. *Food Chem.*, 280, 321–327.

Giusti, A., Armani, A., Sotelo, C.G. (2017). Advances in the analysis of complex food matrices: Species identification in surimi-based products using Next Generation Sequencing technologies. *PLoS ONE*, 12(10), e0185586.

Gopi, K., Mazumder, D., Sammut, J., Saintilan, N. (2019). Determining the provenance and authenticity of seafood: A review of current methodologies. *Trends Food Sci. Technol.*, 91, 294–304.

Griffiths, A.M., Sotelo, C.G., Mendes, R., Perez-Martin, R., Schröder, U., Shorten, M., Silva, H.A., Verrez-Bagnis, V., Mariani, S. (2014). Current methods for seafood authenticity testing in Europe, is there a need for harmonisation? *Food Control*, 45, 95–100.

Grigorakis, K. (2007). Compositional and organoleptic quality of farmed and wild gilthead sea bream (*Sparus aurata*) and sea bass (*Dicentrarchus labrax*) and factors affecting it: A review. *Aquaculture*, 272, 55–75.

Grohmann, L., Barbante, A., Eriksson, R., Gatto, F., Georgieva, T., Huber, I., Hulin, J., Köppel, R., Marchesi, U., Marmin, L. et al. (2021). Guidance document on multiplex real-time PCR methods. JRC Technical Reports [Online]. Available at: http://ec.europa.eu.

Guglielmetti, C., Brusadore, S., Sciuto, S., Esposito, G., Manfredi, M., Marengo, E., Bozzetta, E., Acutis, P.L., Mazza, M. (2021). Wild or farmed gilthead seabream (*Sparus aurata*)? How to distinguish between them by two-dimensional gel electrophoresis. *J. Food Prot.*, 84(4), 592–596.

Güney, O.I. (2019). Consumers' perceived differences between wild and farmed fish: A survey study in Turkey. *J. Aquat. Food Prod. T.*, 28(2), 210–220.

Haiminen, N., Edlund, S., Chambliss, D., Kunitomi, M., Weimer, B.C., Ganesan, B., Baker, R., Markwell, P., Davis, M., Huang, B.C. et al. (2019). Food authentication from shotgun sequencing reads with an application on high protein powders. *NPJ Sci. Food*, 3, 24.

Handy, S.M., Chizhikov, V., Yakes, B.J., Paul, S.Z., Deeds, J.R., Mossoba, M.M. (2014). Microarray chip development using infrared imaging for the identification of catfish species. *Appl. Spectrosc.*, 68(12), 1365–1373.

Hassoun, A., Shumilina, E., Di Donato, F., Foschi, M., Simal-Gandara, J., Biancolillo, A. (2020). Emerging techniques for differentiation of fresh and frozen–thawed seafoods: Highlighting the potential of spectroscopic techniques. *Molecules*, 25, 4472.

He, J. (2018). From country-of-origin labelling (COOL) to seafood import monitoring program (SIMP): How far can seafood traceability rules go? *Mar. Policy*, 96, 163–174.

Heather, J.M. and Chain, B.J. (2016). The sequence of sequencers: The history of sequencing DNA. *Genomics*, 107(1), 1–8.

Hellberg, R.S., Pollack, S., Hanner, R.H. (2016). Seafood species identification using DNA sequencing. In *Seafood Authenticity and Traceability: A DNA Perspective*, Naaum, A.M. and Hanner, R.H. (eds). Academic Press, Amsterdam.

Herrero, B., Madriñán, M., Vieites, J.M., Espiñeira, M. (2010). Authentication of Atlantic cod (*Gadus morhua*) using real time PCR. *J. Agric. Food Chem.*, 58(8), 4794–4799.

Higuchi, R., Fockler, C., Dollinger, G.W.R. (1993). Kinetic PCR analysis: Real-time monitoring of DNA amplification reactions. *Bio/Technology*, 11(9), 1026–1030.

Huang, G.Q., Lau, J.S.K., Mak, K.L. (2003). The impacts of sharing production information on supply chain dynamics: A review of the literature. *Int. J. Prod. Res.*, 41(7), 1483–1517.

Huang, H., Yu, H., Xu, H., Ying, Y. (2008). Near infrared spectroscopy for on/in-line monitoring of quality in foods and beverages: A review. *J. Food Eng.*, 87(3), 303–313.

ISO 20813:2019 (2019). Molecular biomarker analysis – Methods of analysis for the detection and identification of animal species in foods and food products (nucleic acid-based methods) – General requirements and definitions [Online]. Available at: www.iso.org.

ISO 22949-1:2021 (2021). Molecular biomarker analysis – Methods of analysis for the detection and identification of animal species in foods and food products (nucleotide sequencing-based methods) – Part 1: General requirements [Online]. Available at: www.iso.org.

Iwobi, A.N., Huber, I., Hauner, G., Miller, A., Busch, U. (2011). Biochip technology for the detection of animal species in meat products. *Food Anal. Methods*, 4(3), 389–398.

Jilberto, F., Araneda, C., Larraín, M.A. (2017). High resolution melting analysis for identification of commercially-important *Mytilus* species. *Food Chem.*, 229, 716–720.

Kang, T.S. (2019). Basic principles for developing real-time PCR methods used in food analysis: A review. *Trends Food Sic. Technol.*, 91, 574–585.

Kappel, K. and Schröder, U. (2015). Species identification of fishery products in Germany. *J. Consumer Prot. Food Safety*, 10, 31–34.

Kappel, K., Haase, I., Käppel, C., Sotelo, C.G., Schröder, U. (2017). Species identification in mixed tuna samples with next-generation sequencing targeting two short cytochrome *b* gene fragments. *Food Chem.*, 234(1), 212–219.

Kappel, K., Eschbach, E., Fischer, M., Fritsche, J. (2020). Design of a user-friendly and rapid DNA microarray assay for the authentication of ten important food fish species. *Food Chem.*, 311, 125884.

Karoui, R., Hassoun, A., Ethuin, P. (2017). Front face fluorescence spectroscopy enables rapid differentiation of fresh and frozen-thawed sea bass (*Dicentrarchus labrax*) fillets. *J. Food Eng.*, 202, 89–98.

Kent, M. and Daschner, F. (2009). Time domain spectroscopy. In *Fishery Products: Quality, Safety and Authenticity*, Rehbein, H. and Oehlenschlager, J. (eds). Blackwell, Oxford.

Kent, M. and Oehlenschläger, J. (2009). Measuring electrical properties. In *Fishery Products: Quality, Safety and Authenticity*, Rehbein, H. and Oehlenschläger, J. (eds). Blackwell, Oxford.

Kent, M., Oehlenschlager, J., Mierke-Klemeyer, S., Knöchel, R., Daschner, F., Schimmer, O. (2004). Estimation of the quality of frozen cod using a new instrumental method. *Eur. Food Res. Tech.*, 219, 540–544.

Kimiya, T., Sivertsen, A.H., Heia, K. (2013). VIS/NIR spectroscopy for non-destructive freshness assessment of Atlantic salmon (*Salmo salar* L.) fillets. *J. Food Eng.*, 116, 758–764.

Klapper, R. and Schröder, U. (2021). Verification of authenticity: A rapid identification method for commercial scallop species through multiplex real-time PCR. *Food Control*, 121(August), 107574.

Kobus, R., Abuín, J.M., Müller, A., Hellmann, S.L., Pichel, J.C., Pena, T.F., Hildebrandt, A., Hankeln, T., Schmidt, B. (2020). A big data approach to metagenomics for all-food-sequencing. *BMC Bioinform.*, 21(1), 1–15.

Kochzius, M., Nölte, M., Weber, H., Silkenbeumer, N., Hjörleifsdottir, S., Hreggvidsson, G.O., Marteinsson, V., Kappel, K., Planes, S., Tinti, F. et al. (2008). DNA microarrays for identifying fishes. *Mar. Biotechnol.*, 10(2), 207–217.

Kolarova, N. and Napiórkowski, P. (2021). Trace elements in aquatic environment. Origin, distribution, assessment and toxicity effect for the aquatic biota. *Ecohydrol. Hydrobiol.*, 21(4), 655–668.

Launey, S., Ledu, C., Boudry, P., Bonhomme, F., Naciri-Graven, Y. (2002). Geographic structure in the European Flat Oyster (*Ostrea edulis* L.) as revealed by microsatellite polymorphism. *J. Hered.*, 93, 331–351.

Leal, M.C., Pimentel, T., Ricardo, F., Rosa, R., Calado, R. (2015). Seafood traceability: Current needs, available tools, and biotechnological challenges for origin certification. *Trends Biotechnol.*, 33, 331–336.

Lee, D., Lee, S., Lee, C.J., Lee, S.H. (2019). Detection of artificially water-injected frozen octopus minor (Sasaki) using dielectric properties. *J. Chem. (Hindawi. Online)*, 8968351.

Lenas, D.S., Triantafillou, D.J., Chatziantoniou, S., Nathanailides, C. (2011). Fatty acid profile of wild and farmed gilthead sea bream (*Sparus aurata*). *J. Verbrauch. Lebensm.*, 6, 435–440.

Li, L., Boyd, C.E., Sun, Z. (2016). Authentication of fishery and aquaculture products by multi-element and stable isotope analysis. *Food Chem.*, 194, 1238–1244.

Li, L., Han, C., Dong, S., Boyd, C.E. (2019). Use of elemental profiling and isotopic signatures to differentiate Pacific white shrimp (*Litopenaeus vannamei*) from freshwater and seawater culture areas. *Food Control*, 95, 249–256.

Li, H., Xie, R., Yu, W., Wang, N., Chen, A. (2021). Rapid identification of cod and oil fish components based on loop-mediated isothermal amplification. *Aquaculture*, 737209.

Lopez, I. and Pardo, M.A. (2005). Application of relative quantification TaqMan real-time polymerase chain reaction technology for the identification and quantification of *Thunnus alalunga* and *Thunnus albacares*. *J. Agric. Food Chem.*, 53(11), 4554–4560.

Loukovitis, D., Sarropoulou, E., Vogiatzi, E., Tsigenopoulos, C.S., Kotoulas, G., Magoulas, A., Chatziplis, D. (2012). Genetic variation in farmed populations of the gilthead sea bream *Sparus aurata* in Greece using microsatellite DNA markers. *Aquac. Res.*, 43, 239–246.

Mannina, L., Sobolev, A.P., Capitani, D., Iaffaldano, N., Rosato, M.P., Ragni, P., Reale, A., Sorrentino, E., D'Amico, I., Coppola, R. (2008). NMR metabolic profiling of organic and aqueous sea bass extracts: Implications in the discrimination of wild and cultured sea bass. *Talanta*, 77, 433–444.

Matta, M.E., Orland, I.J., Ushikubo, T., Helser, T.E., Black, B.A., Valley, J.W. (2013). Otolith oxygen isotopes measured by high-precision secondary ion mass spectrometry reflect life history of a yellowfin sole (*Limanda aspera*). *Rapid Commun. Mass Spectrom.*, 27, 691–699.

Matthes, N., Pietsch, K., Rullmann, A., Näumann, G., Pöpping, B., Szabo, K. (2020). The barcoding table of animal species (BaTAnS): A new tool to select appropriate methods for animal species identification using DNA barcoding. *Mol. Biol. Rep.*, 47(8), 6457–6461.

Megdal, P.A., Craft, N.A., Handelman, G.J. (2009). A simplified method to distinguish farmed (*Salmo salar*) from wild salmon: Fatty acid ratios versus astaxanthin chiral isomers. *Lipids*, 44(6), 569–576.

Mendes, R., Vieira, H., Pereira, J., Teixeira, B. (2017). Water uptake and cooking losses in *Octopus vulgaris* during industrial and domestic processing. *LWT – Food Sci. Technol.*, 78, 8–15.

Mendes, R., Schimmer, O., Vieira, H., Pereira, J., Teixeira, B. (2018). Control of abusive water addition to *Octopus vulgaris* with non-destructive methods. *J. Sci. Food Agric.*, 98, 369–376.

Moe, T. (1998). Perspectives on traceability in food manufacture. *Trends in Food Science & Technology*, 9, 211–214.

Molkentin, J., Meisel, H., Lehmann, I., Rehbein, H. (2007). Identification of organically farmed Atlantic salmon by analysis of stable isotopes and fatty acids. *Eur. Food Res. Technol.*, 224, 535–543.

Molkentin, J., Lehmann, I., Ostermeyer, U., Rehbein, H. (2015). Traceability of organic fish – Authenticating the production origin of salmonids by chemical and isotopic analysis. *Food Control*, 53, 55–66.

Moreno-Rojas, J.M., Serra, F., Giani, I., Moretti, V.M., Reniero, F., Guillou, C. (2007). The use of stable isotope ratio analyses to discriminate wild and farmed gilthead sea bream (*Sparus aurata*). *Rapid Commun. Mass Spectrom.*, 21, 207–211.

Moreno-Rojas, J.M., Tulli, F., Messina, M., Tibaldi, E., Guillou, C. (2008). Stable isotope ratio analysis as a tool to discriminate between rainbow trout (*O. mykiss*) fed diets based on plant or fish-meal proteins. *Rapid Communications in Mass Spectrometry, An International Journal Devoted to the Rapid Dissemination of Up-to-the-Minute Research in Mass Spectrometry*, 22(23), 3706–3710.

Morrison, D.J., Preston, T., Bron J.E., Hemderson, R.J., Cooper, K., Strachan, F., Bell, J.G. (2007). Authenticating production origin of gilthead sea bream (*Sparus aurata*) by chemical and isotopic fingerprinting. *Lipids*, 42, 537–545.

Naaum, N.M., Warner, K., Mariani, S., Hanner, R.H., Drake Carolin, C. (2016). Seafood mislabeling incidence and impacts. In *Seafood Authenticity and Traceability*, Naaum, A.M. and Hanner, R.H. (eds). Academic Press, Amsterdam.

Noh, E.S., Lee, M.-N., Kim, E.M., Nam, B.-H., Noh, J.K., Park, J.Y., Kim, K.-H., Kang, J.-H. (2020). Discrimination of raw material species in mixed seafood products (surimi) using the next generation sequencing method. *Food Biosci.*, 41, 100786.

Oehlenschläger, J. (2003). Measurement of freshness of fish based on electrical properties. In *Quality of Fish from Catch to Consumer: Labelling, Monitoring and Traceability*, Luten, J.B., Oehlenschläger, J., Olafsdottir, G. (eds). Wageningen Academic Publishers, Wageningen.

Okazaki, E., Yamashita, Y., Uddin, M. (2006). Classification of fresh and frozen-thawed fish – A review. *Refrigeration*, 81, 175–181.

Olsen, P. and Borit, M. (2013). How to define traceability. *Trends Food Sci. Technol.*, 29, 142–150.

Olsen, S.A., Hansen, P.K., Givskud, H., Ervik, A., Samuelsen, O.B. (2015). Changes in fatty acid composition and stable isotope signature of Atlantic cod (*Gadus morhua*) in response to laboratory dietary shifts. *Aquaculture*, 435, 277–285.

Ottavian, M., Facco, P., Fasolato, L., Novelli, E., Mirisola, M., Perini, M., Barolo, M. (2012). Use of near-infrared spectroscopy for fast fraud detection in seafood: Application to the authentication of wild European sea bass (*Dicentrarchus labrax*). *J. Agr. Food Sci.*, 60, 639–648.

Park, Y.J., Noh, E.S., Park, J.Y., Kim, E.M., Noo, J.K., Choi, T-J., Kang, J-H. (2018). Development of a simple, rapid Multiplex PCR method using the 16S rRNA gene to determine the country of origin of brackish water bivalve *Corbicula Japonica*. *Food Anal. Methods*, 11, 3034–3041.

Pimentel, T., Marcelino, J., Ricardo, F., Soares, A.M.V.M., Calado, R. (2017). Bacterial communities 16S rDNA fingerprinting as a potential tracing tool for cultured seabass *Dicentrarchus labrax*. *Sci. Rep.*, 7, 11862.

Pinheiro, L.B., Coleman, V.A., Hindson, C.M., Herrmann, J., Hindson, B.J., Bhat, S., Emslie, K.R. (2012). Evaluation of a droplet digital polymerase chain reaction format for DNA copy number quantification. *Anal. Chem.*, 84(2), 1003–1011.

Piovesana, S., Capriotti, A.L., Caruso, G., Cavaliere, C., La Barbera, G., Chiozzi, R.Z., Laganà, A. (2016). Labeling and label free shotgun proteomics approaches to characterize muscle tissue from farmed and wild gilthead sea bream (*Sparus aurata*). *J. Chromatogr. A*, 1428, 193–201.

Rasmussen, R.S. and Morrissey, M.T. (2009). Application of DNA-based methods to identify fish and seafood substitution on the commercial market. *Compr. Rev. Food Sci. Food Saf.*, 8, 118–154.

Ratnasingham, S. and Hebert, P.D. (2007). Barcoding bold: The barcode of life data system. *Mol. Ecol. Not.*, 7(3), 355–364 [Online]. Available at: http://www.barcodinglife.org.

Rau, J., Eisenberg, T., Wind, C., Huber, I., Pavlovic, M., Becker, R. (2022). Aus der § 64 LFGB-Arbeitsgruppe MALDI-TOF: Leitlinien für die Validierung von Spezies-Identifizierungen mittels MALDI-TOF-MS. *J. Consum. Prot. Food Safety*, 17(1), 97–101.

Regattieri, A., Gamberi, M., Manzini, R. (2007). Traceability of food products: General framework and experimental evidence. *J. Food Eng.*, 81, 347–356.

Rehbein, H. (1992). Physical and biochemical methods for the differentiation between fresh and frozen-thawed fish or fillets. *Ital. J. Food Sci.*, 2, 75–86.

Rehbein, H. and Cakli, S. (2000). The lysosomal enzyme activities of fresh, cooled, frozen and smoked salmon fish (*Onchorhyncus keta* and *Salmo salar*). *Turk. J. Vet. Anim. Sci.*, 24, 103–108.

Ricardo, F., Pimentel, T., Moreira, A., Rey, F., Coimbra, M., Domingues M., Domingues, P., Costa Leal, M., Calado, R. (2015). Potential use of fatty acid profiles of the adductor muscle of cockles (*Cerastoderma edule*) for traceability of collection site. *Sci. Rep.*, 5.

Ripp, F., Krombholz, C.F., Liu, Y., Weber, M., Schäfer, A., Schmidt, B., Köppel, R., Hankeln, T. (2014). All-Food-Seq (AFS): A quantifiable screen for species in biological samples by deep DNA sequencing. *BMC Genom.*, 15, 639.

Sánchez, A., Quinteiro, J., Rey-Mendez, M., Perez-Martin, R.I., Sotelo, C.G. (2009). Identification of European hake species (*Merluccius merluccius*) using real-time PCR. *J. Agric. Food Chem.*, 57(9), 3397–3403.

Sanders, R., Hugget, J.F., Bushell, C.A., Cowen, S., Scott, D.J., Foy, C.A. (2011). Evaluation of digital PCR for absolute DNA quantification. *Anal. Chem.*, 83, 6474–6484.

Sanger, F., Nicklen, S., Coulson, A.R. (1977). DNA sequencing with chain-terminating inhibitors. *Proc. Natl. Acad. Sci. USA*, 74, 5463–5467.

Santaclara, F.J., Velasco, A., Pérez-Martín, R.I., Quinteiro, J., Rey-Méndez, M., Pardo, M.A., Jimenez, E., Sotelo, C.G. (2015). Development of a multiplex PCR-ELISA method for the genetic authentication of *Thunnus* species and *Katsuwonus pelamis* in food products. *Food Chem.*, 180(1), 9–16.

Saull, J., Duggan, C., Hobbs, G., Edwards, T. (2016). The detection of Atlantic cod (*Gadus morhua*) using loop mediated isothermal amplification in conjunction with a simplified DNA extraction process. *Food Control*, 59, 306–313.

Sciuto, S., Esposito, G., Guglielmetti, C., Mazza, M., Cocco, C., Ru, G., Acutis, P.L. (2019). Development of a novel method for rapid discrimination between wild and farmed sea bream. *J. Food Prot.*, 82(11), 1870–1873.

Seafish (2012). Seafish SR654: Review of polyphosphates as additives and testing methods for them in scallops and prawns [Online]. Available at: https://www.seafish.org/document/?id=7cf5af51-1499-47aa-b578-9346f26a95f7 [Accessed 4 August 2021].

Sequid (2021). Added water in fish and meat [Online]. Available at: https://www.sequid.de/en/products/rfq-scan-en/added-water [Accessed 4 August 2021].

Shanahan, C., Kernan, B., Ayalew, G., McDonnell, K., Butler, F., Ward, S. (2009). A framework for beef traceability from farm to slaughter using global standards: An Irish perspective. *Comput. Electr. Agric.*, 66(1), 62–69.

Sharma, P., Kumar, V., Sinha, A.K., Ranjan, J., Kithsiri, H.M., Venkateshwarlu, G. (2010). Comparative fatty acid profiles of wild and farmed tropical freshwater fish rohu (*Labeo rohita*). *Fish Physiol. Biochem.*, 36, 411–417.

Shehata, H.R., Li, J., Chen, S., Redda, H., Cheng, S., Tabujara, N., Li, H., Warriner, K., Hanner, R. (2017). Droplet digital polymerase chain reaction (ddPCR) assays integrated with an internal control for quantification of bovine, porcine, chicken and turkey species in food and feed. *PLoS ONE*, 12(8), e0182872.

Shumilina, E., Møller, I.A., Dikiy, A. (2020). Differentiation of fresh and thawed Atlantic salmon using NMR metabolomics. *Food Chem.*, 314, 126227.

Silva, D.V.T., Baião, D.S., Silva, F.O., Alves, G., Perrone, D., Del Aguila, E.M., Paschoalin, V.M.F. (2019). Betanin, a natural food additive: Stability, bioavailability, antioxidant and preservative ability assessments. *Molecules*, 24(3), 458.

Šimat, V., Bogdanović, T., Krželj, M., Soldo, A., Maršić-Lučić, J. (2012). Differences in chemical, physical and sensory properties during shelf life assessment of wild and farmed gilthead sea bream (*Sparus aurata*, L.). *J. Appl. Ichthyol.*, 28, 95–101.

Spielmann, G., Huber, I., Maggipinto, M., Haszprunar, G., Busch, U., Pavlovic, M. (2017). Comparison of five preparatory protocols for fish species identification using MALDI-TOF MS. *Eur. Food Res. Technol.*, 244(4), 685–694.

Spielmann, G., Ziegler, S., Haszprunar, G., Busch, U., Huber, I., Pavlovic, M. (2019). Using loop-mediated isothermal amplification for fast species delimitation in eels (genus Anguilla), with special reference to the European eel (*Anguilla anguilla*). *Food Control*, 101, 156–162.

Stahl, A. and Schröder, U. (2017). Development of a MALDI–TOF MS-based protein fingerprint database of common food fish allowing fast and reliable identification of fraud and substitution. *J. Agric. Food Chem.*, 65(34), 7519–7527.

Standish, J.D., Sheehy, M., Warner, R.R. (2008). Use of otolith natal elemental signatures as natural tags to evaluate connectivity among open-coast fish populations. *Mar. Ecol. Prog. Ser.*, 356, 259–268.

Stowasser, G., Pond, D.W., Collins, M.A. (2012). Fatty acid trophic markers elucidate resource partitioning within the demersal fish community of South Georgia and Shag Rocks (Southern Ocean). *Mar. Biol.*, 159, 2299–2310.

Strobel, C., Jahreis, G., Kuhnt, K. (2012). Survey of n-3 and n-6 polyunsaturated fatty acids in fish and fish products. *Lipids Health Dis.*, 11, 144.

Sun, Z., Liang, L., Li, J., Liu, X., Sun, J., Zou, X., Zuo, M., Guo, Z. (2020). Rapid detection of Atlantic salmon multi-quality based on impedance properties. *Food Sci., Nutr.*, 8, 862–869.

Taboada, L., Sánchez, A., Pérez-Martín, R.I., Sotelo, C.G. (2017). A new method for the rapid detection of Atlantic cod (*Gadus morhua*), Pacific cod (*Gadus macrocephalus*), Alaska pollock (*Gadus chalcogrammus*) and ling (*Molva molva*) using a lateral flow dipstick assay. *Food Chem.*, 233, 182–189.

Tanner, D.K., Brazner, J.C., Brady, V.J. (2011). Factors influencing carbon, nitrogen, and phosphorus content of fish from a Lake Superior coastal wetland. *Can. J. Fish. Aquat. Sci.*, 57(6), 1243–1251.

Teletchea, F., Maudet, C., Hänni, C. (2005). Food and forensic molecular identification: Update and challenges. *Trends Biotechnol.*, 23, 359–366.

Teletchea, F., Bernillon, J., Duffraisse, M., Laudet, V., Hanni, C. (2008). Molecular identification of vertebrate species by oligonucleotide microarray in food and forensic samples. *J. Appl. Ecol.*, 45(3), 967–975.

Thakur, M. and Gunnlaugsson, V.N. (2018). Information sharing strategies in whitefish supply chains in Norway vs. Iceland: Impact on supply chain decision making. *Int. J. Food Sys. Dyn.*, 9(3), 240–252.

Thomas, C.J. and Cahoon, L.B. (1993). Stable isotope analyses differentiate between different trophic pathways supporting rocky-reef fishes. *Mar. Ecol. Prog. Ser.*, 95, 19–24.

Tulli, F., Moreno-Rojas, J.M., Messina, C.M., Trocino, A., Xiccato, G., Muñoz-Redondo, J.M., Santulli, A., Tibaldi, E. (2020). The use of stable isotope ratio analysis to trace European sea bass (*D. labrax*) originating from different farming systems. *Animals,* 10(11), 2042.

Ulberth, F. (2020). Tools to combat food fraud – A gap analysis. *Food Chem.*, 330, 127044.

US Food and Drug Administration (2002). The Bioterrorism Act of 2002.

Valletta, M., Ragucci, S., Landi, N., Di Maro, A., Pedone, P.V., Russo, R., Chambery, A. (2021). Mass spectrometry-based protein and peptide profiling for food frauds, traceability and authenticity assessment. *Food Chem.*, 365, 130456.

Varunjikar, M.S., Moreno-Ibarguen, C., Andrade-Martinez, J.S., Tung, H.-S., Belghit, I., Palmblad, M., Olsvik, P.A., Reyes, A., Rasinger, J.D., Lie, K.K. (2021). Comparing novel shotgun DNA sequencing and state-of-the-art proteomics approaches for authentication of fish species in mixed samples. *Food Control*, 131, 108417.

Velasco, A., Ramilo-Fernández, G., Sotelo, C.G. (2020). A real-time PCR method for the authentication of common cuttlefish (*Sepia officinalis*) in food products. *Foods*, 9(3), 286.

Velasco, A., Ramilo-Fernández, G., Denis, F., Oliveira, L., Shum, P., Silva, H., Sotelo, C.G. (2021). A new rapid method for the authentication of common octopus (*Octopus vulgaris*) in seafood products using recombinase polymerase amplification (RPA) and lateral flow assay (LFA). *Foods*, 10, 1825.

Viðarsson, J.R., Gregersen, Ó., Olsen, P., Gunnlaugsson, V.N., Djurhuus, D.R., Finnbjörnsson, Þ., Pétursdóttir, A.M. (2015). WhiteFishMaLL: North Atlantic whitefish marine living lab report. *Nordic Innovation Publication*, 86.

Viñas, J. and Tudela, S. (2009) A validated methodology for genetic identification of tuna species (Genus *Thunnus*). *PLOS ONE*, 4(10), e7606.

Vogelstein, B. and Kinzler, K.W. (1999). Digital PCR. *Proc. Natl. Acad. Sci. USA*, 96, 9236–9241.

Vythalingam, L.M., Hossain, M.M., Bhassu, S. (2021). Rapid in-situ detection kit (RisK): Development of loop-mediated isothermal amplification (LAMP) assay for the rapid identification of selected invasive alien fish in Malaysian freshwaters. *Mol. Cell. Probes*, 55, 101683.

Wang, Y., Feng, J., Tian, X. (2019). Application of loop-mediated isothermal amplification (LAMP) for rapid detection of Atlantic cod (*Gadus morhua*), Pacific cod (*Gadus macrocephalus)* and haddock (*Melanogrammus aeglefinus*). *Mol. Cell. Probes*, 47, 101420.

Wang, N., Xing, R.-R., Zhou, M.-Y., Sun, R.-X., Han, J.-X., Zhang, J.-K., Zheng, W.-J., Chen, Y. (2021). Application of DNA barcoding and metabarcoding for species identification in salmon products. *Food Addit. Contam. Part A*, 38(5), 754–768.

Wells, B.K., Thorrold, S.R., Jones, C.M. (2003). Stability of elemental signatures in the scales of spawning weakfish, *Cynoscion Regalis*. *Can. J. Fish. Aquat. Sci.*, 60, 361–369.

Wheeler, D. and Bhagwat, M. (2007). BLAST QuickStart, example-driven Web-based BLAST tutorial. In *Comparative Genomics*, Volumes 1 and 2, Bergman, N.H. (ed.) Humana Press, Totowa [Online]. Available at: https://www.ncbi.nlm.nih.gov/books/NBK1734/.

Wolff, B.A., Johnson, B.M., Breton, A.R., Martinez, P.J., Winkelman, D.L. (2012). Origins of invasive piscivores determined from the strontium isotope ratio (87Sr/86Sr) of otoliths. *Can. J. Fish. Aquat. Sci.*, 69(4), 724–739.

World Economic Forum (2019). Innovation with a purpose: Improving traceability in food value chains through technology innovations [Online]. Available at: https://www3.weforum.org/docs/WEF_Traceability_in_food_value_chains_Digital.pdf.

Xiong, X., Huang, M., Xu, W., Li, Y., Cao, M., Xiong, X. (2021a). Using real time fluorescence loop-mediated isothermal amplification for rapid species authentication of Atlantic salmon (*Salmo salar*). *J. Food Compos. Anal.*, 95, 103659.

Xiong, X., Xu, W., Guo, L., An, J., Huang, L., Qian, H., Cui, X., Li, Y., Cao, M., Xiong, X. et al. (2021b). Development of loop-mediated isothermal amplification (LAMP) assay for rapid screening of skipjack tuna (*Katsuwonus pelamis*) in processed fish products. *J. Food Compos. Anal.*, 104038.

Yamashita, M., Namikoshi, A., Iguchi, J., Takashima, Y., Hossain, M.A., Yabu, T., Yamashita, Y. (2008). Molecular identification of species and the geographic origin of seafood. In *Fisheries for Global Welfare and Environment, 5th World Fisheries Congress*, Tsukamoto, K., Kawamura, T., Takeuchi, T., Beard Jr., T.D., Kaiser, M.J. (eds) [Online]. Available at: https://onlinelibrary.wiley.com/doi/10.1111/j.1095-8649.2009.02441_1.x.

Zhang, X., Liu, Y., Li, Y., Zhao, X. (2017). Identification of the geographical origins of sea cucumber (*Apostichopus japonicus*) in northern China by using stable isotope ratios and fatty acid profiles. *Food Chem.*, 218, 269–276.

Zhou, Y., Zhang, D., Peatman, E., Rhodes, M.A., Liu, J., Davis, D.A. (2017). Effects of various levels of dietary copper supplementation with copper sulfate and copper hydroxychloride on Pacific white shrimp *Litopenaeus vannamei* performance and microbial communities. *Aquaculture*, 476, 94–105.

5

Bacterial Risks and Biopreservation of Seafood Products

Sabrina MACÉ, Delphine PASSERINI and Françoise LEROI

Ifremer, Nantes, France

5.1. Introduction

The flesh of a healthy, living fish is considered sterile because the immune system prevents the growth of microorganisms. When the fish dies, its immune system collapses and the bacteria present can proliferate freely. The poikilothermic nature of fish leads to the growth of bacteria over a wide temperature range including psychrotolerant bacterial species capable of growing at very low temperatures. Fish is one of the most perishable foods on the market (Gram and Huss 1996) due to its relatively high post-mortem pH (> 6.0) and the availability of large amounts of non-protein nitrogenous compounds as a growth substrate (free amino acids, nucleotides, trimethylamine oxide (TMAO)).

The bacterial ecosystem of seafood products, also called microbiota, is composed of endogenous marine bacteria and those from the product processing environment. Each manipulation from slaughter to the consumer's table can influence the composition of the final product's microbiota (Sivertsvik et al. 2002). This microbiota includes different types of bacterial species which can be classified according to their role in food preservation. There are bacteria pathogenic to humans, bacteria responsible for the organoleptic degradation of the matrix and conversely, beneficial bacteria. A good knowledge of the microbiota and its

Current Challenges for the Aquatic Products Processing Industry,
coordinated by Véronique VERREZ-BAGNIS.

evolution during processing and storage is therefore necessary to develop strategies to (i) reduce the development of pathogens and thus limit the microbiological risk and (ii) delay sensory degradation to limit food waste. In this chapter, we will present the current knowledge concerning the study and characterization of seafood microbiota, and then summarize the advances concerning biopreservation, a mild preservation method based on the use of selected protective flora.

5.2. Microbiota, microbial risk and product quality

5.2.1. *Methods for studying the microbiota of seafood*

The microbiota of a given product changes over time and according to storage conditions. In order to follow the dynamics of the microbiota of seafood products and to correlate this information with the quality and spoilage of the product, different types of analyses are used in parallel for a combined or polyphasic approach:

– Classical microbiology, molecular biology and next generation sequencing (NGS) to follow the evolution of different bacterial flora.

– Biochemistry (total volatile basic nitrogen (TVBN), trimethylamine (TMA), biogenic amines, organic acids and volatile compounds) to determine the metabolites produced.

– Sensory analysis, an essential tool to determine the organoleptic quality and the spoilage level of products during storage.

Recently, the use of NGS methods has become widespread. This molecular approach is based on a high-throughput sequencing technology of all genes (metagenomics) or a target gene (metabarcoding). Metabarcoding targeting the 16S rRNA gene has been used in several studies to describe the bacterial genera and microbiota diversity of several seafood products such as salmon, shrimp, cod, yellowfin tuna, red drum, mussels or cuttlefish and blue crab (see Table 5.1). Nevertheless, the combination of cultural and acultural methods is essential. Indeed, although molecular methods allow for an increasingly precise description of bacterial diversity, they are not all quantitative. Moreover, only the isolation of strains enables their spoilage potential, biopreservative effect or other phenotypic characteristics of interest to be tested.

Many bacterial species are present in the microbiota. They co-evolve and interact during storage, impacting the degradation of products. During storage, some bacterial genera become dominant, and thus, the bacterial diversity of the product tends to decrease at the end of storage compared to the first days of production. Some bacterial genera are common to the microbiota of several products: Gram positive bacteria like *Brochothrix* or lactic acid bacteria (*Aerococcus, Carnobacterium*, *Lactobacilli, Lactococcus, Leuconostoc, Vagococcus*) as well as Gram negative bacteria like *Acinetobacter, Photobacterium, Pseudomonas, Psychrobacter, Shewanella, Serratia* and *Vibrio*.

In the coming decades, the influence of the microbiome and its impact on health/safety will be characterized along the food chain, in a continuum from the microbiome of the living fish in its environment, through the microbiome of the processed fish as food, to the microbiome of the consumer's gastrointestinal tract.

	Main bacterial genera (or family) identified during storage	**Storage (days, conditioning, T°C)**	**Reference**
Cod (*Gadus morhua*)	*Fusobacteriaceae* *Brochothrix* *Photobacterium* *Vagococcus* *Carnobacterium* *Clostridium* *Shewanella* *Acinetobacter*	D7; modified atmosphere; 4°C then 8°C	Chaillou et al. (2015)
Blue crab (*Callinectes sapidus*)	*Rhodobacteraceae* *Vibrionaceae*	D6 to D10; under air (live crabs); 4°C or 10°C	Parlapani et al. (2019)
Cooked and peeled shrimp	*Leuconostoc* *Streptococcus* *Vibrio* *Carnobacterium* *Weissela* *Aerococcus* *Trichococcus*	D15 to D49, modified atmosphere; 4°C then 8°C	Chaillou et al. (2015)
Mussels (*Mytilus galloprovincialis*)	*Synechococcus* *Shewanella* *Acidaminococcus* *Psychromonas* *Acinetobacter* *Psychrobacter*	D0 to D15; modified atmosphere, (depurated and not depurated); 4°C	Odeyemi et al. (2019)

Red drum (*Sciaenops ocellatus*)	*Carnobacterium* *Brochothrix* *Lactococcus* *Vagococcus* *Arthrobacter* *Shewanella* *Leuconostoc* *Pseudomonas*	D8 to D29; vacuum, modified atmosphere; 4°C	Silbande et al. (2018a)
Fresh salmon (*Salmo salar*)	*Carnobacterium* Lactobacilli *Photobacterium* *Lactococcus* *Brochothrix* *Vagococcus* *Flavobacterium*	D9-J11; modified atmosphere 4°C then 8°C	Chaillou et al. (2015)
Smoked salmon (*Salmo salar*)	*Photobacterium* *Carnobacterium* *Brochothrix* *Serratia* *Psychrobacter* *Shewanella* Lactobacilli *Lactococcus* *Vibrio* *Salinivibrio* *Aliivibrio* *Pantoea* *Enterococcus* *Staphylococcus* *Pseudomonas*	D0 to D28; D21 to D57; under vacuum; 4°C then 8°C	Chaillou et al. (2015) and Maillet et al. (2021)
Gravlax salmon (*Salmo salar*)	*Photobacterium* *Serratia/Yersinia* *Vibrio* *Lactococcus* *Carnobacterium* lactobacilli *Aerococcus*	D14 to D21; under vacuum at 8°C	Wiernasz et al. (2020)
Cuttlefish (*Sepia officinalis*)	*Psychrobacter* *Pseudomonas*	D0 to D8; under air; 2°C	Parlapani et al. (2018)
Yellowfin tuna (*Thunnus albacares*)	*Brochothrix* *Pseudomonas* *Hafnia*	D6 to D13; on ice; 0°C, vacuum, modified atmosphere; 4°C	Silbande et al. (2016)

Table 5.1. *Microbiota from different seafood products determined by metabarcoding analyses targeting regions of the 16S rRNA gene*

5.2.2. *Pathogenic bacteria and microbiological risks*

The identification of bacterial flora provides information on the presence of pathogenic bacteria and the hygienic quality of the product during processing. Food business operators have to withdraw unsafe food from the market. Harmonized safety criteria for the acceptability of foodstuffs – in particular with regard to the presence of certain pathogenic microorganisms, toxins or metabolites – are defined in the European Regulation (EC) No. 2073/2005. For seafood products, there are only four safety criteria: *Listeria monocytogenes* for ready-to-eat foods, *Salmonella* for cooked crustaceans and cooked or live bivalve mollusks, *Escherichia coli* for live bivalve mollusks, echinoderms, tunicates and gastropods and histamine for fish with a high histidine content (see Table 5.2). However, other pathogenic bacteria can cause food poisoning, and these risks should be known and managed for consumer safety.

5.2.2.1. *Vibrio*

Species of the genus *Vibrio* are Gram-negative, mesophilic bacteria present in marine and estuarine environments. They are also found in the intestine and on the tissues of fish, crustaceans and shellfish. Some species of non-cholera vibrios (*Vibrio parahaemolyticus*, *Vibrio vulnificus* and *Vibrio cholerae* (non-O1/non-O139 serogroups) can cause food poisoning, and *V. cholerae* O1 or O139 can cause cholera. The main cause of transmission is the consumption of fish and seafood, especially shellfish (e.g. oysters) and crustaceans (e.g. tropical shrimp) eaten raw, undercooked or recontaminated after cooking. The concentration of *Vibrio* varies according to the surface temperature of the water, salinity, turbidity, pH or chlorophyll A content. Maximum abundance is observed during the warmer months of the year and in warm seawater (Anses 2019a).

5.2.2.2. *Clostridium botulinum type E*

C. botulinum is a Gram-positive, spore-forming, strict anaerobic bacillus. It is responsible for severe food poisoning in humans and animals (botulism) by ingestion of preformed toxin in food (intoxication) or by intestinal colonization and toxin production in situ (toxi-infection). There are seven types of botulism toxins, from A to G, according to their immunological properties. Four of these forms (types A, B, E and rarely F) can cause human botulism. The foods most often implicated in outbreaks of type E toxin botulism are canned goods and homemade or artisanal products of salted and dried fish, fish marinades or vacuum-packed products (Anses 2019b).

5.2.2.3. *Listeria monocytogenes*

Widely present in water, soil and dust, able to grow at low temperatures (from –2°C), tolerant of acidic and saline conditions and the absence of oxygen (O_2), this pathogenic species can contaminate processing plants (Fonnesbech Vogel et al. 2001). Although cooking is successful in destroying it, it is sometimes found in ready-to-eat foods (Parihar et al. 2008; Gambarin et al. 2012). *L. monocytogenes* toxi-infections can be very serious, even fatal, in immuno-compromised patients and for the fetus in pregnant women (Todd and Notermans 2011; Anses 2020). It is found in the aquatic environment mainly on the skin, in the gills or digestive tract of fish. As a result, contamination of fish meat can occur during the various successive operations of slaughter, evisceration, filleting, etc. It is necessary to qualify suppliers (farms and slaughterhouses) and to monitor the presence of *L. monocytogenes* at reception and throughout the life of the product. Contamination can also occur in the plant, by humans or by contact with soiled surfaces, especially when *L. monocytogenes* forms biofilms which are difficult to remove once settled. *L. monocytogenes* is a regulatory safety criterion for ready-to-eat foods, described in the European regulation (EC) No. 2073/2005. There must be no *L. monocytogenes* in 25 g of product during production. A threshold of less than 100 CFU/g during the whole shelf life of the product can be accepted provided that the manufacturer is able to demonstrate that this limit will be respected. If these values are exceeded, a product withdrawal or recall is necessary.

5.2.2.4. *Histamine*

Histamine is a biogenic amine derived from histidine. The ingestion of foods with a high histamine content causes histamine intoxication (or scombroid intoxication). It is manifested by allergic-type reactions (facio-cervical redness, itching), neurological disorders (headaches, heart palpitations, dizziness) as well as secondary symptoms of a gastrointestinal nature (vomiting, diarrhea) (Union du mareyage français (UMF) 2010). Histamine is responsible for 30–40% of food-borne illnesses related to the consumption of fish products. It is produced in the muscles of fish species containing histidine. The animal is not ill because it is synthesized after the death of the fish through the action of an enzyme: the histidine decarboxylase. This enzyme is produced by mesophilic enterobacteria and also by psychrotolerant species such as *Photobacterium phosphoreum* and *Morganella psychrotolerans* which are able to generate histamine at low temperatures (as low as 2°C). Histamine is a safety criterion of the Regulation (EC) No. 2073/2005 for fishery products made from fish species rich in histidine: an average value of 100 mg/kg must not be exceeded during the conservation of the product. This value increases to 200 mg/kg for the products that have undergone enzymatic maturation in brine and to 400 mg/kg for fish sauces produced by fermentation of fish products.

5.2.2.5. *Other bacterial hazards*

Other bacteria introduced via the environment contaminated by waste, often domestic or industrial, or via human or animal fecal contamination, must also be considered as a microbiological hazard.

5.2.2.5.1. *Salmonella*

Salmonella is a Gram-negative rod-shape bacteria. Salmonellosis is one of the main gastroenteric syndromes of bacterial origin in industrialized countries. The main reservoir of *Salmonella* is the gastrointestinal tract of mammals (pigs, cattle) and birds (domestic poultry). Some strains can also be isolated from other sources, such as cold-blooded animals (reptiles, turtles) and aquatic animals (mollusks, fish). Salmonella in animal feces can contaminate pastures, soils and water and survive for several months; thus, the environment can become a source of danger (Anses 2021). Bivalve mollusks have a concentrating role for bacteria in the environment: by filtering large quantities of water to ensure their respiration and nutrition, they retain particles, including bacteria, whether they are free or associated with absorbed food. The safety criterion of the regulation (EC) No. 2073/2005 indicates the absence of *Salmonella* in 25 g of product (cooked crustaceans or mollusks as well as live bivalves, echinoderms, tunicates and live gastropods).

5.2.2.5.2. *Escherichia coli*

Gram-negative, oxidase-negative rod-shaped bacteria, *Escherichia coli* is normally present in the intestinal microbiota and feces of humans and warm-blooded animals. Its presence in a food indicates a lack of hygiene during a food preparation step and a potential risk regarding the presence of pathogenic organisms. The actual presence of *E. coli* in live shellfish indicates a lack of control of the quality of the water used by establishments or of the decontamination processes implemented or even an initial contamination in the harvesting area. For live bivalve mollusks and live echinoderms, tunicates and gastropods, the limit for this safety criterion is 230 MPN/100 g of meat or intravalvular fluid (MPN = most probable number). For cooked shellfish products, *E. coli* is also a hygiene criterion with a limit of 1/g at the end of the manufacturing process (Regulation (EC) No. 2073/2005).

5.2.2.5.3. *Staphylococcus aureus*

Staphylococcus aureus is a microorganism that is not part of the natural microbiota of fish, but is present on mammalian skin and mucous membranes. Worker handling is the primary source of product contamination (Huss 1995). *S. aureus* produces heat-resistant toxins which can cause severe food poisoning. However, this bacterium does not grow at refrigeration temperatures (0°C–4°C) and cases of poisoning are often related to ready-to-eat products left at too high temperatures. *S. aureus* is a hygiene criterion of Regulation (EC) No. 2073/2005 with a limit of 100 CFU/g at the end of the manufacturing process for shucked and shelled cooked shellfish products.

	Foodstuffs	Limits	Applications
Listeria monocytogenes	Ready-to-eat seafood products in which the growth of *L. monocytogenes* is possible (smoked salmon, cooked peeled shrimp, etc.)	100 CFU/g	Product placed on the market till end of shelf life
		Absence in 25 g	Before the commodity has left the control of the operator who manufactured it
Salmonella	Cooked crustaceans and mollusks	Absence in 25 g	Product placed on the market till end of shelf life
	Live bivalve mollusks and echinoderms, tunicates and gastropods	Absence in 25 g	Product placed on the market till end of shelf life
Escherichia coli	Live bivalve mollusks and echinoderms, tunicates and gastropods	230 MPN/100 g of meat and/or interval fluid	Product placed on the market till end of shelf life
Histamine	Fish products made from fish species rich in histidine (tuna, mackerel)	100 mg/kg	Product placed on the market till end of shelf life
	Fish products that have undergone an enzymatic maturation treatment in brine, made from fish species associated with a high amount of histidine (salted anchovies, salted anchovy products, fish sauce)	200 mg/kg	
	Fish sauces produced by fermentation of fish products	400 mg/kg	

Table 5.2. *Safety criteria for seafood products from Regulation (EC) No. 2073/2005*

5.2.3. *Seafood spoilage bacteria*

Spoilage includes all changes that make a food product unfit for human consumption from a sensory point of view. The first steps of spoilage are from autolytic and/or chemical origin, but are then mainly related to bacterial metabolism responsible for bad odors and flavors. The metabolism of nitrogenous compounds in the meat of marine products by bacteria often generates the production of unpleasant odors. For example, the degradation of urea causes ammonia odors, and the reduction of TMAO to TMA contributes to the amine odors typical of rotten fish. TMAO is a molecule present in the flesh of marine fish that plays a role in osmoregulation (Huss 1995). In anaerobic conditions, the use of TMAO as a final electron acceptor by some bacteria facilitates their growth in the absence of oxygen. The degradation of sulfur amino-acid like cysteine and methionine causes sulfurous odors.

5.2.3.1. *The specific spoilage bacteria*

During product storage, some bacteria grow faster depending on various parameters such as temperature, protective atmosphere and microbial interactions. At the time of spoilage, the microbiota of the product is composed of: (i) Specific Spoilage Organisms (SSO) that produce metabolites responsible for unpleasant odors and (ii) bacteria that grow on the product but do not cause unpleasant changes (Gram and Huss 1996; Gram and Dalgaard 2002; Lyhs 2009). The growth of specific spoilage bacteria and the production of their metabolites can be used as indicators for determining the shelf life of a product.

Determining the bacteria responsible for spoilage of a given product is not an easy task and requires an approach combining several types of assays: sensory, chemical and microbiological analyses. The identification of these spoilers is based on the comparison of the spoilage characteristics of the naturally contaminated product with those induced on the product inoculated by spoilage microbiota isolates. These analyses can be performed on sterile media which are as close as possible to the food, such as fish juice, or during challenge tests on sterile or pauci-microbial products (containing as few microorganisms as possible). The final method is the one that comes closest to the naturally contaminated product. It has been shown that within the same species, different strains may not reveal the same spoilage potential. Hence, it is preferable to inoculate several strains of the same species during these challenge tests.

5.2.3.2. *Influence of conditioning on spoilage bacteria*

Food preservation methods can alter the microbiota naturally present on the product and influence the development of specific spoilage bacteria.

Spoilage of fresh marine fish from temperate waters, stored under air (e.g. on ice), is mainly caused by bacteria belonging to the "Shewanella putrefaciens-like" group, including *S. putrefaciens*, *S. baltica* or other related species. *Shewanella* produce H_2S and other sulfur compounds as well as hypoxanthine and TMA. Spoilage of tropical freshwater fish is often related to *Pseudomonas*. Many species produce typical sensory characteristics such as foul odors with "fruity and rotten" characteristics. *Pseudomonas* produce various spoilage metabolites such as esters, aldehydes, sulfur compounds like methylmercaptan, but no H_2S or TMA. These bacteria are indeed not able to reduce TMAO, and therefore, their growth is limited in the absence of oxygen. *Shewanella* spp. and *Pseudomonas* are also considered the spoilage bacteria of warm-water marine fish (Gram and Huss 1996; Koutsoumanis and Nychas 1999; Leroi 2014).

For fresh vacuum-packed fish containing TMAO, such as cod, *S. putrefaciens* and *P. phosphoreum* have been identified as the specific spoilage bacteria (Gram and Huss 1996).

The protective atmosphere enriched with CO_2 inhibits the growth of *Pseudomonas* and *S. putrefaciens*, while *P. phosphoreum* and lactic acid bacteria are more resistant to CO_2 (Gram and Dalgaard 2002). For fresh marine fish preserved with this type of conditioning, *P. phosphoreum* thus appears to be the main specific spoilage bacterium. It produces high amount of TMA and has been characterized as a specific spoilage bacterium in cod (Dalgaard et al. 1997) and salmon (Emborg et al. 2002; Macé et al. 2013). For cooked peeled shrimp or warm water fish packed under modified atmosphere, lactic acid bacteria such as *Carnobacterium maltaromaticum* associated with *Brochothrix thermosphacta* are responsible for spoilage. For cooked whole tropical shrimp, *C. maltaromatcum and S. baltica* have been identified as the main spoilers (Laursen et al. 2005, 2006; Mejlholm et al. 2005; Leisner et al. 2007; Jaffrès et al. 2011; Macé et al. 2014). Some lactic acid bacteria are responsible for the production of ammonia, H_2S and biogenic amines (Leroi 2010). *B. thermosphacta* is capable of producing ammonia, aldehydes (acetaldehyde, isovaleraldehyde) and ketones (diacetyl displays cheese-odor) (Dalgaard 2000; Jaffrès et al. 2011).

The addition of salt, slight acidification and vacuum storage for lightly preserved products, such as cold-smoked salmon, will inhibit aerobic Gram-negative bacteria. Under these conditions, the microbiota is dominated by lactic acid bacteria (lactobacilli and *Carnobacterium*) or other Gram-positive bacteria such as *B. thermosphacta* associated with fermentative Gram-negative bacteria such as *P. phosphoreum* and psychrotrophic enterobacteria (Leroi and Joffraud 2011; Chaillou et al. 2015; Wiernasz et al. 2021). *Latilactobacillus sakei*, *Companilactobacillus farciminis*, *Latilactobacillus curvatus*, *Serratia liquefaciens*-like, as well as some strains of *B. thermosphacta* and *P. phosphoreum* have been found to be highly detrimental to smoked salmon (Stohr et al. 2001). *L. sakei* was associated with the production of H_2S, acetic acid and esters and *S. liquefaciens* with the production of TMA and sulfur compounds. *P. phosphoreum* has been identified as the most significant producer of biogenic amines in smoked salmon, although enterobacteria and lactic acid bacteria are also capable of producing them.

Increasing the preservative pressure – for example, by acidifying with preservatives such as sorbate and benzoate for pickled herring – leads to the growth of lactic acid bacteria, including lactobacilli and yeast (Gram and Dalgaard 2002; Lyhs and Björkroth 2008).

5.2.3.3. *Influence of bacterial associations*

The observed variation of the microbiota diversity during storage testifies of numerous bacterial interactions. The dynamic of these interactions may influence the spoilage of a given matrix by impacting the nature and quantity of the metabolites synthesized. The interaction of certain species can be beneficial. In smoked salmon, co-inoculation of *L. sakei* and *S. liquefaciens-like*, two species that induce high spoilage when inoculated separately, slows the spoilage process (Joffraud et al. 2006). In tuna, the presence of *B. thermosphacta* or *P. phosphoreum* decreases the undesirable odor caused by *Hafnia alvei* in pure culture (sulfur/cabbage, pyrrolidine, fecal), probably due to a 100 times lower concentration of the latter when co-inoculated (Silbande et al. 2018b). Other interactions have little or no impact on the sensory evolution of the matrix. This is the case for *C. maltaromaticum/P. phosphoreum* or *C. maltaromaticum/B. thermosphacta* pairs in smoked salmon. Finally, it has been shown that the presence of two species can give rise to new odors. For example, simultaneous inoculation of *C. maltaromaticum* and *B. thermosphacta* generates an unpleasant "wet dog" odor not identified in the presence of the independently inoculated species on shrimp (Laursen et al. 2006). Similarly, two non-spoiling species such as *Vibrio* sp. and *B. thermosphacta* produce unpleasant odors in co-culture in smoked salmon (Joffraud et al. 2006).

5.3. Biopreservation of seafood products

Biopreservation is a gentle food preservation technology that consists of preventing the development of undesirable microorganisms (pathogens or spoilers) by introducing into the product harmless microorganisms selected for their antimicrobial properties (Pilet and Leroi 2011). Unlike fermentation where microorganisms have a technological role (milk is transformed into yogurt, cabbage into sauerkraut), biopreservation aims to preserve delicate products without changing their organoleptic and nutritional characteristics. It is a natural method that can replace the addition of additives or allow the scales of technological treatments to be reduced, such as heat, salt and smoke.

5.3.1. *Bacterial interactions*

In order to propose effective biopreservation strategies, it is necessary to understand the dynamics of bacterial communities in seafood. This aims to better predict the effect of the addition of bioprotective strains on the microbiota and thus on the sensory profile of the matrix. Bacterial interactions are driven by several mechanisms.

5.3.1.1. *Bacterial competition*

Some scientists suggest that competitions are prevalent in the natural environment (Foster and Bell 2012). Indeed, microorganisms wage a constant battle for access to resources. This is especially true in seafood, where carbon substrates are scarce and undiverse. Glucose, lactate and free amino acids are predominantly found (Jaaskelainen et al. 2019). The preponderance of certain species is determined by the needs of each individual and the access and rate of consumption of available nutrients. Therefore, microorganisms develop active strategies that allow them to be more competitive and reach resources faster. For example, they can increase their mobility, adhere to specific surfaces or synthesize antimicrobial compounds (Hibbing et al. 2010).

Gram-negative bacteria are a rich source of antimicrobial compounds (Hu et al. 2015; Ng et al. 2015; Masschelein et al. 2017; Carroll et al. 2020). They belong to different chemical classes such as polyketides, non-ribosomal peptides, lipopeptides, alkaloids, terpenes and quinones (Ng et al. 2015). The molecules produced can be associated with a bacterial genus or species, but are often strain-specific. Various examples have been described in the literature among species colonizing seafood. *Shewanella* sp. MTCC 12715 produces macrocyclic polyketides with anti-*Staphylococcus* and anti-*Vibrio* activities (Chakraborty et al. 2021). *Pseudomonas fluorescens* produces mupirocin, a polyketide active against some Gram-positive bacteria, and *Serratia* sp. produces the broad-spectrum antibiotic andremide (Masschelein et al. 2017). Many bacteria also synthesize siderophores, iron-chelating molecules involved in bacterial competition. Indeed, iron is an essential element for the life of microorganisms and is generally present in low quantities in food matrices. Some bacterial species, such as *Shewanella* sp. or *Pseudomonas* sp. are known to synthesize siderophores in order to internalize the iron necessary for their survival (Gram 1994; Soe et al. 2012). Indeed, the competition for iron by these two bacterial genera has been studied in cod tissue (Gram and Melchiorsen 1996).

Lactic acid bacteria isolated from seafood produce numerous bacteriocins (Gómez-Sala et al. 2015). These ribosome-synthesized peptides have long been studied for their antimicrobial properties (Arnison et al. 2013; Silva et al. 2018; Trejo-González et al. 2021). There is a wide diversity of bacteriocins in terms of their structure, size and modes of action. In bacteria isolated from seafood, some strains have been extensively studied for their application in biopreservation (see section 5.3.3). In addition to bacteriocins, lactic acid bacteria synthesize other antimicrobial substances such as organic acids, hydrogen peroxide, diacetyl or carbon dioxide. These molecules and favorable product packaging conditions may explain the preponderance of lactic acid bacteria in certain products. But despite all the knowledge on this bacterial group, some species still hide some mysteries. This is the case of *Lactococcus piscium*, a species often found in the microbiota of seafood. The CNCM I-4031 strain, which does not produce bacteriocins, is currently being studied to understand the mechanism involved in preventing the growth of *L. monocytogenes* (Fall et al. 2010; Saraoui et al. 2016).

Bacterial viruses also affect the dynamics of microbial communities. Phages can both lie dormant in a bacterial host (then called a prophage) and lyse it under conditions of stress in order to infect other targets. A recent study found over 109 phages per gram of cheese matrix (Dugat-Bony et al. 2020). The prevalence of phages in seafood is not known, yet several bacterial genera represented host them. This is the case of lactic acid bacteria or the genera *Listeria* or *Bacillus* (Canchaya et al. 2003). The isolation of phages from seafood processing plant samples (Arachchi et al. 2013) and the activation of prophage gene expression in *L. piscium* in coculture with other lactic acid bacteria (Andreevskaya et al. 2018) also point in this direction. The use of phage therapy as a means of biopreservation is proving to be an interesting strategy. Virulent phages have been used to target *Shewanella* in catfish fillets (Yang et al. 2019).

5.3.1.2. *Bacterial cooperation*

In food, a particular mechanism called the "Jameson effect" was described in 1962 (Jameson 1962). It is defined as the cessation of population growth when a species has reached its maximum concentration. Several studies describe this phenomenon in seafood (Ross 2000). These observations could be explained today by quorum sensing, a bacterial communication system that is the subject of numerous studies (Waters and Bassler 2005). Quorum sensing induces the production of autoinducers in response to cell density. These autoinducers can regulate the expression of specific genes such as those involved in the synthesis of antimicrobial compounds, bioluminescence, virulence or biofilm formation. In Gram-positive bacteria, small peptides called AIPs (Autoinducing Peptides) serve as signal molecules. In Gram-negative bacteria, the best known are the acyl homoserine lactones (AHLs) encoded by the lux genes. The presence of AHLs of different classes has been detected in seafood (Gram et al. 2002). Enterobacteria strains isolated from fish and *Pseudomonas*, *Photobacterium* and *Aeromonas* strains have also shown the ability to produce these types of molecules (Ravn et al. 2001; Gram et al. 2002). The knowledge gained on quorum sensing molecules in seafood allows for the development of new biopreservation strategies. Recently, the use of two strains of *Lactiplantibacillus plantarum* and *Lacticaseibacillus casei* communicating through the LuxS/AI-2 system was successful for the biopreservation of chilled shrimp and decreased the concentration of TVBN (Li et al. 2019). Another strategy was to treat diced salmon with selected quorum sensing inhibitors called 5'-AMP (5'-adenylic acid) and 5'-CMP (Cytidine-5'-monophosphate) to combat *S. baltica* and *P. fluorescens-related* spoilage (Wang et al. 2021).

Another process of bacterial cooperation called metabiosis has been described in seafood since the 2000s. It corresponds to interspecies sequential metabolic interactions (Jørgensen et al. 2000; Gram and Dalgaard 2002), i.e. the development of certain microorganisms through the use of substrates derived from the metabolism of other microorganisms. This phenomenon has helped explain the accumulation of high concentrations of biogenic amines (putrescine and agmatine) in smoked salmon. These are believed to be produced from arginine through the combined activity of enzymes synthesized by *P. phosphoreum, S. liquefaciens, H. alvei, Carnobacterium* sp. or *L. sakei* (Jørgensen et al. 2000). This phenomenon is considered harmful, but from a biological point of view, it allows microorganisms to survive in a rather hostile environment.

5.3.2. Selection of bioprotective microorganisms

The selection of biopreservation microorganisms, also called protective cultures (PCs), requires collections to be screened. The microorganisms, often bacteria for seafood applications, can come from various biotopes. However, it is interesting to have candidates isolated from the food to be biopreserved, which guarantees a good re-implantation in the product. Therefore, knowledge of microbial ecosystems is a prerequisite for successful biopreservation. Prioritizing the many criteria for selecting PCs is important to minimize the number of tests.

5.3.2.1. *Antimicrobial activity*

The first step in selecting a PC is to screen their antimicrobial activity against target strains. Since not all strains of the same species are equally susceptible to PCs (Brillet et al. 2004), it is important to test multiple strains when possible.

There are many type strains of pathogens which can be used as targets; however, laboratories that are not equipped to work with class 2 pathogens often perform an initial screening on surrogate species, such as *Listeria innocua* instead of *L. monocytogenes*. A final validation on the pathogen species is always necessary. The target bacteria for spoilage are rather specific to each product and must be carefully selected according to the matrix.

The tests are generally completed under model conditions, in liquid or agar culture media, sometimes in fish juice. Different methods exist. We can mention the spot on lawn method which consists of spotting drops of bacterial culture on agar plates previously inoculated with a target strain. After incubation at the growth temperature of the target (or at the storage temperature of the fish to be as close as possible to the real conditions), the presence of clear halos around the spots indicates an inhibition whose intensity is stronger as the diameter of the halo increases (Matamoros et al. 2009a). Inhibitions specifically related to the excretion of antimicrobial compounds can be observed by depositing sterilized culture supernatants. In case of activity, different rapid tests are implemented for a pre-characterization of active molecules, including organic acids, hydrogen peroxide or bacteriocins. All of these methods can be miniaturized to allow rapid screening of several hundred bacteria under various conditions. If no collections of PCs are available for screening, another strategy is to enumerate the bacteria present in the food to be biopreserved by the classical Petri dish technique, and then to apply a thin overlay of agar inoculated with the target strain on the surface of the dish. The use of selective enumeration media and their incubation conditions (temperature, anaerobiosis) then allows the screening to be directed towards more adapted PCs. A specific medium for lactic bacteria can be used, for example. Lactic acid bacteria are, indeed, good candidates for biopreservation because they are often present in foods, carry little risk for health and have a wide range of antimicrobial activities.

Omics tools are now also used for pre-selection of PCs. Biosynthetic clusters, often involved in the production of non-ribosomal antimicrobial peptides, terpenes and polyketides or bacteriocins, can be searched in silico in available genomes. The presence of such gene clusters can predict potential antimicrobial activity. The genomic comparison and de-replication process can then determine whether the molecules for which they code are novel. For example, Begrem et al. (2020) showed the high potential diversity of antimicrobial metabolites in several *Carnobacterium* species.

5.3.2.2. *Sensory acceptability*

One of the big challenges of biopreservation is to find bacteria which have antimicrobial capabilities but do not degrade the sensory quality of the product. Unfortunately, many bacteria are responsible for bad odors, flavors or textures. Some strains belonging to the lactic acid bacteria group, in particular produce organic acids which, although necessary in fermentation, are not compatible with a biopreservation application. Some produce bitter flavors, CO_2 responsible for the swelling of the packaging, H_2S and other malodorous molecules. The low acidification and the absence of bad odors can be tested in a first step in a model liquid medium. To mimic the food, some use fish juices enriched with malodorous metabolite precursor molecules, such as TMAO and sulfur amino acids (Wiernasz et al. 2017). The more efficient tool for defining sensory profiles remains human perception. Without involving a huge panel, a few people trained in sensory analysis can sniff out different cultures to select the most neutral candidates possible. It is important to know that the ability of a bacterial species to spoil a food is strain dependent, although broad signatures can emerge.

5.3.2.3. *Validation of antimicrobial and organoleptic effects in foodstuffs in a pauci-microbial matrix*

The results observed in liquid medium must be validated in food. The microbiota, the composition of the product (sugars, proteins, vitamins, etc.) as well as the technological parameters, such as salt, smoke, atmosphere and storage temperature, have an impact on the implantation of PCs and their metabolism.

The growth and antimicrobial activity of PCs are verified by carrying out challenge tests. These consist of co-inoculating the PC and the target strain (pathogen, spoilage) in the food, containing if possible a very low quantity of endogenous bacteria. This is particularly important when we wish to test the effect of PCs on target spoilage bacteria, whose natural presence in food could make interpretation difficult. Sterile or *pauci*-microbial matrices can be obtained either by processing the product under excellent hygienic conditions or by applying a non-destructive decontamination process such as ionization. The ISO 20976-1 (2019) standard for performing challenge tests specifies the importance of using cocktails of several target strains from the food rather than collection strains, and of performing pre-cultures of the targets under stress conditions similar to those they may encounter in the matrix (low temperatures, salt, etc.). In addition, the AFNOR standard (NF V01-003, 2018) specifies the protocol for validating the shelf lives of refrigerated products. The scenario is a conservation for 1/3 of the time of the total shelf life at 4°C and 2/3 at 8°C, mimicking the temperature profile observed from distribution to conservation at the consumer.

The use of a *pauci*-microbial matrix also makes it possible to control that the PC, even when inoculated at high concentration, does not cause organoleptic changes in the product until the end of the shelf life. For example, a strain of *L. sakei*, very promising to prevent the development of histaminogenic bacteria such as *Photobacterium* sp. and *Morganella* sp., has been found acceptable in tuna but responsible for strong sulfurous odors in other products (salmon, shrimp, cod). Acidification of a PC in a model liquid medium is often considered a rejection criterion, whereas some strains may be acceptable when introduced into a low-sugar fish meat with a high buffering capacity. For example, a bioprotective strain of *L. gelidum* performed poorly in gravlax because the presence of sugar in the recipe led to acidification and production of sticky exopolysaccharides on the surface of the product, whereas it was very effective in extending the shelf life of fresh salmon and cooked peeled shrimp (Brillet 2011).

5.3.2.4. *Safety of protective bacterial culture*

Biopreservation bacteria must not represent a risk to human health. The first criterion is the detailed taxonomic knowledge of the strain. New sequencing tools enable excellent identification of bacteria and total genome sequencing is now easily accessible, allowing some important metabolic pathways to be specified in order to guarantee the safety of bacteria. Although not mandatory, the membership of the species to the QPS list (Quality Presumption of Safety) of the European Food Safety Agency (EFSA) facilitates its acceptance (EFSA Panel on Biological Hazards (BIOHAZ) 2022). Numerous lactic acid bacteria are listed, which often directs the selection towards this taxonomic group. Phenotypic and physiological knowledge is also important, as well as data regarding their origin, their ability to colonize certain niches and their traditional use, or not, in food processing.

The antibiotic resistance profile is sought to avoid selecting PCs that would allow horizontal transfer of resistance genes, one of the critical public health issues identified by the World Health Organization. The physiological tests recommended by EFSA include the determination of minimum inhibitory concentrations (MICs) of several antibiotics which can be performed according to ISO 10932 (2010). However, acceptable MIC thresholds are not defined for all bacterial species. The study of a large number of strains for a given species may define an average resistance threshold, specific to the species. The in silico study of genes involved in antibiotic resistance completes the phenotypic tests and specifies if they are located on mobile elements (plasmids), generating a higher risk of transmission.

The production of toxic molecules such as biogenic amines is also a criterion. Histamine is a molecule responsible for allergic type syndromes and is regulated in some seafood products. The assays can be performed in liquid medium rich in histidine or directly in product. The presence of gene coding for histidine decarboxylase, detected by genome analysis or by specific PCR, now completes these analyses. The production of tyramine, which is not regulated, is sometimes a rejection criterion for some starter producers because this molecule can be responsible for migraines, at very high concentrations (> 5,000 mg/kg).

Although a few toxicity studies of PCs are performed on animals, cell culture tests are sometimes done. A first screening can be achieved by applying bacterial cultures or sterile filtered culture supernatant to Vero or Caco-2 cells. Lactate dehydrogenase (LDH) released from the cytosol of cells is an indicator of necrotic cell death when released. In addition, the integrity of the epithelial barrier in the presence of PCs can be measured by transepithelial electrical resistance on Caco-2 cell monolayers. Hemolytic activity is investigated by phenotypic tests on blood agar plates, the results of which depend on the conditions and the blood used. These tests can be complemented by genomic analysis of the presence of genes coding for hemolysins and also for other endo or exotoxins or virulence factors involved in adhesion, invasion or resistance to phagocytosis by lymphocytes.

5.3.2.5. *Technological skills*

To consider an industrial application, it is important that PCs can be easily produced in a fermenter and at a reasonable cost. Bacteria which grow well on simple culture media and at temperatures which do not require energy-intensive heating or refrigeration are obviously more interesting. The ability to grow on media with vegetable rather than animal proteins can also be important for vegan productions, for example. Moreover, some bacteria require a salt content that can damage the fermenters. The ability of PCs to withstand freeze-drying is also important. The strains are usually marketed as freeze-dried powder to be rehydrated and must be easily revivified.

5.3.2.6. *Validation of results in naturally contaminated products*

The final step in the selection of PCs is the application in naturally contaminated foods. We have seen that the effects of a PC can vary depending on the products to be treated, their composition and their endogenous flora. Each plant has its own specificities (origin of the raw material, technological parameters of transformation, cleaning/disinfection process) and its own signature in terms of environmental microbiota. To validate the effect of PCs on spoilage, PCs are introduced into products recovered in the factory just after manufacturing, and the control is a non-biopreserved batch. During preservation, classical analyses are performed (bacterial counts, sensory fingerprints, physico-chemical measurements), and completed by omic methods to study finely the microbiota (metabarcoding, metagenomics) and the volatilome. Specific data processing tools are developed to link all these results.

The validation of the effect of a PC on pathogenic bacteria is more complicated to show. Their prevalence in seafood products is (fortunately) rarely sufficient to acquire statistically exploitable data. Hence, the effect of PCs against pathogens, in the presence of endogenous flora, is assessed in the laboratory on pathogen-artificially contaminated commercial foods.

5.3.3. *Examples of the application of protective cultures in seafood*

The bacteria mostly used in fermented food processing are generally lactic acid bacteria, which are difficult to implant in the fresh fish flesh. Because of the increase of lightly processed products consumption such as salted, smoked, acidified, lightly cooked or preserved fish and shellfish under protective atmosphere, the importance of lactic acid bacteria has been highlighted and thereby, most of the biopreservation examples concern these products. Studies on the biopreservation of seafood products began in the 1990s, with a particular focus on the control of pathogenic bacteria, especially *L. monocytogenes*. It is only 20 years later that the first trials to combat organoleptic spoilage took place. There are many studies on the selection of PCs with antimicrobial activities, but we will only present here those that have led to product applications, and more particularly those for which the use of live PCs is used (culture supernatants and bacteriocins causing a regulatory problem).

5.3.3.1. *Control of the Listeria monocytogenes risk*

The genus *Carnobacterium* probably has the most examples of application against *Listeria* sp. in seafood. This genus includes 12 species, two of which, *C. divergens* and *C. maltaromaticum*, are very frequently isolated from marine products. The percentage of strains with anti-listeria activity is much higher than in other lactic acid bacteria species (Passerini et al. 2021). Beyond their antimicrobial properties, *Carnobacterium* sp. are particularly interesting because they multiply well at low temperatures and resist the pH, salinity and packaging conditions (vacuum or modified atmosphere CO_2/N_2) encountered in seafood products, giving them an important ecological advantage. Their sensory acceptability depends on the strain and the food but *C. divergens* and *C. maltaromaticum* species are often quite neutral, not very acidifying and do not produce histamine.

Several authors have isolated *C. divergens* strains (V41, M35, etc.) that slow down or completely prevent the growth of *L. monocytogenes* in vacuum-packed smoked salmon or in cooked peeled shrimp tails packaged in a protective atmosphere throughout the product's shelf life. This effect has been validated on a large number of *L. monocytogenes* strains (Brillet et al. 2004, 2005; Tahiri et al. 2009; Saraoui et al. 2017). We find other numerous examples of *L. monocytogenes* control in smoked salmon and salmon gravlax by *C. maltaromaticum* strains (Nilsson et al. 1999; Vescovo et al. 2006; Wiernasz et al. 2020). The activity is often due to the production of class IIa bacteriocins. The culture supernatant, even though it has an interesting listericidal activity just after its introduction into the food, often loses its efficacy over time. However, some carnobacteria also inhibit *L. monocytogenes* without any evidence of bacteriocin production. Nutritional competition for glucose could partly explain this phenomenon (Nilsson et al. 2005). As with all lactic acid bacteria species, the effect of the strains is matrix dependent. *C. maltaromaticum* CTC1741, for example, shows an inhibitory effect on *L. monocytogenes* in only one of the three batches of smoked salmon tested, with no explanation given (Aymerich et al. 2019).

While the anti-listeria effect of *Carnobacterium* sp. is often proven, studies do not always consider their impact on food quality and safety. Many strains of *Carnobacterium* sp. do not appear to affect the organoleptic properties of seafood, although they may produce mild buttery/caramel odors. However, other strains result in more unpleasant, chlorinated, malty, chemical or sour odors, enhanced in the presence of other species such as *B. thermosphacta* or *Vibrio* sp. (Joffraud et al. 2006; Leisner et al. 2007; Jaffrès et al. 2011). *C. divergens* V41, *C. maltaromaticum* V1, SF668 and SF1944 selected for their anti-listeria activity do not degrade the quality of smoked salmon and salmon gravlax. They also do not produce histamine. Tyramine concentrations, although varying between studies, rarely exceed 350 mg/kg (Brillet et al. 2005). Given the evidence of their benefits for seafood biopreservation and the minimal risk associated with tyramine production, some European and Canadian producers of ferments are already commercializing *Carnobacterium* ssp. strains and other producers are considering marketing them in the very near future.

Lactobacilli have also been extensively studied. Strains of *L. sakei* isolated from seafood have been shown to inhibit *L. monocytogenes* in smoked salmon (Katla et al. 2001; Weiss and Hammes 2006; Lacumin et al. 2021). The bacteriocin-producing strain CTC494 has been robustly studied in numerous batches of vacuum-packed cold-smoked salmon, hot-smoked sea bass and sea bream under protective atmosphere (Aymerich et al. 2019; Bolivar et al. 2021). The anti-listeria effect varies with the storage temperature profile and efficacy is optimal with an inoculum of 10^4

L. sakei/g. Beyond that, the *L. sakei* strain is no longer sensorially acceptable. Other lactobacillus species (*L. casei, L. curvatus, L. plantarum, Lactiplantibacillus pentosus, Lactobacillus delbrueckii*) show efficacy in various fresh or processed products.

Lactococci are widely used in the food industry, particularly *L. lactis* in dairy products. However, its mesophilic nature does not give it an ecological advantage in seafood, although a successful trial was conducted against *L. innocua* on cooked shelled shrimp (Hwanhlem et al. 2015). The species *L. piscium/carnosum*, on the contrary, is quite often found in processed fish. Two strains have been particularly studied and are still the subject of numerous works by INRAE/Oniris and Ifremer laboratories which have notably shown that the species does not present a risk a priori for human consumption. *L. piscium* is able to inhibit the growth of *L. monocytogenes* by 2–4 log (CFU/g) in cooked peeled shrimp stored under vacuum or protective atmosphere (Matamoros et al. 2009b; Fall et al. 2010). All *L. monocytogenes* strains tested are susceptible to *L. piscium* CNCM I-4031, and this strain also decreases the virulence of *L. monocytogenes* (Saraoui et al. 2018). Regardless of the PC/target ratios, *Listeria* growth always stops when *L. piscium* has reached its early stationary phase (Saraoui et al. 2018). The mechanism of inhibition is not related to the production of antimicrobial peptides, organic acids, hydrogen peroxide or other molecules excreted massively into the medium, nor to nutritional competition. Contact between the bioprotective strain and the pathogen is required to observe inhibition (Saraoui et al. 2016).

Other strains of *L. gelidum*, *Pediococcus acidilactici* and *Streptococcus phocae* species have also been shown to reduce the growth of *L. monocytogenes* or *L. innocua* in shrimp, Indian sardinella and smoked salmon, but the work is less complete.

5.3.3.2. *Control of other pathogens*

There is little work on the control of *Salmonella* or *S. aureus*. These pathogens are less often implicated in seafood-related illnesses. In addition, biopreservation is less suitable for bacterial decontamination and complete elimination than for inhibition of growth. This is the requirement for *Salmonella* in some products, and even *L. monocytogenes* in the United States (zero tolerance). However, a study by Tasaku et al. (2017) presents the reduction of *Salmonella typhimurium* in Tilapia by-products by *Pediococcus pentosaceus* and of *S. aureus* in fresh whole shrimp by *L. rhamnosus* or *L. plantarum* and by *L. piscium* and *L. gelidum* in cooked peeled shrimp.

Oyster decontamination trials have also been conducted (Sorée et al. 2022). The use of *Lactobacillus* sp. strains or other lactic acid bacteria in the purification baths resulted in a 2-log reduction in *V. parahaemolyticus* contamination. This pathogen was also reduced by 1 log when bacteriocin from a *L. lactis* subps. *lactis* strain was applied on Nile perch paté.

We have seen that histamine poisoning is linked to the production of histamine by bacteria, particularly *Enterobacteriaceae*. Some authors have tried to select PCs capable of degrading histamine. However, waiting until the molecule is formed and then degrading it is not the most desirable strategy. It is better to prevent its production, for example, by limiting the growth of the producing bacteria. There is a patent on the use of a strain of *L. sakei* that prevents the growth of *Morganella morganii* and *P. phosphoreum* in fresh vacuum-packed tuna, and very significantly decreases the concentration of histamine at abuse temperatures (Podeur et al. 2016).

5.3.3.3. *Improvement of the sensory quality*

As early as 1995, it was shown that lactic acid bacteria strains could increase the (sensory) shelf-life of some products (Kim et al. 1995; Leroi et al. 1996). However, the microbiota and metabolic interactions involved in spoilage are complex, so explanations and reproducibility of results are quite rare. A clear example is a strain of *Vagococcus pennaei* that significantly decreases organoleptic spoilage of gravlax salmon during preservation while it does not develop well on the product, and little changes are observed on the composition of the endogenous microbiota (Wiernasz et al. 2020).

Many species (*L. casei, L. paracasei, L. curvatus, L. plantarum, L. lactis,* etc.) have been used to delay spoilage of vacuum-packed fresh salmon fillets (+15 days), hake and megrim on ice, swordfish fillets and tuna burgers (Passerini et al. 2021). The most extensive work is on shrimp. *C. divergens* V41, known for its anti-listeria activity, increases the use-by date of cooked peeled tropical shrimp tails packaged in a protective atmosphere by at least 10 days. One explanation is the inhibition of *Enterobacteriaceae* and *Shewanella* sp. (Saraoui et al. 2017). *L. piscium* also increases the shelf-life of cooked peeled shrimp by at least 15 days (Matamoros et al. 2009b). The remarkable effect could be explained by the inhibition of *B. thermosphacta* (Fall et al. 2010, 2012). Moreover, the combination of *C. divergens* V41 and *L. piscium* CNCM I-4030 both inhibited *L. monocytogenes* growth and increased the use-by date of cooked peeled shrimp (Saraoui et al. 2017). To our knowledge, there is at least one industrial application in France involving the use of a *L. gelidum* strain (LLO ferment, Biocéane company) for the preservation of peeled cooked tropical shrimp.

5.3.4. *Regulatory aspects*

The use of microbial food cultures (ferments) in food manufacturing is not specifically regulated in Europe. There is no legal definition of ferments, so the EFFCA (European Food and Feed Culture Association) has proposed the following: "Microbial food cultures are live bacteria, yeasts and moulds used for in food production". Currently, traditionally used ferments are considered as ingredients and must be clearly listed as such on the packaging. The precise name of the species does not necessarily have to be listed, and the terms "ferments" or "lactic ferments" are often used. The status of PCs used for preservation purposes is currently not clearly defined (ingredients or additives), but their use must be mentioned on the label.

Although there is no specific regulation, a food in which biopreservative strains are added must meet, as any food, the conditions of the hygiene package and not be harmful to health or unfit for human consumption (Regulation (EC) No. 178/2002, 2002). The added bacteria must therefore meet the safety criteria listed in the previous paragraph. In 2007, EFSA introduced the concept of "Qualified Presumption of Safety" (QPS) of microorganisms used in food and feed. This system allows the safety assessment procedures of microorganisms to be harmonized within the different sample groups of EFSA but is not intended to evaluate all microorganisms. The evaluation is done on species for which sufficient knowledge is available to affirm that all strains are assumed to be safe. A list of species has been established and is updated regularly. However, strains that are not on the list can be used, either because they have not been the subject of any application or because work at the strain level has proven its safety, even though knowledge at the species level is considered insufficient to put it on the QPS list. In addition, the International Dairy Federation (IDF) has compiled an inventory of microorganisms with technological benefits in dairy products, now extended to other foods, based on extensive knowledge and bibliographic studies, and publishes a regularly updated list used worldwide.

The use of a PC for the biopreservation of food is therefore essentially based on its proven efficacy, its safety and the history of traditional use of this species in food without adverse health consequences, followed over several generations and on a wide range of populations. If this is the case, no application for pre-marketing authorization is required. If, on the contrary, the product is considered to be novel on the European market (i.e. not consumed to any significant degree in the Community before May 14, 1997), then the food must comply with Novel Food Regulation (EC) No. 258/97. A novel food must undergo a pre-market assessment and an authorization procedure including a risk assessment by EFSA and a risk management by the European Commission before it can be placed on the market.

Beside the regulations, certain codes of practice and the specifications of products with a quality sign must be taken into account. In addition, for some foods, specifications may limit the use of PCs. In France, the AFNOR V45-065 standard for smoked salmon stipulates, for example, that only salt and sugar are authorized as ingredients. On the contrary, the use of PCs can lead to non-compliance with microbiological criteria relating to aerobic mesophilic flora and lactic flora (e.g. criteria of the *Fédération du Commerce et de la Distribution*, French Commerce, Services and Distribution Federation).

The United States also has no specific regulations for the use of ferments in food. The FDA (Food and Drug Administration) has introduced the GRAS (Generally Recognized as Safe) status that concerns microbial food cultures and also ingredients in the broad sense. Unlike QPS, GRAS status is granted for the use of a microorganism (a strain in a food) and not for a microbial species. Many ferments have been used in products for centuries and those with documented use prior to 1958 automatically have GRAS status. In other cases, a substantiated notification can be made to the FDA by a person or industry demonstrating GRAS status, and a receipt from the FDA stating that they have no objection is sufficient.

5.4. Conclusion

Biopreservation is a natural food preservation technique that meets the growing demand for safe, convenient and authentic food without chemical preservatives. It is consistent with the European objectives of promoting the "Clean label". Although industrial applications already exist for seafood products, certain regulatory and technological barriers remain. Biopreservation is, indeed, quite complex to implement since the selected microorganisms must meet the dual challenge of limiting undesirable bacteria (pathogens, spoilers) and not degrading the quality of the product. A better understanding of bacterial interactions will therefore allow better control of its evolution during food preservation and more efficient biopreservation strategies to be proposed. Today, genome analysis of isolated bacteria or complex samples (metagenomics), as well as the numerous bioinformatics tools available, makes it possible to predict the presence of genes coding for molecules involved in bacterial interactions. However, the predicted molecules largely exceed those expressed under laboratory conditions and probably active in the matrix. Therefore, accessing the in situ metabolic functionalities of microbial communities and studying the role of individual organisms in the microbiota remains a major challenge. New strategies such as metatranscriptomics should in the coming years lead to a better understanding of the dynamics of seafood microbiota.

5.5. References

Andreevskaya, M., Jääskeläinen, E., Johansson, P., Ylinen, A., Paulin, L., Björkroth, J., Auvinen, P. (2018). Food spoilage-associated *Leuconostoc*, *Lactococcus*, and *Lactobacillus* species display different survival strategies in response to competition. *Applied and Environmental Microbiology*, 84(13), e00554-18. https://doi.org/10.1128/AEM.00554-18.

ANSES (2019a). Fiche de description de danger biologique transmissible par les aliments : vibrions entéropathogènes : *Vibrio parahaemolyticus*, *Vibrio cholerae* non-O1 / non-O139 et *Vibrio vulnificus*. Report, 2016-SA-0272, 5.

ANSES (2019b). Fiche de description de danger biologique transmissible par les aliments : *Clostridium botulinum*, *Clostridium neurotoxinogènes*. Report.

ANSES (2020). Fiche de description de danger biologique transmissible par les aliments : *Listeria monocytogenes*. Report, 2016-SA-0081, 4.

ANSES (2021). Fiche de description de danger biologique transmissible par les aliments : *Salmonella* spp. Report, 2016-SA-0080, p. 4.

Arachchi, G.J.G., Mutukumira, A.N., Dias-Wanigasekera, B.M., Cruz, C.D., McIntyre, L., Young, J., Flint, S.H., Hudson, A., Billington, C. (2013). Characteristics of three listeriaphages isolated from New Zealand seafood environments. *Journal of Applied Microbiology*, 115(6), 1427–1438. https://doi.org/10.1111/jam.12332.

Arnison, P.G., Bibb, M.J., Bierbaum, G., Bowers, A.A., Bugni, T.S., Bulaj, G., Camarero, J.A., Campopiano, D.J., Challis, G.L., Clardy, J. et al. (2013). Ribosomally synthesized and post-translationally modified peptide natural products : Overview and recommendations for a universal nomenclature. *Natural Product Reports*, 30(1), 108–160. https://doi.org/10.1039/C2NP20085F.

Aymerich, T., Rodriguez, M., Garriga, M., Bover-Cid, S. (2019). Assessment of the bioprotective potential of lactic acid bacteria against *Listeria monocytogenes* on vacuum-packed cold-smoked salmon stored at 8°C. *Food Microbiology*, 83, 64–70. https://doi.org/10.1016/j.fm.2019.04.011.

Begrem, S., Ivaniuk, F., Gigout-Chevalier, F., Kolypczuk, L., Bonnetot, S., Leroi, F., Grovel, O., Delbarre-Ladrat, C., Passerini, D. (2020). New insight into antimicrobial compounds from food and marine-sourced *Carnobacterium* species through phenotype and genome analyses. *Microorganisms*, 8(7), 1093. https://doi.org/10.3390/microorganisms8071093.

Bolivar, A., Peres Costa, J.C.C., Posada-Izquierdo, G.D., Bover-Cid, S., Zurera, G., Perez-Rodriguez, F. (2021). Quantifying the bioprotective effect of *Lactobacillus sakei* CTC494 against *Listeria monocytogenes* on vacuum packaged hot-smoked sea bream. *Food Microbiology*, 94, 103649. https://doi.org/10.1016/j.fm.2020.103649.

Brillet, A. (2011). Towards a sustainable food chain: Food process, bioprocessing and food quality management. *6th International Technical Symposium*. https://cigr.org/node/686.

Brillet, A., Pilet, M.F., Prevost, H., Bouttefroy, A., Leroi, F. (2004). Biodiversity of *Listeria monocytogenes* sensitivity to bacteriocin-producing *Carnobacterium* strains and application in sterile cold-smoked salmon. *Journal of Applied Microbiology*, 97(5), 1029–1037. https://doi.org/10.1111/j.1365-2672.2004.02383.x.

Brillet, A., Pilet, M.F., Prevost, H., Cardinal, M., Leroi, F. (2005). Effect of inoculation of *Carnobacterium divergens* V41, a biopreservative strain against *Listeria monocytogenes* risk, on the microbiological, chemical and sensory quality of cold-smoked salmon. *International Journal of Food Microbiology*, 104(3), 309–324. https://doi.org/10.1016/j.ijfoodmicro.2005.03.012.

Canchaya, C., Proux, C., Fournous, G., Bruttin, A., Brüssow, H. (2003). Prophage genomics. *Microbiology and Molecular Biology Reviews: MMBR*, 67(2), 238–276. https://doi.org/10.1128/MMBR.67.2.238-276.2003.

Carroll, A.R., Copp, B.R., Davis, R.A., Keyzers, R.A., Prinsep, M.R. (2020). Marine natural products. *Natural Product Reports*, 37(2), 175–223. https://doi.org/10.1039/C9NP00069K.

Chaillou, S., Chaulot-Talmon, A., Caekebeke, H., Cardinal, M., Christieans, S., Denis, C., Desmonts, M.H., Dousset, X., Feurer, C., Hamon, E. et al. (2015). Origin and ecological selection of core and food-specific bacterial communities associated with meat and seafood spoilage. *The ISME Journal*, 9(5), 1105–1118. https://doi.org/10.1038/ismej.2014.202.

Chakraborty, K., Kizhakkekalam, V.K., Joy, M. (2021). Chemical mining of heterotrophic *Shewanella algae* reveals anti-infective potential of macrocyclic polyketides against multidrug-resistant pathogens. *Bioorganic Chemistry*, 108, 104533. https://doi.org/10.1016/j.bioorg.2020.104533.

Dalgaard, P. (1995). Qualitative and quantitative characterization of spoilage bacteria from packed fish. *International Journal of Food Microbiology*, 26(3), 319–333. https://doi.org/10.1016/0168-1605(94)00137-U.

Dalgaard, P. (2000). *Freshness, Quality and Safety in Seafoods*. Teagasc, The National Food Centre, Dunsinea, Castleknock, Dublin.

Dalgaard, P., Mejlholm, O., Christiansen, T.J., Huss, H.H. (1997). Importance of *Photobacterium phosphoreum* in relation to spoilage of modified atmosphere-packed fish products. *Letters in Applied Microbiology*, 24(5), 373–378. https://doi.org/10.1046/j.1472-765X.1997.00152.x.

Dugat-Bony, E., Lossouarn, J., De Paepe, M., Sarthou, A.-S., Fedala, Y., Petit, M.-A., Chaillou, S. (2020). Viral metagenomic analysis of the cheese surface: A comparative study of rapid procedures for extracting viral particles. *Food Microbiology*, 85, 103278. https://doi.org/10.1016/j.fm.2019.103278.

EFSA Panel on Biological Hazards (BIOHAZ) (2022). Updated list of QPS-recommended biological agents for safety risk assessments carried out by EFSA. Zenodo. https://doi.org/10.5281/zenodo.5905365.

Emborg, J., Laursen, B.G., Rathjen, T., Dalgaard, P. (2002). Microbial spoilage and formation of biogenic amines in fresh and thawed modified atmosphere-packed salmon (*Salmo salar*) at 2°C. *Journal of Applied Microbiology*, 92(4), 790–799. https://doi.org/10.1046/j.1365-2672.2002.01588.x.

European Parliament (1997). Règlement (CE) n° 258/97 du Parlement européen et du Conseil du 27 janvier 1997 relatif aux nouveaux aliments et aux nouveaux ingrédients alimentaires. Document, European Parliament.

European Parliament (2002). Règlement (CE) n° 178/2002 du Parlement européen et du Conseil du 28 janvier 2002 établissant les principes généraux et les prescriptions générales de la législation alimentaire, instituant l'Autorité européenne de sécurité des aliments et fixant des procédures relatives à la sécurité des denrées alimentaires. Document, European Parliament.

European Parliament (2005). Règlement (CE) n° 2073/2005 de la Commission du 15 novembre 2005 concernant les critères microbiologiques applicables aux denrées alimentaires. Document, European Parliament.

Fall, P.A., Leroi, F., Chevalier, F., Guerin, C., Pilet, M.-F. (2010). Protective effect of a non-bacteriocinogenic *Lactococcus piscium* CNCM I-4031 strain against *Listeria monocytogenes* in sterilized tropical cooked peeled shrimp. *Journal of Aquatic Food Product Technology*, 19(2), 84–92. https://doi.org/10.1080/10498850.2010.486910.

Fall, P.A., Pilet, M.F., Leduc, F., Cardinal, M., Duflos, G., Guérin, C., Joffraud, J.-J., Leroi, F. (2012). Sensory and physicochemical evolution of tropical cooked peeled shrimp inoculated by *Brochothrix thermosphacta* and *Lactococcus piscium* CNCM I-4031 during storage at 8°C. *International Journal of Food Microbiology*, 152(3), 82–90. https://doi.org/10.1016/j.ijfoodmicro.2011.07.015.

Fonnesbech Vogel, B., Huss, H.H., Ojeniyi, B., Ahrens, P., Gram, L. (2001). Elucidation of *Listeria monocytogenes* contamination routes in cold-smoked salmon processing plants detected by DNA-based typing methods. *Applied and Environmental Microbiology*, 67(6), 2586–2595. https://doi.org/10.1128/AEM.67.6.2586-2595.2001.

Foster, K.R. and Bell, T. (2012). Competition, not cooperation, dominates interactions among culturable microbial species. *Current Biology: CB*, 22(19), 1845–1850. https://doi.org/10.1016/j.cub.2012.08.005.

Gambarin, P., Magnabosco, C., Losio, M.N., Pavoni, E., Gattuso, A., Arcangeli, G., Favretti, M. (2012). *Listeria monocytogenes* in ready-to-eat seafood and potential hazards for the consumers. *International Journal of Microbiology*, 1–10. https://doi.org/10.1155/2012/497635.

Gómez-Sala, B., Muñoz-Atienza, E., Sánchez, J., Basanta, A., Herranz, C., Hernández, P.E., Cintas, L.M. (2015). Bacteriocin production by lactic acid bacteria isolated from fish, seafood and fish products. *European Food Research and Technology*, 241(3), 341–356. https://doi.org/10.1007/s00217-015-2465-3.

Gram, L. (1994). Siderophore-mediated iron sequestering by *Shewanella putrefaciens*. *Applied and Environmental Microbiology*, 60(6), 2132–2136. https://doi.org/10.1128/aem.60.6.2132-2136.1994.

Gram, L. and Dalgaard, P. (2002). Fish spoilage bacteria – Problems and solutions. *Current Opinion in Biotechnology*, 13(3), 262–266. https://doi.org/10.1016/S0958-1669(02)00309-9.

Gram, L. and Huss, H.H. (1996). Microbiological spoilage of fish and fish products. *International Journal of Food Microbiology*, 33(1), 121–137. https://doi.org/10.1016/0168-1605(96)01134-8.

Gram, L. and Melchiorsen, J. (1996). Interaction between fish spoilage bacteria *Pseudomonas* sp. and *Shewanella putrefaciens* in fish extracts and on fish tissue. *Journal of Applied Bacteriology*, 80(6), 589–595. https://doi.org/10.1111/j.1365-2672.1996.tb03262.x.

Gram, L., Ravn, L., Rasch, M., Bruhn, J.B., Christensen, A.B., Givskov, M. (2002). Food spoilage – Interactions between food spoilage bacteria. *International Journal of Food Microbiology*, 78(1–2), 79–97. https://doi.org/10.1016/S0168-1605(02)00233-7.

Hibbing, M.E., Fuqua, C., Parsek, M.R., Peterson, S.B. (2010). Bacterial competition: Surviving and thriving in the microbial jungle. *Nature Reviews Microbiology*, 8(1), 15–25. https://doi.org/10.1038/nrmicro2259.

Hu, Y., Chen, J., Hu, G., Yu, J., Zhu, X., Lin, Y., Chen, S., Yuan, J. (2015). Statistical research on the bioactivity of new marine natural products discovered during the 28 years from 1985 to 2012. *Marine Drugs*, 13(1), 202–221. https://doi.org/10.3390/md13010202.

Huss, H.H. (1995). Quality and quality changes in fresh fish. FAO Fisheries, Technical Paper 348.

Hwanhlem, N., Jaffres, E., Dousset, X., Pillot, G., Choiset, Y., Haertle, T., H-Kittikun, A., Chobert, J.-M. (2015). Application of a nisin Z-producing *Lactococcus lactis* subsp *lactis* KT2W2L isolated from brackish water for biopreservation in cooked, peeled and ionized tropical shrimps during storage at 8°C under modified atmosphere packaging. *European Food Research and Technology*, 240(6), 1259–1269. https://doi.org/10.1007/s00217-015-2428-8.

ISO 10932 (2010). Milk and milk products – Determination of the minimal inhibitory concentration (MIC) of antibiotics applicable to bifidobacteria and non-enterococcal lactic acid bacteria (LAB). Standard, ISO.

ISO 20976-1 (2019). Microbiology of the food chain – Requirements and guidelines for conducting challenge tests of food and feed products – Part 1: Challenge tests to study growth potential, lag time and maximum growth rate. Standard, ISO.

Jaaskelainen, E., Jakobsen, L.M.A., Hultman, J., Eggers, N., Bertram, H.C., Bjorkroth, J. (2019). Metabolomics and bacterial diversity of packaged yellowfin tuna (*Thunnus albacares*) and salmon (*Salmo salar*) show fish species-specific spoilage development during chilled storage. *International Journal of Food Microbiology*, 293, 44–52. https://doi.org/10.1016/j.ijfoodmicro.2018.12.021.

Jaffrès, E., Lalanne, V., Macé, S., Cornet, J., Cardinal, M., Sérot, T., Dousset, X., Joffraud, J.-J. (2011). Sensory characteristics of spoilage and volatile compounds associated with bacteria isolated from cooked and peeled tropical shrimps using SPME–GC–MS analysis. *International Journal of Food Microbiology*, 147(3), 195–202. https://doi.org/10.1016/j.ijfoodmicro.2011.04.008.

Jameson, J.E. (1962). A discussion of the dynamics of salmonella enrichment. *Journal of Hygiene*, 60(2), 193–207. https://doi.org/10.1017/S0022172400039462.

Joffraud, J.-J., Cardinal, M., Cornet, J., Chasles, J.-S., Léon, S., Gigout, F., Leroi, F. (2006). Effect of bacterial interactions on the spoilage of cold-smoked salmon. *International Journal of Food Microbiology*, 112(1), 51–61. https://doi.org/10.1016/j.ijfoodmicro.2006.05.014.

Jørgensen, L.V., Huss, H.H., Dalgaard, P. (2000). The effect of biogenic amine production by single bacterial cultures and metabiosis on cold-smoked salmon. *Journal of Applied Microbiology*, 89(6), 920–934. https://doi.org/10.1046/j.1365-2672.2000.01196.x.

Katla, T., Moretro, T., Aasen, I.M., Holck, A., Axelsson, L., Naterstad, K. (2001). Inhibition of *Listeria monocytogenes* in cold smoked salmon by addition of sakacin P and/or live *Lactobacillus sakei* cultures. *Food Microbiology*, 18(4), 431–439. https://doi.org/10.1006/fmic.2001.0420.

Kim, C., Hearnsberger, J., Vickery, A., White, C., Marshall, D. (1995). Sodium-acetate and bifidobacteria increase shelf-life of refrigerated catfish fillets. *Journal of Food Science*, 60(1), 25–27. https://doi.org/10.1111/j.1365-2621.1995.tb05598.x.

Koutsoumanis, K. and Nychas, G.-J.E. (1999). Chemical and sensory changes associated with microbial flora of Mediterranean Boque (*Boops boops*) stored aerobically at 0, 3, 7, and 10°C. *Applied and Environmental Microbiology*, 65(2), 698–706. https://doi.org/10.1128/AEM.65.2.698-706.1999.

Lacumin, L., Cappellari, G., Pellegrini, M., Basso, M., Comi, G. (2021). Analysis of the bioprotective potential of different lactic acid bacteria against *Listeria monocytogenes* in cold-smoked sea bass, a new product packaged under vacuum and stored at 6 ± 2°C. *Frontiers in Microbiology*, 12, 796655. https://doi.org/10.3389/fmicb.2021.796655.

Laursen, B.G., Bay, L., Cleenwerck, I., Vancanneyt, M., Swings, J., Dalgaard, P., Leisner, J.J. (2005). *Carnobacterium divergens* and *Carnobacterium maltaromaticum* as spoilers or protective cultures in meat and seafood: Phenotypic and genotypic characterization. *Systematic and Applied Microbiology*, 28(2), 151–164. https://doi.org/10.1016/j.syapm.2004.12.001.

Laursen, B.G., Leisner, J.J., Dalgaard, P. (2006). *Carnobacterium* species: Effect of metabolic activity and interaction with *Brochothrix thermosphacta* on sensory characteristics of modified atmosphere packed shrimp. *Journal of Agricultural and Food Chemistry*, 54(10), 3604–3611. https://doi.org/10.1021/jf053017f.

Leisner, J.J., Laursen, B.G., Prévost, H., Drider, D., Dalgaard, P. (2007). *Carnobacterium*: Positive and negative effects in the environment and in foods. *FEMS Microbiology Reviews*, 31(5), 592–613. https://doi.org/10.1111/j.1574-6976.2007.00080.x.

Leroi, F. (2010). Occurrence and role of lactic acid bacteria in seafood products. *Food Microbiology*, 27(6), 698–709. https://doi.org/10.1016/j.fm.2010.05.016.

Leroi, F. (2014). Role of bacteria in seafood products. In *Seafood Science*, Kim, S.-K. (ed.). CRC Press, Boca Raton.

Leroi, F. and Joffraud, J.-J. (2011). Microbial degradation of seafood. In *Aquaculture Microbiology and Biotechnology*, Volume 2, Montet, D. and Ray, R.C. (eds). CRC Press, Boca Raton.

Leroi, F., Arbey, N., Joffraud, J.-J., Chevalier, F. (1996). Effect of inoculation with lactic acid bacteria on extending the shelf-life of vacuum-packed cold smoked salmon. *International Journal of Food Science and Technology*, 31(6), 497–504. https://doi.org/10.1046/j.1365-2621.1996.00366.x.

Li, J., Yang, X., Shi, G., Chang, J., Liu, Z., Zeng, M. (2019). Cooperation of lactic acid bacteria regulated by the AI-2/LuxS system involve in the biopreservation of refrigerated shrimp. *Food Research International*, 120, 679–687. https://doi.org/10.1016/j.foodres.2018.11.025.

Lyhs, U. (2009). Microbiological methods. In *Fishery Products*, Rehbein, H. and Oehlenschlger, J. (eds). Wiley-Blackwell, Hoboken. https://doi.org/10.1002/9781444322668.ch15.

Lyhs, U. and Björkroth, J.K. (2008). *Lactobacillus sakei/curvatus* is the prevailing lactic acid bacterium group in spoiled maatjes herring. *Food Microbiology*, 25(3), 529–533. https://doi.org/10.1016/j.fm.2007.11.007.

Macé, S., Joffraud, J.-J., Cardinal, M., Malcheva, M., Cornet, J., Lalanne, V., Chevalier, F., Sérot, T., Pilet, M.-F., Dousset, X. (2013). Evaluation of the spoilage potential of bacteria isolated from spoiled raw salmon (*Salmo salar*) fillets stored under modified atmosphere packaging. *International Journal of Food Microbiology*, 160(3), 227–238. https://doi.org/10.1016/j.ijfoodmicro.2012.10.013.

Macé, S., Cardinal, M., Jaffrès, E., Cornet, J., Lalanne, V., Chevalier, F., Sérot, T., Pilet, M.-F., Dousset, X., Joffraud, J.-J. (2014). Evaluation of the spoilage potential of bacteria isolated from spoiled cooked whole tropical shrimp (*Penaeus vannamei*) stored under modified atmosphere packaging. *Food Microbiology*, 40, 9–17. https://doi.org/10.1016/j.fm.2013.11.018.

Maillet, A., Denojean, P., Bouju-Albert, A., Scaon, E., Leuillet, S., Dousset, X., Jaffrès, E., Combrisson, J., Prévost, H. (2021). Characterization of bacterial communities of cold-smoked salmon during storage. *Foods*, 10(2), 362. https://doi.org/10.3390/foods10020362.

Masschelein, J., Jenner, M., Challis, G.L. (2017). Antibiotics from gram-negative bacteria: A comprehensive overview and selected biosynthetic highlights. *Natural Product Reports*, 34(7), 712–783. https://doi.org/10.1039/c7np00010c.

Matamoros, S., Pilet, M.F., Gigout, F., Prevost, H., Leroi, F. (2009a). Selection and evaluation of seafood-borne psychrotrophic lactic acid bacteria as inhibitors of pathogenic and spoilage bacteria. *Food Microbiology*, 26(6), 638–644. https://doi.org/10.1016/j.fm.2009.04.011.

Matamoros, S., Leroi, F., Cardinal, M., Gigout, F., Chadli, F.K., Cornet, J., Prevost, H., Pilet, M.F. (2009b). Psychrotrophic lactic acid bacteria used to improve the safety and quality of vacuum-packaged cooked and peeled tropical shrimp and cold-smoked salmon. *Journal of Food Protection*, 72(2), 365–374. https://doi.org/10.4315/0362-028X-72.2.365.

Mejlholm, O., Bøknaes, N., Dalgaard, P. (2005). Shelf life and safety aspects of chilled cooked and peeled shrimps (*Pandalus borealis*) in modified atmosphere packaging. *Journal of Applied Microbiology*, 99(1), 66–76. https://doi.org/10.1111/j.1365-2672.2005.02582.x.

NF V01-003 (2018). Traçabilité et sécurité des aliments – management et hygiène – lignes directrices pour la réalisation de tests de vieillissement microbiologique – aliments périssables réfrigérés. Standard, Afnor Editions.

Ng, T.B., Cheung, R.C.F., Wong, J.H., Bekhit, A.A., Bekhit, A.E.-D. (2015). Antibacterial products of marine organisms. *Applied Microbiology and Biotechnology*, 99(10), 4145–4173. https://doi.org/10.1007/s00253-015-6553-x.

Nilsson, L., Gram, L., Huss, H.H. (1999). Growth control of *Listeria monocytogenes* on cold-smoked salmon using a competitive lactic acid bacteria flora. *Journal of Food Protection*, 62(4), 336–342. https://doi.org/10.4315/0362-028X-62.4.336.

Nilsson, L., Hansen, T.B., Garrido, P., Buchrieser, C., Glaser, P., Knochel, S., Gram, L., Gravesen, A. (2005). Growth inhibition of *Listeria monocytogenes* by a nonbacteriocinogenic *Carnobacterium piscicola*. *Journal of Applied Microbiology*, 98(1), 172–183. https://doi.org/10.1111/j.1365-2672.2004.02438.x.

Odeyemi, O.A., Burke, C.M., Bolch, C.C.J., Stanley, R. (2019). Spoilage microbial community profiling by 16S rRNA amplicon sequencing of modified atmosphere packaged live mussels stored at 4°C. *Food Research International*, 121, 568–576. https://doi.org/10.1016/j.foodres.2018.12.017.

Parihar, V.S., Barbuddhe, S.B., Danielsson-Tham, M.L., Tham, W. (2008). Isolation and characterization of *Listeria* species from tropical seafoods. *Food Control*, 19(6), 566–569. https://doi.org/10.1016/j.foodcont.2007.06.009.

Parlapani, F.F., Michailidou, S., Anagnostopoulos, D.A., Sakellariou, A.K., Pasentsis, K., Psomopoulos, F., Argiriou, A., Haroutounian, S.A., Boziaris, I.S. (2018). Microbial spoilage investigation of thawed common cuttlefish (*Sepia officinalis*) stored at 2°C using next generation sequencing and volatilome analysis. *Food Microbiology*, 76, 518–525. https://doi.org/10.1016/j.fm.2018.08.004.

Parlapani, F.F., Michailidou, S., Anagnostopoulos, D.A., Koromilas, S., Kios, K., Pasentsis, K., Psomopoulos, F., Argiriou, A., Haroutounian, S.A., Boziaris, I.S. (2019). Bacterial communities and potential spoilage markers of whole blue crab (*Callinectes sapidus*) stored under commercial simulated conditions. *Food Microbiology*, 82, 325–333. https://doi.org/10.1016/j.fm.2019.03.011.

Passerini, D., Kolypczuk, L., Macé, S., Pilet, M.F., Leroi, F. (2021). Biopréservation des produits de la mer avec des bactéries marines. *Techniques de l'Ingénieur*, BIO9240 v1, 1–7.

Pilet, M.-F. and Leroi, F. (2011). Applications of protective cultures, bacteriocins and bacteriophages in fresh seafood and seafood products. In *Protective Cultures, Antimicrobial Metabolites and Bacteriophages for Food and Beverage Biopreservation*, Lacroix, C. (ed.). Woodhead Publishing, Sawston. https://doi.org/10.1533/9780857090522.3.324.

Podeur, G., Leroi, F., Pilet, M.F., Prévost, H. (2016). Use of *Lactobacillus sakei* for the biopreservation of products from the sea. United States Patent US20160143302A1. https://patents.google.com/patent/US20160143302A1/fr [Accessed 23 April 2022].

Ravn, L., Christensen, A.B., Molin, S., Givskov, M., Gram, L. (2001). Methods for detecting acylated homoserine lactones produced by Gram-negative bacteria and their application in studies of AHL-production kinetics. *Journal of Microbiological Methods*, 44(3), 239–251. https://doi.org/10.1016/S0167-7012(01)00217-2.

Ross, T. (2000). Predictive modelling of the growth and survival of *Listeria* in fishery products. *International Journal of Food Microbiology*, 62(3), 231–245. https://doi.org/10.1016/S0168-1605(00)00340-8.

Saraoui, T., Fall, P.A., Leroi, F., Antignac, J.-P., Chéreau, S., Pilet, M.F. (2016). Inhibition mechanism of *Listeria monocytogenes* by a bioprotective bacteria *Lactococcus piscium* CNCM I-4031. *Food Microbiology*, 53, 70–78. https://doi.org/10.1016/j.fm.2015.01.002.

Saraoui, T., Cornet, J., Guillouet, E., Pilet, M.F., Chevalier, F., Joffraud, J.-J., Leroi, F. (2017). Improving simultaneously the quality and safety of cooked and peeled shrimp using a cocktail of bioprotective lactic acid bacteria. *International Journal of Food Microbiology*, 241, 69–77. https://doi.org/10.1016/j.ijfoodmicro.2016.09.024.

Saraoui, T., Leroi, F., Chevalier, F., Cappelier, J.-M., Passerini, D., Pilet, M.-F. (2018). Bioprotective effect of *Lactococcus piscium* CNCM I-4031 against *Listeria monocytogenes* growth and virulence. *Frontiers in Microbiology*, 9, 1564. https://doi.org/10.3389/fmicb.2018.01564.

Silbande, A., Adenet, S., Smith-Ravin, J., Joffraud, J.-J., Rochefort, K., Leroi, F. (2016). Quality assessment of ice-stored tropical yellowfin tuna (*Thunnus albacares*) and influence of vacuum and modified atmosphere packaging. *Food Microbiology*, 60, 62–72. https://doi.org/10.1016/j.fm.2016.06.016.

Silbande, A., Adenet, S., Chopin, C., Cornet, J., Smith-Ravin, J., Rochefort, K., Leroi, F. (2018a). Effect of vacuum and modified atmosphere packaging on the microbiological, chemical and sensory properties of tropical red drum (*Sciaenops ocellatus*) fillets stored at 4°C. *International Journal of Food Microbiology*, 266, 31–41. https://doi.org/10.1016/j.ijfoodmicro.2017.10.015.

Silbande, A., Cornet, J., Cardinal, M., Chevalier, F., Rochefort, K., Smith-Ravin, J., Adenet, S., Leroi, F. (2018b). Characterization of the spoilage potential of pure and mixed cultures of bacterial species isolated from tropical yellowfin tuna (*Thunnus albacares*). *Journal of Applied Microbiology*, 124(2), 559–571. https://doi.org/10.1111/jam.13663.

Silva, C.C.G., Silva, S.P.M., Ribeiro, S.C. (2018). Application of bacteriocins and protective cultures in dairy food preservation. *Frontiers in Microbiology*, 9, 594. https://doi.org/10.3389/fmicb.2018.00594.

Sivertsvik, M., Jeksrud, W.K., Rosnes, J.T. (2002). A review of modified atmosphere packaging of fish and fishery products – Significance of microbial growth, activities and safety. *International Journal of Food Science and Technology*, 37(2), 107–127. https://doi.org/10.1046/j.1365-2621.2002.00548.x.

Soe, C.Z., Pakchung, A.A.H., Codd, R. (2012). Directing the biosynthesis of putrebactin or desferrioxamine B in *Shewanella putrefaciens* through the upstream inhibition of ornithine decarboxylase. *Chemistry Biodiversity*, 9(9), 1880–1890. https://doi.org/10.1002/cbdv.201200014.

Sorée, M., Kolypczuk, L., Hadjiev, E., Lozach, S., Verrez-Bagnis, V., Delbarre-Ladrat, C., Hervio Heath, D., Passerini, D. (2023). Screening of marine lactic acid bacteria for *Vibrio parahaemolyticus* inhibition and application to depuration in Pacific oysters (Crassostrea gigas). *Journal of Applied Microbiology*, 134(2), lxac081. https://doi.org/10.1093/jambio/lxac081.

Stohr, V., Joffraud, J.J., Cardinal, M., Leroi, F. (2001). Spoilage potential and sensory profile associated with bacteria isolated from cold-smoked salmon. *Food Research International*, 34(9), 797–806. https://doi.org/10.1016/S0963-9969(01)00101-6.

Tahiri, I., Desbiens, M., Kheadr, E., Lacroix, C., Fliss, I. (2009). Comparison of different application strategies of divergicin M35 for inactivation of *Listeria monocytogenes* in cold-smoked wild salmon. *Food Microbiology*, 26(8), 783–793. https://doi.org/10.1016/j.fm.2009.05.003.

Tasaku, S., Siripornadulsil, S., Siripornadulsil, W. (2017). Inhibitory activity of food-originated *Pediococcus pentosaceus* NP6 against *Salmonella enterica* serovar typhimurium in Nile tilapia by-products. *Chiang Mai Journal of Science*, 44(2), 383–393.

Todd, E.C.D. and Notermans, S. (2011). Surveillance of listeriosis and its causative pathogen, *Listeria monocytogenes*. *Food Control*, 22(9), 1484–1490. https://doi.org/10.1016/j.foodcont.2010.07.021.

Trejo-González, L., Gutiérrez-Carrillo, A.-E., Rodríguez-Hernández, A.-I., del Rocío López-Cuellar, M., Chavarría-Hernández, N. (2021). Bacteriocins produced by LAB isolated from cheeses within the period 2009–2021: A review. *Probiotics and Antimicrobial Proteins*. https://doi.org/10.1007/s12602-021-09825-0.

Union du mareyage français (UMF) (2010). Guide des bonnes pratiques et application de l'HACCP. *Editions des journaux officiels*, 5941/2010.

Vescovo, M., Scolari, G., Zacconi, C. (2006). Inhibition of *Listeria innocua* growth by antimicrobial-producing lactic acid cultures in vacuum-packed cold-smoked salmon. *Food Microbiology*, 23(7), 689–693. https://doi.org/10.1016/j.fm.2005.12.002.

Wang, Y., Wang, Y., Chen, J., Koseki, S., Yang, Q., Yu, H., Fu, L. (2021). Screening and preservation application of quorum sensing inhibitors of *Pseudomonas fluorescens* and *Shewanella baltica* in seafood products. *LWT*, 149, 111749. https://doi.org/10.1016/j.lwt.2021.111749.

Waters, C.M. and Bassler, B.L. (2005). Quorum sensing: Cell-to-cell communication in bacteria. *Annual Review of Cell and Developmental Biology*, 21(1), 319–346. https://doi.org/10.1146/annurev.cellbio.21.012704.131001.

Weiss, A. and Hammes, W.P. (2006). Lactic acid bacteria as protective cultures against *Listeria* spp. on cold-smoked salmon. *European Food Research and Technology*, 222(3–4), 343–346. https://doi.org/10.1007/s00217-005-0116-9.

Wiernasz, N., Cornet, J., Cardinal, M., Pilet, M.-F., Passerini, D., Leroi, F. (2017). Lactic acid bacteria selection for biopreservation as a part of hurdle technology approach applied on seafood. *Frontiers in Marine Science*, 4, 119. https://doi.org/10.3389/fmars.2017.00119.

Wiernasz, N., Leroi, F., Chevalier, F., Cornet, J., Cardinal, M., Rohloff, J., Passerini, D., Skırnisdóttir, S., Pilet, M.-F. (2020). Salmon gravlax biopreservation with lactic acid bacteria: A polyphasic approach to assessing the impact on organoleptic properties, microbial ecosystem and volatilome composition. *Frontiers in Microbiology*, 10, 3103. https://doi.org/10.3389/fmicb.2019.03103.

Wiernasz, N., Gigout, F., Cardinal, M., Cornet, J., Rohloff, J., Courcoux, P., Vigneau, E., Skírnisdottír, S., Passerini, D., Pilet, M.-F. et al. (2021). Effect of the manufacturing process on the microbiota, organoleptic properties and volatilome of three salmon-based products. *Foods*, 10(11), 2517. https://doi.org/10.3390/foods10112517.

Yang, Z., Tao, X., Zhang, H., Rao, S., Gao, L., Pan, Z., Jiao, X. (2019). Isolation and characterization of virulent phages infecting *Shewanella baltica* and *Shewanella putrefaciens*, and their application for biopreservation of chilled channel catfish (*Ictalurus punctatus*). *International Journal of Food Microbiology*, 292, 107–117. https://doi.org/10.1016/j.ijfoodmicro.2018.12.020.

6

Fish Parasites and Associated Risks

Mélanie GAY[1] and Véronique VERREZ-BAGNIS[2]
[1] *ANSES, Boulogne-sur-Mer, France*
[2] *Ifremer, Nantes, France*

6.1. Introduction

With nearly 35,000 recorded species, fish is a very diverse group of aquatic vertebrates that participate in many parasitic cycles, as definitive hosts (hosting the sexual or adult parasitic stages), intermediate hosts (hosting the asexual or larval parasitic stages) or paratenic or transport hosts (that may not participate in the parasitic cycle, but often effectively carry a larval stage to the definitive host) (Dei-Cas 1996).

Most fish are predators of other fish and/or invertebrates. They are therefore exposed to infection by numerous parasitic protists and metazoans through their food, some of which represent a risk for human health. They can also be infested by active penetration of parasites at different stages (trematode cercariae, actinospores of Myxosporidia). The risk for the consumer comes either from the actual parasitism of the fish by infective stages of parasites pathogenic to humans or from the contamination of external surfaces or nets by protists (e.g. *Cryptosporidium* sp. oocysts) (Roberts et al. 2007) or similar organisms (e.g. Microsporidia spores) contaminating fresh water or oceans.

In addition to the description of the main parasites responsible for zoonoses and present in fishery products, and the description of induced pathologies, biological cycles and distribution areas, this chapter[1] will address various other aspects related to the presence of these pathogens such as induced allergic phenomena, detection and identification methods, a particular case of aquaculture fish, and finally existing regulations and prevention methods.

6.2. Prevalence and significance of parasitosis for the sector

The presence of parasites in fishery and aquaculture products induces two main types of risks: economic and health-related. On the one hand, the presence of zoonotic parasites induces a health risk. Indeed, some of the parasites present can induce more or less severe and frequent human pathologies (Audicana and Kennedy 2008; Dupouy-Camet et al. 2020b). On the other hand, the presence of parasites can induce different economic costs, directly or indirectly. Beyond the regulatory aspect, prohibiting the marketing of obviously parasitized products (see section 6.7), the consumer may be put off by the presence of parasite during consumption, leading to a global rejection of this category of fish products. Furthermore, for aquaculture products, the use of veterinary treatments, the loss of yield or the mortalities induce additional costs impacting the profitability of the production (Costello 2009; Frenette et al. 2022; Zhong et al. 2022). During processing stage, the time spent sorting the wild or farmed products (careful evisceration, trimming, manual elimination, observation on a candling table or by eye) also induces additional costs without guaranteeing a total elimination of parasites (Levsen et al. 2005; Mercken et al. 2022).

Some species of parasites and/or in some geographic areas may have very high prevalences. For example, Anisakidae have been described as occurring in more than 200 species of fish with prevalences as high as 100% (Fiorenza et al. 2020; Mercken et al. 2020). In some Asia regions, Opisthorchiidae are present in wild and farmed freshwater fish at very high intensities (Long et al. 2021).

6.2.1. *Protists*

Representatives of *Apicomplexa*, amoebae, flagellates and ciliates frequently parasitize the internal organs, gills or body surface of fish. *Apicomplexa* (Coccidia, Hemosporina) have been reported as significant pathogens for fish, especially farmed fish. There are about 250 species of coccidia parasites of fish, belonging to

1. Some of the aspects of fish parasites have been mentioned in the chapter titled "Minimally Processed Seafood Products and Bacterial and Parasitic Hazards" by Brauge et al. (2022) in the book *Hazards in the Food Processing and Distribution Chain*, edited by N. Haddad, published by ISTE, London, and John Wiley & Sons, New York.

eight genera. Sometimes pathogenic to their hosts (Bruno et al. 2006), the parasitic specificity of the protists explains, at least partially, the lack of risk of protists from fish, to consumer in general.

The coccidia of the genus *Cryptosporidium* deserve, however, special consideration. Two species have been described: *C. molnari* in gilt-head sea bream (*Sparus aurata*) and European bass (*Dicentrarchus labrax*) (Alvarez-Pellitero and Sitjà-Bobadilla 2002) and *C. scophthalmi* in turbot (*Scophthalmus maximus*) (Alvarez-Pellitero et al. 2004). Several species of *Cryptosporidium* from various mammals have been shown to transgress the species barrier to infect humans (Guyot et al. 2001). However, we have no data on the potential infectivity of fish-infested *Cryptosporidium* to humans or other mammals. These aquatic vertebrates may, however, passively transport human-infectious oocysts of other *Cryptosporidium* species that frequently contaminate fresh, estuarine or marine waters (Guyot et al. 2005). In their epidemiological study, Certad et al. (2019) identified *Cryptosporidium* spp. in 46 of 1,853 fish collected from the Northeast Atlantic Ocean. The most infested species was saithe. Thus, for fishermen, who are highly exposed to infection through direct daily handling of their fishery products, the risk of infection is much higher than that of the general population (Roberts et al. 2007). In addition, edible bivalves are quite frequently infected with *Cryptosporidium* oocysts (Li et al. 2006); however, human infection from this source has not been reported, to our knowledge.

In addition, fish are hosts of several species of amphizoic amoebae (e.g. free-living amoebae capable of invading their tissues). However, apart from the genera *Naegleria* and *Acanthamoeba*, responsible for amoebic meningoencephalitis and keratitis in humans, other amoebae which can be isolated from fish have not been reported for this host. As for *Naegleria* and *Acanthamoeba* infections, isolated from freshwater fish organs on several occasions (Dykova and Lom 2004), the circumstances of occurrence of these infections in humans (bathing in fresh water for *Naegleria*, contamination of duckweed or other mechanisms for *Acanthamoeba*) have not been associated with fish consumption.

Among other parasitic protists, ecto- or endoparasitic flagellates (e.g. trypanosomes) and the abundant fauna of ciliates, sometimes highly specialized, which parasitize fish, do not, however, lead to any risk for the consumer.

6.2.2. *Microsporidia and mesomycetozoea*

These organisms, with strong fungal affinities, constitute a very large group including more than 100 genera with more than 1,000 different species, of which only a small percentage (11 species) are responsible for infection in humans. While many species infect the animal world, four species share human pathology: *Enterocytozoon bieneusi*, *Encephalitozoon intestinalis*, *E. hellem* and *E. cuniculi*. Opportunistic parasitosis, microsporidiosis is mainly observed in patients undergoing profound immunodeficiency. However, cases of intestinal microsporidiosis have been reported in immunocompetent subjects returning from tropical areas and asymptomatic carriage has also been reported.

Fish can be the source of microsporidiosis through two mechanisms. On the one hand, spores of Microsporidia pathogenic to humans can be found on the coastline. Those of *Pleistophora* sp. have been detected in the Seine, the Loire, in irrigation waters (Fournier et al. 2000) and in freshwater bivalves in Ireland (Graczyck et al. 2004). Thus, the considerations made above about *Cryptosporidium* spp. passively contaminating fish for human consumption can be applied to Microsporidia spores as well. On the other hand, Microsporidia very similar to those parasitizing fish have been identified in AIDS patients (Didier et al. 2004; Cali et al. 2005). The hypothesis of fish-to-human transmission has therefore been formulated (Curry 1999), but no evidence in this sense has been provided for the moment.

Finally, it is necessary to mention the Mesomycetozoea, which are organisms phylogenetically placed at the boundary of Fungi and Animalia (Mendoza et al. 2005). They include *Rhinosporidium seeberi*, a human pathogen that causes rhinosporidiosis, a chronic disease characterized by the presence of polypous lesions in the mucous membranes, mainly in the nasal, ocular and pharyngeal regions. The genera *Dermocystidium*, *Ichthyophonus* and the "rosette agent" are recognized pathogens of fish. *Ichthyophonus hoferi*, in haddock, salmon, plaice and herring, produces rounded formations ("sporangia") of variable size that generate endospores within invaded tissues. *Dermocystidium* sp. and the "rosette agent", renamed *Sphaerothecum destruens* in 2004 (Arkhush et al. 2003), attack gills and other tissues of salmonids and other fish. Mesomycetozoea also parasitize amphibians, but other than *R. seeberi*, these microorganisms do not appear to parasitize mammals.

6.2.3. *Stramenopiles*

Our notions on the taxonomy, phylogeny and pathogenicity of members of the genus *Blastocystis* have evolved considerably in recent years. It is now known that they belong to the stramenopile group (brown algal lineage) (Lecointre and Le Guyader 2001) and that they show considerable genetic diversity with at least 12 species with low host specificity (Noël et al. 2005). Because of their interspecies transmissibility and pathogenicity, the zoonotic potential of *Blastocystis* spp. is recognized (Leelayoova et al. 2004; Hoevers and Snowden 2005). Gantois et al.

(2020) described the presence of *Blastocystis* in 3.5% of 345 fish analyzed (herring, mackerel, whiting and saithe). The *Blastocystis* subtypes identified belongs either to subtypes described in human pathologies or to subtypes isolated from poikilothermic animals. These results showed the possibility of transmission of *Blastocystis* by fish, but, to date, no human case from fish products has been described.

6.2.4. *Metazoans*

Myxozoa, plathelminths, acanthocephalans, nematodes, pentastomes, hirudines and arthropods frequently parasitize fish at all latitudes. In addition to their role as definitive host, fish, especially teleosts, play essential roles as intermediate hosts and paratenic hosts in the life cycle of trematodes, cestodes, nematodes and even pentastomes whose definitive hosts are the large predators placed at the top of the pyramid of aquatic ecosystems (carnivorous sharks, mammals, crocodiles, piscivorous birds). Some of these parasites can be responsible for diseases in fish and/or be potentially at the origin of zoonoses (e.g. opisthorchiidoses, diphyllobothhriases, anisakidoses) (Dupouy-Camet and Peduzzi 2004; Chai et al. 2005; Dupouy-Camet et al. 2020a). The main parasites responsible for zoonoses and present in fishery products are described below. The case of Myxozoa, recently identified as inducing human pathologies, is identified in the section on emerging zoonoses below.

6.2.4.1. *Trematodes*

The trematode class consists of three subclasses: Monogenea, Aspidogastrea and Digenea. Only trematodes in the Digenea subclass can infest humans. These trematodes are responsible for the infection of more than 40 million people worldwide with a higher prevalence in the South East Asian and Western Pacific regions (World Health Organization Regional Office for the Western Pacific 2002 (WHO 2015)). Monogenea and Aspidogastrea frequently parasitize fish but will not be considered in this book because they do not pose a health risk to humans.

Digenea are non-segmented endoparasites, hermaphrodites, generally obligate parasites (Hyman 1951). Adult Digenea are generally located either in the digestive tract and associated organs (liver, biliary organs, pancreas, lungs) or in the urinary tract. Their biological cycle is heterotopic, involving the final host, a first intermediate host, usually a gastropod mollusk, and then, frequently, a second intermediate host. This one is actively infested by a free-swimming larval stage, the cercaria.

Although 33 species of digenetic trematodes are known to be transmissible to humans through the consumption of fish and shellfish, only a few species have been implicated in zoonotic diseases with high public health impact (Chai et al. 2005; Chai and Jung 2020). Among these, the family Opisthorchiidae, which includes distomas (or flukes) responsible for hepatic distomatosis in the Far East, Europe and America, is the one with the highest health impact due to the frequency of infections and their severity (Chai et al. 2005). In addition to these hepatic affections, trematodes can be responsible for digestive pathologies, mainly Heterophyidae, or pulmonary, such as *Paragonimus*.

6.2.4.1.1. Opisthorchiidae – hepatic pathologies

This family includes the species *Clonorchis sinensis* ("Chinese liver fluke"), *Opisthorchis viverrini*, *O. felineus* ("cat liver fluke") and *Metorchis bilis* in Europe and Asia, and *O. guayaquilensis* and *M. conjunctus* in the Americas. Opisthorchiidae trematodes reach sexual maturity in the bile ducts of domestic and wild mammals including humans, into which the adults release their embryonated eggs. The eggs are released with the feces. The next stage, the miracidium, emerges from these eggs and actively infests a gastropod mollusk. The next stage, the cercaria, free and swimmer, is emitted in the external environment. It swims to find, in less than 24–48 hours, a fish, the second intermediate host. It actively penetrates and encysts in the underlying skin or muscle where it becomes a metacercariae. More rarely, the cercariae may infest fins, scales or internal organs. In the case of *C. sinensis,* approximately 40 species of freshwater fishes in the family Cyprinidae may act as intermediate hosts (Neva and Brown 1994).

Humans contract the infection by eating raw, undercooked, pickled, dried or smoked freshwater or brackish fish containing encysted metacercariae. After ingestion, these larvae encyst in the duodenum, releasing young flukes that reach the liver by ascending the bile duct. The longevity of these parasites is significant, varying from 8 to more than 20 years.

Opisthorchiidoses are zoonoses whose agents circulate widely in ecosystems and have many mammalian reservoirs. Their geographic distribution, like that of other trematodoses, is conditioned by the distribution of host mollusks. These parasitoses are present in a significant way in Eastern and Central Europe, especially in Russia, as well as in Asia. Numerous studies report 35 million people infected with *C. sinensis*, including 15 million in China (Lun et al. 2005; Furst et al. 2012). The prevalence of opisthorchidosis is currently increasing due to the success of raw fish dishes. In addition, clonorchiasis is sometimes reported in non-endemic areas such as the USA or Western Europe, either carried by Asian emigrants or after ingestion of imported marinated or smoked fish preparations containing live metacercariae. In the early 1980s, a prevalence of clonorchiasis of 14% was found in a population of 1,170 Indochinese refugees who arrived in Northern France (Schmit et al. 1982). Human cases of opisthorchiasis are occasionally described in Italy (Wunderink et al. 2014; Scaramozzino et al. 2018).

Probably, the most striking aspect of the pathological changes induced by Opisthorchiidae, particularly *C. sinensis* and *O. viverrini*, is the frequent association of chronic opisthorchidosis with the development of (usually well-differentiated) cholangiocarcinoma of the intra- or extra-hepatic bile ducts (Furst et al. 2012; Pengput and Schwartz 2020). The mechanism of this neoplasia remains unknown. At present, the malignant degeneration associated with hepato-biliary opisthorchidosis, which darkens the prognosis of these trematodoses, constitutes the strongest association of an eukaryotic pathogen with a cancerous lesion.

6.2.4.1.2. Heterophyidae and other trematodes – intestinal pathologies

Human intestinal distomatoses are caused by trematodes belonging mainly to the families Heterophyidae and Echinostomatidae. The family Heterophyidae is the most significant. It includes small trematodes of which about 35 species are responsible for zoonotic intestinal distomatosis. The taxonomic position of the genera belonging to this family is, however, controversial. Several authors defend the existence of this family (Chai and Jung 2020) while others include these genera in the family Opisthorchiidae (Sokolov et al. 2021). In this chapter, these parasites are treated separately from the Opisthorchiidae because of the differences in the pathologies induced.

The life cycle of intestinal distomatosis agents follows the general pattern of that of the Opisthorchiidae. The human host becomes accidentally infected by ingestion of infective metacercariae contained in intermediate second hosts, mainly Mugilidae, Cyprinidae and Salmonidae for Heterophyidae, Cyprinidae for Echinostomatidae and Salmonidae for *N. salmincola*.

Generally considered to be of lesser importance than hepato-biliary distomatoses, intestinal distomatoses do induce less severe symptomatology, centered on diarrhea and abdominal pain. However, heterophyidoses, especially those caused by *Stellantchasmus falcatus*, *Haplorchis* sp. and *Procerovum* sp., can often cause fatal lesions of the brain, spinal cord and heart (especially in the Philippines), probably as a result of the embolization of egg clusters during massive infections (Chai et al. 2005). As for *N. salmincola*, whose infection is generally contracted by eating insufficiently cooked infected salmon, the parasite is itself a host for the bacterium *Neorickettsia helminthoeca*. This prokaryote causes the disease called “salmon poisoning” in dogs, which is fatal without treatment. The preferred site of metacercariae in salmon is the kidney, so careful filleting eliminates the risk of infection (Macpherson 2005). To date, 74 species of the family Heterophyidae have been described as zoonotic, with an estimated number of 40–50 million affected patients (Chai and Jung 2020). It should be noted that these data are based solely on estimates, as diagnosis relies on egg identification. Furthermore, the distribution of these parasites is global, and little data are available on their impact in some cases. Thus, some genera have been described as minor zoonotic, with little data available on their distribution, and the pathologies induced (Chai and Jung 2020).

6.2.4.2. *Cestodes*

Cestodes are found in adult or larval stages in aquatic vertebrates. It is the larval cestodes that pose a risk to the consumer, either by developing in the intestine to reach adult state (dibothriocephalosis) or by remaining in the larval state and invading tissues (sparganosis). The agents of dibothriocephalosis are species of the genera *Dibothriocephalus*, intestinal parasites of birds and fish-eating mammals. On the contrary, sparganosis is essentially determined by the plerocercoid larvae (or spargana) of cestodes of the genus *Spirometra*, adult parasites of carnivorous mammals. The cestodes responsible for these two diseases belong to the family Diphyllobothriidae. The life cycle of the cestodes of these two genera follows essentially the same pattern, although the ecological context differs considerably depending on the species. The eggs are laid without embryos, their development continuing in fresh or sea water, depending on the species, until the formation of a ciliated embryo. Once hatched and ingested by copepods, this embryo develops into a procercoid larva. The procercoid larva continues its development in fish in the case of *Dibothriocephalus* spp. or in frogs in the case of *Spirometra* spp. These second intermediate hosts become infested by ingesting the infested copepods. The procercoid larva passes through the digestive wall, encysts in various organs and evolves into a plerocercoid larva or sparganum, a whitish, non-segmented worm. It may remain alive in the tissues for months or years, waiting for this second intermediate host to be ingested by the final host. In reality, these fish, in the case of *Dibothriocephalus*, or frogs, in the case of *Spirometra*, are often ingested either by other fish (perch or pike) in the former case or, in the latter case, by snakes (Dei-Cas et al. 1976), other reptiles or other predatory vertebrates. The plerocercoid larvae then pass through the digestive wall of these paratenic hosts (holding hosts) and encyst alive in their muscles or in other locations. They will evolve into the adult stage in the intestine of their definitive host (dogs, other carnivores, bears, humans, birds, for *Dibothriocephalus*; wild or domestic carnivores for *Spirometra*).

Twenty million people are estimated to be infested with adult Diphyllobothriidae worldwide (Chai et al. 2005). Humans and other piscivorous mammals become infected by eating raw or undercooked fish meat infested with plerocercoid larvae. The zoonosis is most prevalent in countries where consumption of raw or marinated fish is common.

D. latus is the largest of the human parasites (from two to eight or even up to 20 m) and can live for several years. This cestode is apparently responsible for the largest number of cases of dibothriocephalosis. Perch, pike or lake trout are frequent hosts in Europe. *D. latus is* apparently restricted to the Palearctic (northern Eurasia) and Nearctic (North America, especially Alaska, Great Lakes) areas, and according to Dupouy-Camet and Peduzzi (2004), its prevalence is reported to be decreasing in the Baltic and Scandinavian areas, while it seems to persist in the French and Italian Alpine areas (Dupouy-Camet et al. 2015). Outbreaks of dibothriocephalosis in South America are either due to human introduction or due to other species of the genus *Dibothriocephalus*, such as *D. dendriticum*, especially since the responsible plerocercoid larvae seemed to be present in salmonids (rainbow trout), who are not the typical hosts of *D. latus*.

Other species of the genus *Dibothriocephalus* can infest also humans as adults but apparently less frequently. The most significant are: *D. pacificum*, *D. cordatum*, *D. ursi*, *D. dendriticum*, *D. lanceolatum*, *D. dalliae*, *D. nihonkaiense*, *D. klebanovskii* and *D. yonagoensis*. Secondary hosts are freshwater fish, anadromous or catadromous, with the exception of species from marine ecosystems, such as *D. pacificum*, *D. cameroni*, *D. cordatum* and *Diplogonoporus* sp. whose second intermediate host is a marine fish. *D. pacificum* is a parasite of sea lions transmitted to humans by marine fish in coastal areas of Chile, Peru, Ecuador and Japan, where the consumption of raw fish dishes (ceviche, sushi and others) is widespread (Chai et al. 2005). Cases of dibothriocephalosis can also be observed outside of specific geographic areas. For example, a case of human infection with *Diplogonoporus balaenopterae*, probably contracted by eating fish imported from Japan, has been reported in Spain (Clavel et al. 1997). Recent studies (Gustinelli et al. 2016; Menconi et al. 2020b; Menconi et al. 2021) have shown the presence of *D. latus* larvae in perch caught from Italian lakes. Prevalences showed spatial variations probably related either to the presence of human fecal contamination or to environmental differences.

Between 2016 and 2020, 11 human cases of *Dibothriocephalus nihonkaiense* infestation were described in France. The main route of infestation for this species is salmon. The authors who described these cases assumed the consumption of imported salmon that had not undergone sanitizing freezing (Autier et al. 2019; Robert-Gangneux et al. 2019; Greigert et al. 2020; Paugam and Yéra 2021).

Infestation with *Dibothriocephalus* sp. is usually caused by a single worm, but intestinal obstructions caused by several specimens are sometimes observed (Neva and Brown 1994). The infestation may have few or no symptoms or may cause a variety of digestive symptoms (abdominal pain, gastric heaviness, diarrhea, nausea, vomiting, proglottid emissions), associated with weight loss, asthenia, neurological disorders, malnutrition and anemia. Diagnosis is made by detection of *Dibothriocephalus* sp. eggs in the stool or with vomit, sometimes associated with proglottids.

6.2.4.3. *Nematodes*

Nematodes are round worms of very variable sizes with separate sexes and a remarkably uniform structure. They are non-segmented and have a pseudocelomic body cavity. Nematodes are one of the richest zoological groups in terms of number of species (about 20,000 known species but estimated to be several million). Many lineages have evolved to parasitize vertebrates, invertebrates and even plants. Among the numerous parasitic nematodes of fishes, only a few are of risk for the consumer. The most significant ones belong to the orders Ascaridida (families: Anisakidae, with the genera *Anisakis*, *Pseudoterranova*, *Contracaecum* and Dioctophymatidae, with the species *Dioctophyma renale* and *Eustrongylides* sp.) and Spirurida (family Gnathostomatidae: *Gnathostoma* sp.).

6.2.4.3.1. Anisakidosis

The genus *Anisakis* was described by Dujardin in 1845, but these worms were known as fish parasites since the 13th century. Anisakiasis (or anisakidosis) is a digestive parasitosis resulting from ingestion of raw or undercooked fish, or cephalopods, contaminated with larvae of ascarid nematodes of the genera *Anisakis*, *Contracaecum* and *Pseudoterranova* (*Phocanema*). Human infestation may remain asymptomatic or have a clinical expression of variable significance depending on larval localization, the extent of the infestation and the intensity of the tissue response. Diagnosis is difficult and often only endoscopic or surgical treatment is effective (Dei-Cas 1996).

The definitive hosts of *Anisakis* sp., including *A. simplex*, "herring worm", *Pseudoterranova* including *P. decipiens*, called "cod worm" and *Contracaecum* sp., are, respectively, whales, pinnipeds and fish-eating birds. Third-stage larvae belonging to *A. simplex*, *A. pegreffii*, *P. decipiens* and *P. azarasi* have been described as causing human anisakidosis (Yera et al. 2016; Brunet et al. 2017; Suzuki et al. 2021). On the contrary, *Hysterothylacium* sp., belonging to the Anisakidae or Raphidascaridae families depending on the authors, whose adults are found in teleost fish, would not be able to determine anisakidosis in endothermic hosts, due, at least in part, to their susceptibility to high temperatures (they do not tolerate temperatures > 30°C), which has been shown experimentally (Gonzalez 1998).

Anisakidosis is a cosmopolitan disease, most common in temperate and cold regions around the world. Its prevalence is higher in regions where raw fish meals are consumed, such as Japan (1,000 cases per year) where 95% of anisakidosis cases have been reported (Sterling 2006) or the Netherlands. In the latter country, since 1965, legislation has required all fish intended for human consumption to be frozen at −20°C for more than 24 hours; this measure has led to a very significant reduction in the incidence of anisakidosis, which was previously 300 cases per year (Dei-Cas 1996). Sporadic cases of anisakidosis have been reported in France, Belgium, Germany, England, USA, Canada and Chile (Adroher-Auroux and Benitez-Rodriguez 2020). The prevalence in the Mediterranean basin and Africa remains poorly known, as also in Uruguay where *Anisakis* spp. have been found in sea lions and larvae in marine fish (Botto 1974; Botto et al. 1976). In France, anisakidosis is rather rare. Between January 1985 and September 1987, 21 cases were diagnosed (Hubert et al. 1989). Between 2010 and 2014, a similar study counted 37 cases of anisakidosis, this pathology thus showing no significant increase between the two studies (Yera et al. 2018). However, parasitism rates in consumed fish are very high. Two major research programs have assessed the distribution of these larvae in consumed fish. Between 2010 and 2014, the Fish-Parasites program (ANR-10-ALIA-004) determined the prevalence and intensity of Anisakidae larvae in 1,768 marine fish belonging to 18 species. Overall, 57% of the fish analyzed had at least one larva. However, significant differences were observed according to fish species, size or geographical origin. The distributions between fillets and viscera were also very variable depending on the fish species. The European program PARASITE (Grant Agreement (GA) no. 312068) studied the distribution of Anisakidae larvae in European fishing waters with the analysis of more than 17,000 individuals belonging to 16 species. Beyond the differences in distribution observed between fish species, this large-scale study has highlighted geographical "hotspots" with very high values, differences in the distribution of Anisakidae species, a scale of risk for the consumer according to the species of fish consumed, and also the possibility of considerably reducing this risk by steps such as trimming the fillets (Levsen et al. 2018). In 2017, a surveillance plan was initiated and coordinated by the General Directorate of Food of the French Ministry of Agriculture on the presence of Anisakidae larvae in fish at the distribution stage. The samples were analyzed by non-destructive detection methods (used by professionals) and by a destructive and exhaustive detection method. Seven commercially important fish species were selected and 205 samples were analyzed. Overall, the rate of samples infested with parasites visible to the naked eye was 42.9% and that obtained by the destructive and exhaustive detection method was

62.9%. The prevalences of infestation observed by the latter method ranged from 29.7% (saithe) to 88.9% (whiting) and were significantly different between fish species (Gay et al. 2019). Thus, many studies on the distribution of Anisakidae larvae could be cited and analyzed, each with different approaches and/or objectives (Timi et al. 2014; Gomes et al. 2020; Mille et al. 2020; Smaldone et al. 2020; Bao et al. 2021). Recently, three teams have pooled these studies to establish meta-analyses of the distribution of Anisakidae larvae in fishery products (Fiorenza et al. 2020; Mercken et al. 2020; Rahmati et al. 2020). These studies have highlighted the influence of different factors on this distribution, such as the distribution and biology of the hosts (fish and marine mammals), the improvement of detection methods, the biology of the parasites, the pollution or the overfishing.

Adult Anisakidae nematodes parasitize the stomach of cetaceans (whales, dolphins, porpoises) and pinnipeds (seals, sea lions, walruses). After mating, the female lays eggs which, eliminated with the host's feces, will embryonate and hatch in the bottom of the sea. These larvae undergo two intra-ovular molts (L1-L2-L3) and keep the cuticle of the second larval stage (Koie and Fagerholm 1993). These are ingested by euphausiid crustaceans (krill in the North Sea, *Meganyctiphanes norvegica*, other species), decapods, copepods or amphipods, and will continue their evolution by locating in the hemocoel of these first intermediate hosts (Dei-Cas 1996; Chai et al. 2005). Marine mammal definitive hosts could become directly infested with L3s during ingestion of infested crustaceans (a whale can consume two to three tons of euphausiids per meal). These are also the source of infestation for many marine fish and squid that act as paratenic hosts. In fish, L3 larvae are found in the digestive tract, abdominal cavity, plated on the gonads, in the liver or in the muscles (Mudry et al. 1986). These larvae are infectious to marine mammal definitive hosts, to other piscivorous animals, such as larger fishes, potential paratenic hosts (Smith and Wootten 1978) or to humans, accidental hosts. Larvae can thus be transmitted from fish to fish through predation relationships because they are able to resettle in a new host. Carnivorous fishes and squids corresponding to a high level in the food chain act as parasite concentrators. When infested fish or squid are ingested by the final host marine mammals, the third stage larva molts twice and becomes an adult. On the contrary, in humans, which constitute a parasite dead end, the larvae will dig itself into the digestive wall and may cause, in its thickness or in the neighboring tissues, the formation of a parasite granuloma rich in eosinophils.

Humans become infected by ingesting raw, undercooked, smoked or marinated fish. Sushi, a typical Japanese dish, is often a source of infestation. The larvae are particularly resistant. They keep their mobility intact for several days at 4°C in tap water, do not die at −20°C for less than 20 hours and resist for more than 30 minutes at 45°C (Dei-Cas 1996).

Human infestation may remain asymptomatic or manifest itself with non-specific symptoms. The significance of the infestation depends on the importance of the infestation, the location of the larval granulomas in the gastrointestinal tract and the host response. Depending on the location of the larval damage, three groups of clinical forms can be distinguished: (i) gastric (8% in the Netherlands, 65% in Japan): pseudo-ulcerous syndromes, sometimes pseudo-cancerous with impairment of the general condition and digestive hemorrhages (hematemesis, melena); (ii) intestinal (90% in Holland, 30% in Japan): predominantly occlusive or pseudo-occlusive signs; (iii) other localizations: esophageal, colonic, rectal, mesenteric, pancreatic and hepatic. Pulmonary forms, with fever, chest pain, hypoxemia, lung lesions and pleural effusion are rare (Matsuoka et al. 1994). Diagnosis of anisakidosis is often difficult. The history (abdominal pain with or without digestive bleeding after ingestion of raw fish or squid) and epidemiological data (patient living in a fishing area, frequent consumption of raw or undercooked, marinated, smoked or lightly salted fish meals) sometimes lead to the diagnosis of anisakidosis. If the topography of the pain suggests gastric involvement, endoscopy may be used to confirm the diagnosis and directly remove the larvae. The medical imaging techniques may be useful. Parasitological diagnosis consists of identifying the larvae, either those removed during endoscopy or those visible in the histological sections of the operating rooms. This identification is the only current means of providing a diagnosis of certainty. However, when larvae are not found in the sections examined, but the clinico-epidemiological context, the hypereosinophilia, the results of the serodiagnosis and the histopathological examination are concordant, the diagnosis of anisakiasis can be retained. Serodiagnosis is useful and many techniques have been proposed (Huang and Bussieras 1988).

6.2.4.3.2. Gnathostomosis

Gnathostomosis is a zoonotic parasitosis determined by larvae or immature adults of spirurid nematodes of the genus *Gnathostoma*. This disease, contracted by ingestion of raw or undercooked food prepared from freshwater fish, and also from frogs, snakes, birds and mammals, is frequent in Asia, particularly in Japan and Thailand. Its severity depends on the location of the worms, the neurological forms being particularly dreadful. Two species are generally incriminated: *G. spinigerum*, a parasite of wild and domestic felids, and *G. hispidum*, a parasite of pigs and boars (Dei-Cas 1996; Liu et al. 2020). However, two other species have an increasing impact in Japan (Nawa 1991): *G. nipponicum* and *G. doloresi*. Human cases of gnahostomosis reported in France and Europe seem to correspond to people having eaten raw fish in infested areas (Ménard et al. 2003). To date, 5,000 human cases have been described worldwide, mainly in Japan, Thailand and China (Liu et al. 2020).

Not only domestic cats but also tigers, leopards and dogs are hosts of *G. spinigerum* (Acha and Szyfre 1989). In rural areas of Thailand, 30% of cats and 26% of dogs have been parasitized by *G. spinigerum* (Jaroonvesama 1988). In markets in Thailand, infestation rates were 37% in bony fish, 80% in eels and 90% in frogs (Daensvang 1972, 1982). Sporadic cases have been reported in Mexico (Leon-Regagnon et al. 2005; Ligon 2005), California, Australia and Israel (Acha and Szyfres 1989). *G. hispidum* is relatively common in pigs but rare in humans.

G. spinigerum is a small nematode that inhabits the gastric wall of felids or canids, definitive hosts. The eggs reach the outside environment with the feces. In fresh water, they give birth to a free swimming larva (L1). It is ingested by a *Cyclops* copepod where it molts into L2. When the parasitized copepods are ingested by a freshwater fish or other receptive aquatic vertebrates, such as frogs or snakes, the L2 continues its development by leaving their digestive tract and gaining their skeletal muscles and reaches the L3 stage and encysts. This larva is infective to the definitive host (Acha and Szyfres 1989).

Humans usually contract the disease by ingesting raw, undercooked, marinated or smoked fish infested with L3. Ingestion of undercooked food prepared from frogs, snakes, chicken, pork or other animals contaminated with these larvae may also cause the disease. Finally, humans may become infected by drinking water contaminated with infested *Cyclops* (Jaroonvesama 1988).

A single larva can determine symptoms induced by its migrations (Dei-Cas 1996). The worm does not reach gastric localization or sexual maturity in humans, but immature young adults (Hira et al. 1989) are sometimes found in this accidental host. Clinical manifestations usually begin one or two days after ingestion of the infesting meal: nausea, ptyalism, urticaria, pruritus and gastralgia associated with leukocytosis and hyper-eosinophilia (Acha and Szyfres 1989). Later, the patient experiences acute pain, associated with the migration of larvae into the abdominal or thoracic organs. A month later, the chronic phase begins, characterized by an attenuation of symptomatology and a tendency to normalization of the eosinophilia. In endemic countries, any syndrome including erratic and recurrent edema with leukocytosis and hypereosinophilia is considered pathognomonic of gnathostomosis (Acha and Szyfres 1989). The parasitologist should identify worms obtained during surgery or observed in skin or other organ biopsies. Specific antibodies (IgG) are detectable by enzyme-linked immunosorbent assay (ELISA). High levels of specific IgE have also been reported (Jaroonvesam 1988). When possible, surgical removal of the larvae should be performed. The only effective prophylactic measure is not to ingest raw or undercooked fish, poultry or other meats.

6.2.4.3.3. Nematodoses due to Dioctophymatidae nematodes

There are probably fewer than two dozen known human cases of dioctophyma worldwide. The agent is *Dioctophyma renale*, an adult parasite of the kidney of piscivorous carnivores and dogs, with larvae found in freshwater teleost fish and frogs (Dei-Cas 1996). Dioctophymosis is a rather cosmopolitan parasitosis. It exists in Europe, Asia and America, but is thought to be absent from Africa and Oceania (Euzéby 1964). Humans and dogs are accidental hosts that contract dioctophymosis by ingesting raw or poorly cooked paratenic fish or frog hosts. They usually harbor a single parasite. *D. renale* causes renal colic and hematuria in humans. In the absence of treatment, this picture evolves towards a fatal cachexia. The diagnosis is made either at autopsy, or by extraction of worms that have spread to the urethra or given rise to renal abscesses that fistulate on the skin, or by the discovery of eggs in the urine (Garcia and Bruckner 1988). Prevention is achieved by thoroughly cooking or freezing at −20°C for more than 24 hours any fish intended for consumption.

Other Dioctophymatidae nematodes can parasitize humans. For example, *Eustrongylides* sp., as adults, parasitize the digestive tract of aquatic birds. Their larvae develop in the tissues of fish, amphibians and reptiles, which can cause various pathologies in humans (Guardone et al. 2021). The zoonotic potential of these larvae, which have been identified in other vertebrates, including crocodiles (Junker et al. 2006), is real. The sources of infection for both paratenic hosts and humans are raw ingested freshwater fish (Dei-Cas 1996). Thus, two human cases reported in the USA contracted this parasitosis by ingesting live minnows (Eberhard et al. 1989; Narr et al. 1996). The genus *Eustrongylides* is present in Asia, Africa, North and South America. In Europe, it has been described in the Danube basin (Juhásová et al. 2019) and in Italy (Guardone et al. 2021; Menconi et al. 2020c). To date, no human case has been described in Europe (Guardone et al. 2021).

6.2.4.3.4. Intestinal capillariosis

The nematodes of the genus *Capillaria* belong to the order Trichocephalida. They are included in the same family (Trichuridae) as the genus *Trichuris* and are relatively close taxonomically to the genus *Trichinella* (family Trichinellidae). Of the three zoonotic species, two are monoxenous and one, *C. philippinensis*, is heteroxenous (Dei-Cas 1996). Cannibalism between rats plays a significant role in the transmission of *C. hepatica* in the wild. *C. aerophila*, an agent of pulmonary capillariasis, which is rare in humans, releases infective eggs into the upper respiratory tract of its vertebrate hosts. *C. philippinensis* causes intestinal capillariasis, an often severe condition with significant mortality (over 6%).

Human cases of this parasitosis have been reported in the Philippines, Thailand, Taiwan (Lu et al. 2006), the Middle East, Egypt and Colombia. A notable episode occurred in the Philippines in the 1960s with over 1,000 clustered cases and 100 deaths (Betson et al. 2020). People usually become infested by ingesting small infested fish, raw or undercooked, especially those that can be swallowed whole (Neva and Brown 1994). Infestation induces abdominal pain and diarrhea. Diagnosis is made by finding eggs, larvae and/or adults in the stool. Prevention involves not ingesting raw, undercooked, smoked or marinated fish (Dei-Cas 1996).

6.2.4.4. Acanthocephalus

Acanthocephalans are non-metamerized parasitic worms, with a pseudocoeloma and without a digestive tract. Parasites of fish, amphibians, reptiles, birds and mammals, these worms can rarely parasitize humans. The two most significant species in human pathology are *Moniliformis moniliformis* (*M. dubius*), from rats, and *Macracanthorhynchus hirudinaceus*, from pigs, which are cosmopolitan in distribution (Prociv et al. 1990). However, although more rarely, acanthocephalans from marine vertebrates have been reported in humans (Schmidt 1971; Beaver et al. 1983; Tada et al. 1983; Adams et al. 1997; Kobayashi et al. 2002). Human infestation, source and mechanism of which remain uncertain, is thought to be contracted through ingestion of either intermediate crustacean hosts or fish or other paratenic hosts. Data are scarce for aquatic vertebrate acanthocephalans.

6.3. Emerging parasitosis

Despite much work on fish parasites, the status of many parasites is still uncertain regarding their zoonotic potential. This section describes some examples of parasites which are not well known but for which studies are needed to determine their potential effect on human health, and for which there has been a renewed interest.

Long considered to be unicellular organisms, Myxozoa are, in the phylogeny of life, microscopic metazoans close to Cnidaria (Lecointre and Le Guyader 2001). The relationship with cnidarians is also illustrated by the structure of the polar capsule of Myxozoa, which evokes, on morphogenetic and ultrastructural levels, the typical nematocyst of the former. It is now known that most Myxozoa are dixenic (two hosts in their cycle), with a life cycle including a vertebrate host and an invertebrate host. The fish is the vertebrate host in which the asexual process that leads to the generation of the typical Myxosporidia spores (myxospores) takes place. Oligochaete annelids (less often polychaetes) or bryozoans constitute the invertebrate host, in which sexualization and actinospore generation takes place

(Kent et al. 2001). In fact for a long time, Myxosporidia and actinosporidia were considered as two separate zoological groups. It is now known that they correspond to two parasitic stages of the same organisms (Myxozoa). Myxozoa parasitize mainly fish and are responsible for various pathologies in these vertebrates, which have a significant economic impact (Kent et al. 2001). Myxozoa have also been rarely reported as parasites of plathelminths, amphibians and reptiles. A few recent reports have described the presence of myxosporidial spores in the brains of moles, in inflammatory lesions of waterfowl, and in the feces of an AIDS patient (Weiss 2003). In the last decade, human cases have been reported in Korea and Japan following the consumption of farmed fish infested with *Kudoa septempunctata* (Yahata et al. 2015; Kim et al. 2018). This parasite induces digestive pathologies following the consumption of raw fish. Recently, a study conducted on sushi and sashimi showed the presence of this parasite at the distribution stage in South Korea (Koo et al. 2021). In addition, recent data have shown the frequent presence of other species of Myxosporidia of the genus *Kudoa* in tuna (Hoai et al. 2022). These authors concluded their study by questioning the impact of the presence of this parasite both from an economic point of view (repugnant aspect and alteration of the fish flesh) and from a sanitary point of view. Indeed, they recall that to date, two species of *Kudoa* have been associated with human cases. Despite the great diversity and the important presence of this group of parasites in fish, wild or farmed, of economic importance, few data are available on their zoonotic potential.

Many species of trematodes are present in fish at different stages (metacercariae or adults). Some families, such as Opisthorchiidae, have been extensively studied because they induce many human cases (Chai et al. 2005). Within this same family, some genera are still supposed to be zoonotic with however few human cases and/or little data on their distribution (Chai and Jung 2017, 2020). In addition, the genus *Clinostomum*, belonging to the family Clinostomidae, has been described as zoonotic since 1938, inducing pathologies of the upper airways (laryngitis, pharyngitis). This parasite is present in the metacercal stage in many species of fish, mainly Perciformes and Cypriniformes, and also Salmoniformes and Siluriformes. Its presence has been described in the Danube basin and in Italy (Juhásová et al. 2019; Menconi et al. 2020a) but without associated human cases. In late 2019 and 2020, this parasite was observed for the first time in France on river perch in the Jura. Since this first observation, it has also been observed in Doubs, Aisne and Indre. The presence of this parasite could be due either to its transport by piscivorous birds, definitive hosts, or by fish during restocking campaigns. No human case has been described in France. However, the presence of this parasite associated with the culinary traditions of raw or marinated perch fillets could represent an increased risk.

These two examples show the need to maintain active surveillance on the presence of new potentially zoonotic parasites and also on the modification of distribution areas of zoonotic parasites and/or consumption patterns of infested products.

6.4. The special case of aquaculture species

Aquaculture is one of the most important sectors of the agri-food industry with an estimated turnover of 250 billion, and representing 52% of the supply of fish for human consumption (FAO 2020). Among the countries producing aquaculture products, Asia is in the lead, producing 69% of farmed fish (FAO 2020). Asia has the highest prevalence of clonorchiasis (*Clonorchis sinensis* distomatosis or Chinese fluke) (currently, it is estimated that the infection affects about 30 million humans). In addition, more than 50 other zoonotic trematodes parasitizing fish belonging to the Heterophyidae and Echinostomatidae have been described in major Asian countries (China, Thailand, Korea, Vietnam). Until recently, the transmission of trematode metacercariae to humans was mainly attributed to the consumption of fish products. However, a study in Vietnam showed that traditional integrated farming/fish farming systems are particularly at risk of zoonotic trematode infection (Chi et al. 2008). Surveys in other countries have also linked clonoriasis to aquaculture fish consumption (dos Santos and Howgate 2011).

Freshwater and marine aquaculture fish are generally raised in controlled environments and fed with formulated feeds whose manufacturing processes exclude the presence of viable parasites. Therefore, there is little risk of trophic infestation of aquaculture fish by either Anisakidae or Diphyllobrothidae. However, larvae of *Dibothriocephalus* spp. have been found in farmed salmonids, whether in Atlantic salmon (cases reported in the United States and Ireland), trout (cases reported in Finland) or rainbow trout (cases reported in Scotland, Russia and Chile) (Cabello 2007; dos Santos and Howgate 2011). Despite a diet free of viable larval cestodes, farmed salmonids are likely to ingest zooplankton infected with procercoids that occur naturally in lakes. The likelihood of larval cestode infection in farmed salmon is low, but possible because smolts are sometimes reared in cages in lakes before being reared at sea.

Regarding the presence of Anisakidae in farmed marine fish, numerous studies have been conducted, particularly in Europe, which have demonstrated the absence of Anisakidae in farmed salmon (which represents the most produced marine farmed fish). This negligible risk led the European Commission in 2011 to exempt farmed salmon products from the requirement for sanitizing freezing if they are to be consumed raw. Since this decision, two studies have shown the possible presence of

Anisakidae in farmed salmon, but only in salmon that had been discarded (Mo et al. 2014; Levsen and Maage 2016).

However, it should be noted that the risk of infestation of marine aquaculture fish by *Anisakis* sp. exists for species reared after capture of juveniles at sea as may be the case for farmed cod and tuna in Europe (Žilić and Mladineo 2006; Mladineo et al. 2011).

Despite the incidence of gnathostomosis, there are very few studies on the rate of infection of farmed fish by *Gnathostoma* sp. nematodes. However, a 2011 study in Thailand showed that *G. spinigerum* L3 larvae infected farmed eels at a high rate, but this was only during winter and the rainy season (Saksirisampant and Thanomsub 2012).

6.5. Allergies related to the presence of parasites

In humans, Anisakidae, can be responsible for allergic responses with clinical signs including urticaria, rhinitis, coughing or gastrointestinal disorders. Allergies to *Anisakis simplex* are particularly well known and studied. *Anisakis* represents the only case known so far of being both a parasitic worm and a food allergen (Fitzsimmons et al. 2014). The larva of *Anisakis simplex* after ingestion can indeed produce an immediate allergic reaction, the signs of which can range from urticaria to angioedema and anaphylactic shock (Moreno-Ancillo et al. 1997). Asthma and conjunctivitis have also been reported after sensitization to *Anisakis simplex* as well as contact dermatitis and arthralgia. Most commonly, these symptoms occur in patients within minutes or hours of eating raw or lightly cooked fish. Some patients experience symptoms of allergic reactions (urticaria and/or angioedema) associated with abdominal pain and vomiting, which usually appear 4 or 5 hours before and after eating raw or undercooked fish. In this case, it is called gastroallergic anisakiasis. The question of the necessity of an initial sensitization to *Anisakis* with the live parasite (ingestion of viable L3 larvae causing anisakidosis) is not resolved at present. Indeed, some patients have allergic reactions to *Anisakis* antigens present in foods which have been cooked or frozen, or by skin contact, inhalation or skin testing (Audicana 2022).

A total of 14 allergens have so far been identified in *Anisakis simplex* (Ani s1 to Ani s14) and listed by the WHO/IUIS Allergen Nomenclature Sub-Committee (2021). Monéo et al. (2017), in a literature review, indicate that eight of them are excreted-secreted (ES) allergens. These proteins are synthesized in the dorsal esophageal gland and in the secretory cells of the digestive tract. Three other allergens, paramyosin, tropomyosin and a third protein whose function is unknown at this time, are somatic proteins. The last three allergens (Ani s11, Ani s12 and Ani s14) are of unknown types. Paramyosin (Ani s2) and tropomyosin (Ani s3) are responsible for cross-reactions with invertebrates such as mites, crustaceans, mollusks, midges and cockroaches. Some of these allergens (in particular Ani s1, Ani S4 (a cystatin) and Ani s5, s8 and s9 which are proteins of the SXP/RAL family) are relatively resistant to digestive enzymes and heat treatments.

Many asymptomatic patients have elevated IgE levels against *A. simplex*. In fact, identification of *Anisakis simplex* allergy is often difficult due to significant cross-reactions with other allergens. This diagnosis is more difficult to make in children than in adults. It is usually made by skin tests. Studies have shown that secreted/excreted antigens are the most specific in identifying parasitized patients. It is recommended to use isolated allergens for diagnosis to avoid false positive responses due to cross-reactions. Furthermore, their use in a mixture for diagnosis is highly recommended to enable an accurate diagnosis. Monéo et al. (2017) were thus able to show, using these diagnostic tools, that: (i) the non-digestive route of sensitization is also relevant, (ii) *Anisakis simplex* seems to be a significant causative agent of chronic urticaria and (iii) the number of highly sensitized people can reach 7% of the general population in endemic countries.

A recent study also showed that exposing mice to a crude extract of *Anisakis pegreffi* larvae by air could cause them to develop asthma (Choi et al. 2021). The authors conclude that *A. pegreffi* has the potential to cause similar effects in humans. Similarly, another recent study compared the proteomes and "allergomes" of type 3 larvae of *A. simplex*, *Pseudoterranova decipiens* and *Contracaecum osculatum* by liquid chromatography-tandem mass spectrometry (LC/MS) analyses (Kochanowski et al. 2020). The results showed that there were 28 probable allergens in *A. simplex* and *P. decipiens* and 25 in *C. osculatum*. These results suggest that, in addition to *A. simplex*, the latter two species can also be considered as potential sources of allergens which can lead to hypersensitivity.

The potential presence of *Anisakis* allergens in food products containing fish implies the recommendation of a total eviction of these products in the diet of people sensitized to these allergens in order to decrease their specific IgE level and consequently the occurrence of chronic or acute episodes induced by the parasite. Indeed, it is almost impossible to totally eliminate the allergens that may be present, even in the case of surimi for which a washing of the fish mince is a part of the industrial manufacturing process. The only way to avoid allergic reactions in sensitive people who still want to continue to eat fish would be to replace the wild fish by farmed fish as many studies indicate that they are virtually free of Anisakidae. However, a recent study has shown that feeding farmed fish with processed fishmeal (therefore having undergone a significant heat treatment) did not prevent the Ani s4 allergen from being found in the muscle of the fish (Polimeno et al. 2021).

In order to directly measure the presence of allergens in products, immunological or proteomic methods (usually by chromatography coupled with mass spectrometry) can be used. For example, a method using the ELISA (enzyme-linked immunosorbent assay) technique has been developed by Werner et al. (2011) and validated by Fæste et al. (2015). However, the ELISA technique cannot be applied to highly processed food products. In this case, mass spectrometry analysis could be used.

6.6. The legislative framework for consumer protection or regulatory requirements for parasites in aquatic products

Several texts of the "Hygiene Package" mention the parasites and the control measures to be implemented (EC 853/2004, EC 2074/2005, EC 852/2004). The main elements are:

– The need to set up visual controls for the detection of visible parasites by qualified personnel on a representative number of samples according to the nature of the fishery products, their geographical origin and the use for which they are intended: the products obviously parasitized must not be put on the market.

– Mandatory sanitizing freezing (–20°C at all points of the product for at least 24 hours or −35°C for at least 15 hours) for products to be consumed raw or almost raw, for marinated, salted and/or products having undergone other inadequate treatment.

– This freezing is not necessary for products (1) that have undergone sufficient heat treatment (60°C for 1 min through) or (2) that have been kept frozen for a sufficiently long period of time or (3) from wild catches if there are epidemiological data indicating that the fishing grounds do not present a health hazard with regard to the presence of parasites or if the competent authority authorizes it or (4) from farmed fish, reared from embryos and exclusively fed with a diet free of viable parasites and either reared in an environment free of viable parasites or the products do not present a health hazard with respect to the presence of viable parasites.

– The definition of certain terms such as: visible parasite (any parasite or group of parasites having a size, color or texture that can be clearly distinguished from fish tissue), visual inspection (non-destructive examination of fish or fishery products, performed with or without optical means of magnification and under good lighting conditions for the human eye, including candling if necessary) and candling (in the case of flatfish or fish fillets, observation of the fish illuminated by a light source in a darkened room against the light for the purpose of detecting parasites).

– Food business operators must not accept ingredients or raw materials (suspected) of being contaminated with parasites in such a way that, even after sorting and/or preparation or processing, the final product would be unfit for human consumption.

6.7. Methods for the detection of parasites and mainly nematode larvae

Since European legislation requires that any fishery product intended for human consumption that is obviously parasitized be withdrawn from the market, there is a need for effective detection methods to identify the presence of parasites in fish fillets. This European regulation is mainly aimed at detecting the presence of nematodes, the most common parasites found in the edible parts of fishery products. Anisakidae are in the list of the 10 most significant parasites in Europe according to a study using multi-criteria analysis based on the recommendations of the Food and Agriculture Organization of the United Nations (FAO) and the World Health Organization (WHO) (Bouwknegt et al. 2018). Anisakidae larvae can be found in almost all marine fish species. Neither early gutting, nor specific trimming of fillets to remove ventral parts, nor the choice to market only fish caught in less infested areas can guarantee a parasite-free product.

The majority of the larvae of these nematodes found in fillets are usually coiled in a spiral shape, whitish in color or even transparent. These coils are a few millimeters in diameter and are very difficult to detect with the naked eye once the larvae have entered the fillets. However, in order to comply with the regulation, professionals are required to carry out visual inspections at all stages of product processing. The term "visual inspection" is defined as a non-destructive examination of fish or fishery products, with or without the use of optical means to enlarge the sample, and under proper light, including candling if necessary. The use of candling tables that allow for non-destructive assessment of fishery products can, of course, only be applied to fillets. In the case of eviscerated whole fish, only a visual inspection to estimate if there are parasites left in the abdominal cavity can be performed. However, this simple and inexpensive technique does not provide quantitative results due to the difficulty in locating larvae in nets. The disadvantages of this method are that the results are very dependent on the thickness of the fillets (beyond a depth of 4 mm, larvae are generally not detectable), the presence of skin on the fillet, the fat content and pigmentation (and thus the species of fish) and finally the operator's level of experience, as well as their fatigue (Levsen et al. 2005;

Mercken et al. 2022). Two methods, whose standards were published in 2021, enable the estimation of the number of nematode larvae in the fish tissues; however, their use implies the destruction of the tissues. These are, on the one hand, the so-called artificial digestion method (using pepsin in an acid medium) (ISO 23036-2:2021), and, on the other hand, a method requiring the use of a hydraulic press and UV light (ISO 23036-1:2021). The latter method uses the natural fluorescence of the Anisakidae larvae previously frozen. Briefly, the sample (fillet/viscera) is flattened or pressed and then frozen and visualized under UV. Anisakidae larvae contain lipofuscin in their cells which is an auto-fluorescent pigment (Pippy 1970). When the cells, under the action of freezing, destructure, the lipofuscin is released. And when the sample is put under UV, the larvae fluoresce much more than the fish flesh.

However, these current techniques are often time-consuming as they require either digestion or prior pressing and freezing. There is still a need to develop new techniques that are faster and non-destructive and allow for product control throughout the production line. Klapper et al. (2021) have listed in a recent bibliographic review, all the works already carried out using imaging techniques in order to rapidly detect Anisakidae larvae. However, to our knowledge, none of these techniques have yet been implemented in industries.

Methods	Advantages	Disadvantages
Observation with the naked eye	Quick No equipment Non-destructive	Not exhaustive Need for expertise Eye strain
Observation on a candling table	Quick Non-destructive	Not exhaustive Need for expertise Eye tiredness Material needed
Peptic digestion	Exhaustive Practice on specific samples (liver, highly parasitized flanks)	Material needed Loss of damaged larvae Chemical risk Destruction of the product
Press/UV	Exhaustive Precise location of larvae No chemical products	Difficult on specific samples (liver, highly parasitized flanks) Physical risk (injury to operators) Destruction of the product
PCRq	Sensitivity and specificity Speed	Cost Selection of the test outlet

Table 6.1. *Advantages and disadvantages of methods for detecting Anisakidae larvae in fishery and aquaculture products*

6.8. Identification of parasites

6.8.1. *Identification of nematode larvae*

Anisakidae larvae are not identifiable to the species level upon visual examination because related species are very similar morphologically and the larvae at all stages lack the organs and sexual characteristics necessary for identification. Therefore, it is recommended to use molecular methods to define the nematode species (Mattiucci and Nascetti 2008). In general, nuclear markers such as ribosomal DNA intergenic transcribed regions (ITS) and mitochondrial markers such as cytochrome c oxidase subunit I (COI) and cytochrome oxidase subunit II (COII) genes are used. Two methods are currently recommended by the European Reference Laboratory for Parasites (EURLP: https://www.iss.it/web/iss-en/eurlp), to identify Anisakidae larvae to species. They both use ITS markers either with a PCR/RFLP method (combination of a classical polymerase chain reaction (PCR) followed by fragment size analysis of the PCR products digested with one or more restriction enzymes (Kijewska et al. 2000; Espiñeira et al. 2010) or with a multiplex PCR method (standard PCR reaction using two or more primer pairs for the same reaction) (Umehara hara et al. 2008; Fang et al. 2010).

Methods for rapid identification of the presence and species of Anisakidae larvae in food products have been very recently developed, based on loop-mediated isothermal amplification (LAMP) or real-time PCR (qPCR) (Lopez and Pardo 2010; Godinez-Gonzalez et al. 2017). The LAMP method developed by Cammilleri et al. (2020) is used to detect the presence of *Anisakis* spp. in a food product. The LAMP method is a simple, inexpensive because it is an isothermal, and rapid method that requires only the Bst DNA polymerase, and a set of four specially designed internal (FIP/BIP) and external (F3/B3) primers that recognize a total of six distinct sequences on the target DNA. Compared to conventional PCR methods, the LAMP method is less affected by PCR inhibitors, making it particularly well suited for analysis of transformed products. According to this study, the detection limit of this method applied to the detection of *Anisakis* spp. is two times lower than the limit of a qPCR detection method.

6.8.2. *Identification of other parasite species*

Kudoa spores can be identified to species by morphological analysis under a microscope. The morphometric characters to be considered according to the guide written by Lom and Arthur (1989) are length, width and thickness of spores, as well as length and width of polar capsules. Different allometric ratios are also defined as additional morphological descriptors to identify kudoid myxospores to species.

Determination of *Kudoa* species by molecular analysis is usually achieved by amplifying and then sequencing the gene encoding the small subunit (SSU) of the 18S ribosomal RNA and comparing this sequence to the GenBank database (program BLAST (Altschul et al. 1997)).

The 18S ribosomal RNA gene sequence is also used as a species marker for the genus *Cryptosporidium* either after nested PCR (Certad et al. 2015) or by metabarcoding using specific primers for the genus *Cryptosporidium* (Zahedi et al. 2019).

Specific identification of adult tapeworms of the genus *Dibothriocephalus*, usually based on morphological observation of capped eggs and adult worm segments, allow identification to the genus level but are unreliable for assessing species identity, because the different taxa are extremely similar to each other. Species differentiation is based on the characteristics of the scolex or the reproductive system. These are often unavailable in human infections. In fish, direct examination of dissected fillets enables the plerocercoid larvae to be observed and to be identified according to their morphology. Examination on a candling table leads to an efficient detection of plerocercoid larvae. In order to identify the species of these parasites, different molecular methods based, in general, on the cytochrome oxidase c subunit I gene have been developed.

The procedure standardized by the European Reference Laboratory for the identification of metacercariae of trematode species of Opisthorchiidae and Heterophyidae, which are generally smaller than 400 µm, requires digestion with pepsin-HCl solution of fish muscle tissue, or compression of the muscle between two glass plates, both followed by observation under a microscope. Metacercariae isolated in this way are identified by an expert at the genus level on morphological criteria. Multiplex PCRs based on ITS and COI amplification have been developed to rapidly identify *Opisthorchis felineus*, *Pseudamphistomum truncatum*, *Metorchis* spp., *Metagonimus* spp. and *Apophallus* spp. which are the major zoonotic or potentially zoonotic trematodes in Europe (Caffara et al. 2020). Other multiplex PCRs have also been developed to identify other metacercariae such as *Clonorchis sinensis* and *Opisthorchis viverrini* (Le et al. 2006).

6.9. Prevention and control of parasite risk

Consumer prevention to the parasitic risk induced by the consumption of contaminated aquatic products can be achieved: (i) by controlling the quality of the raw material by favoring the least affected fishing areas or parasite-free farms,

(ii) by the eviction of the contaminated parts or (iii) by the use of physicochemical treatments killing the parasites.

To date, most of the data and research on preventive treatments has focused on Anisakidae (especially *A. simplex*). Although research has been carried out on trematodes, especially *Clonorchis sinensis*, which is one of the most common foodborne zoonoses, gaps in knowledge, for example, about the biology (habits and life history characteristics) as well as the mechanism of infection of intermediate hosts prevent effective measures to block their transmission on farms. Indeed, the control through technological or biological means (often more effective means of control) of freshwater snails which are the primary intermediate hosts of *C. sinensis* represents a feasible control strategy (Tang et al. 2016). In addition, to date, there are little data in the literature on time–temperature schedules to be applied to products susceptible to infestation by trematode metacercariae, which are much more heat resistant than nematodes and most other multicellular parasites (EFSA 2010). Thus, the treatment schedules (sanitizing freezing and cooking of the product) defined by the different regulatory bodies (European Commission, Food and Drug Administration and the CODEX Alimentarius) are essentially anisakiasis prevention treatments.

Among the chemical treatments most commonly used to inactivate Anisakidae larvae, salting and pickling (usually including vinegar and salt) kill these larvae after three to six weeks (these data are different depending on the studies). However, the survival of the larvae depends on many factors such as the size of the fish and its lipid content, as well as the ingredients of the marinade. Such treatments, if not combined with freezing beforehand, need to be optimized for each type of product. Similarly *Opisthorchi viverrini* metacercariae are removed by soaking fish fillets (carp) in 50% salt brines according to Prasongwatana et al. (2013). These authors also note that traditional fermentation of these freshwater fish as it is performed in Thailand kills all metacercariae.

Regarding physical treatments, apart from freezing and cooking, whose effectiveness depends not only on the species of parasite and its stage of development, but also on the time–temperature scales, some processes such as the use of high pressure, drying, hot smoking, irradiation and low-voltage currents, have shown their effectiveness in killing Anisakidae larvaes; however, there is still little information on the conditions to be used in order to preserve the taste qualities of the treated products.

For example, Brutti et al. (2010) showed that using 300 MPa for 5 minutes killed all *Anisakis simplex* larvae present in mackerel.

According to compilation work by Franssen et al. (2019), irradiation between 0.1 and 0.5 kGy is effective in inactivating all parasites that can infest fish, except for Anisakidae for which a gamma radiation dose of more than 10 kGy would be required (which is double the dose allowed in France for meat and fish treatment).

A patent has been taken in Spain for the use of low voltage current on whole fish to inactivate Anisakidae larvae. However, studies proving the interest of this method are not available.

The effect of smoking on parasite survival depends on whether the fish is smoked hot (temperature above 60°C) or cold. In the case of cold smoking, this treatment must imperatively be associated with a prior treatment of inactivation of Anisakidae larvae (e.g. freezing).

The international standards of good practice related to the different fish processing techniques have been developed by the Codex Alimentarius. However, they have been developed on the basis of the experiences gained with fish contamination by parasites in Europe and North America. They therefore mainly address nematode contamination. In general, regulations can help to control certain aspects of the products, but they cannot replace basic cooking and freezing rules to be respected by the consumer. And while European regulations and compliance with prophylactic rules can effectively protect the consumer from infection with live parasites, they cannot guarantee a parasite-free product. It should be noted that whatever the standard, heating or freezing procedures do not inactivate the allergens of these larvae.

Finally, it seems difficult to establish preventive guidelines that can be effectively applied to all species of parasites that may infest fish. Rather, preventive approaches should be defined taking into account the dietary habits (targeted prevention campaigns for consumers), the species of fish consumed, their parasites and the susceptibility of the latter to available treatments.

6.10. Future perspectives and approaches

There are still many unknowns regarding the distribution of zoonotic parasites in fishery products, their life cycle, their impacts on human health, the factors influencing their distribution or their pathogenicity. Among these factors, some are biotic, such as fish species, size or presence of hosts and others are abiotic, such as temperature or geographical area.

In addition, some elements could induce changes in the broad distribution of parasites (host spectrum, geographic distribution, prevalence and intensity) while others could alter consumer exposure.

Thus, climate change is described as potentially impacting parasite populations. However, it is difficult to predict the direction in which climate change will affect parasite populations (Pozio 2020). Indeed, the increase in temperature will accelerate the life cycle of parasites having stages in the external environment or in poikilothermic hosts but will reduce the life span of parasites (Pozio 2020). However, climate change will lead to a modification of the distribution of the hosts, sometimes leading to the disappearance of one of the links of the biological cycle or on the contrary, to the introduction of a new host in this cycle (Marcogliese 2001; Dudley et al. 2015).

With regard to consumer exposure, beyond changes in the distribution of parasites, elements at the consumption stage, such as culinary habits and knowledge of the hazard, will modify the risk. The development of home cooking of raw, undercooked or minimally processed products is a significant risk factor that increases the risk of contracting a zoonotic parasite (Slifko et al. 2000). However, improved consumer knowledge with the integration of good practices (sufficient sanitizing freezing, consumption of fillets rather than viscera) will reduce the risk related to the presence of zoonotic parasites in fishery products.

In conclusion, the risk related to the presence of zoonotic parasites in fishery products is often neglected and/or not well known by professionals and consumers. However, this risk could increase in the future years due to the emergence of new parasites, environmental modifications due to climate change or changes in dietary habits. Therefore, surveillance of these parasites should be maintained and developed, preferably with a holistic approach to take into account the factors of change.

A tribute in memory of Dr Eduardo Dei-Cas

The two authors of this chapter would like to dedicate this chapter to Dr. Eduardo Dei-Cas. Eduardo was a hospital practitioner, lecturer and researcher in parasitology known and recognized, among other things, for his studies on the opportunistic fungus *Pneumocystis*. In the years leading up to his death in 2014, he coordinated a project funded by the Agence Nationale pour la Recherche in France on fish parasites to identify risks and define prevention strategies for the consumer. This work has given a new impetus to the theme of zoonotic parasites in fish products in France and has led to the development of many new research projects.

He had, on this occasion, written a chapter with one of the authors of this chapter, a book that unfortunately could not be published.

6.11. References

Acha, P.N. and Szyfres, B. (1989). *Zoonoses et maladies transmissibles communes à l'Homme et aux animaux*. Office International des Epizooties, Paris.

Adams, A.M., Murrell, K.D., Cross, J.H. (1997). Parasites of fish and risks to public health. *Rev. Sci. Tech.*, 16, 652–660.

Adroher-Auroux, F.J. and Benitez-Rodriguez, R. (2020). Anisakiasis and *Anisakis*: An underdiagnosed emerging disease and its main etiological agents. *Res. Vet. Sci.*, 132, 535–545.

Altschul, S.F., Madden, T.L., Schäffer, A.A., Zhang, J., Zhang, Z., Miller, W., Lipman, D.J. (1997). Gapped BLAST and PSI-BLAST: A new generation of protein database search programs. *Nucleic Acids Res.*, 25, 3389–3402.

Alvarez-Pellitero, P. and Sitja-Bobadilla, A. (2002). *Cryptosporidium molnari* n. sp. (Apicomplexa: Cryptosporidiidae) infecting two marine fish species, *Sparus aurata* L. and *Dicentrarchus labrax* L. *Int. J. Parasitol.*, 32, 1007–1021.

Alvarez-Pellitero, P., Quiroga, M.I., Sitja-Bobadilla, A., Redondo, M.J., Palenzuela, O., Padros, F., Vazquez, S., Nieto, J.M. (2004). *Cryptosporidium scophthalmi* n. sp. (Apicomplexa: Cryptosporidiidae) from cultured turbot Scophthalmus maximus. Light and electron microscope description and histopathological study. *Dis. Aquat. Organ.*, 62, 133–145.

Arkush, K.D., Mendoza, L., Adkison, M.A., Hedrick, R.P. (2003). Observations on the life stages of *Sphaerothecum destruens* n. g., n. sp., a mesomycetozoean fish pathogen formerly referred to as the rosette agent. *J. Eukaryot. Microbiol.*, 50, 430–438.

Audicana, M.T. (2022). *Anisakis*, something is moving inside the fish. *Pathogens*, 11(3), 326.

Audicana, M.T. and Kennedy, M.W. (2008). *Anisakis simplex*: From obscure infectious worm to inducer of immune hypersensitivity. *Clin. Microbiol. Rev.*, 21(2), 360–379.

Autier, B., Belaz, S., Degeilh, B., Gangneux, J.P., Robert-Gangneux, F. (2019). *Dibothriocephalus nihonkaiensis*: An emerging foodborne parasite in Brittany (France)? *Parasites & Vectors*, 12(1), 267.

Bao, M., Cipriani, P., Giulietti, L., Drivenes, N., Levsen, A. (2021). Quality issues related to the presence of the fish parasitic nematode *Hysterothylacium aduncum* in export shipments of fresh Northeast Arctic cod (*Gadus morhua*). *Food Control*, 121, 107724.

Beaver, P., Otsuji, T., Otsuji, A., Yoshimura, H., Uchikawa, R., Sato, A. (1983). Acanthocephalan, probably *Bolbosoma*, from the peritoneal cavity of man in Japan. *Am. J. Trop. Med. Hyg.*, 32, 1016–1018.

Betson, M., Alonte, A.J., Ancog, R.C., Aquino, A.M., Belizario, V.Y., Bordado, A.M., Clark, J., Corales, M.C., Dacuma, M.G., Divina, B. et al. (2020). Zoonotic transmission of intestinal helminths in southeast Asia: Implications for control and elimination. *Adv. Parasitol.*, 106, 47–131.

Botto, C. (1974). Sobre une nueva especie desl género *Filocapsularia* Deslongchamps, 1924 (Nematoda, Ascarididea). *Neotropica*, 20, 133–139.

Botto, C., Osimani, J.J., Mane Garzon, F. (1976). Sobre la presencia de larves de *Anisakis sp* en peces de la costa atlantica uruguaya y su patogenicidad experimental para el perro y el gato. *Revista Uruguaya de Patologia Clinica y Microbiologia*, 14, 49–62.

Bouwknegt, M., Devleesschauwer, B., Graham, H., Robertson, L.J., van der Giessen, J.W. the Euro-BP workshop participants (2018). Prioritisation of food-borne parasites in Europe, 2016. *Eurosurveillance*, 23(9), 17.

Brauge, T., Copin, S., Gay, M., Midelet, G. (2022). Minimally processed seafood products and bacterial and parasitic hazards. In *Hazards in the Food Processing and Distribution Chain*, Haddad, N. (ed.). ISTE Ltd, London, and John Wiley & Sons, New York.

Brunet, J., Pesson, B., Royant, M., Lemoine, J.P., Pfaff, A.W., Abou-Bacar, A., Year, H., Fréalle, E., Dupouy-Camet, J., Merino-Espinosa, G. et al. (2017). Molecular diagnosis of *Pseudoterranova decipiens* s.s in human, France. *BioMed Central Infectious Diseases*, 17(1), 397.

Bruno, D.W., Nowak, B., Elliott, D.G. (2006). Guide to the identification of fish protozoan and metazoan parasites in stained tissue sections. *Dis. Aquat. Organ.*, 70, 1–36.

Brutti, A., Rovere, P., Cavallero, S., D'Amelio, S., Danesi, P., Arcangeli, G. (2010). Inactivation of *Anisakis simplex* larvae in raw fish using high hydrostatic pressure treatments. *Food Control*, 21(3), 331–333.

Cabello, F.C. (2007). Salmon aquaculture and transmission of the fish tapeworm. *Emerg. Infect. Dis.*, 13(1), 169.

Caffara, M., Gustinelli, A., Mazzone, A., Fioravanti, M.L. (2020). Multiplex PCR for simultaneous identification of the most common European Opisthorchiid and Heterophyid in fish or fish products. *Food and Waterborne Parasitolo.*, 19, e00081.

Cali, A., Weiss, L.M., Takvorian, P.M. (2005). A review of the development of two types of human skeletal muscle infections from microsporidia associated with pathology in invertebrates and cold-blooded vertebrates. *Folia Parasitol. (Praha)*, 52, 51–61.

Cammilleri, G., Ferrantelli, V., Pulvirenti, A., Drago, C., Stampone, G., Del Rocio Quintero Macias, G., Drago, S., Arcoleo, G., Costa, A., Geraci, F. et al. (2020). Validation of a commercial loop-mediated isothermal amplification (LAMP) assay for the rapid detection of *Anisakis* spp. DNA in processed fish products. *Foods*, 9(1), 92.

Certad, G., Dupouy-Camet, J., Gantois, N., Hammouma-Ghelboun, O., Pottier, M., Guyot, K., Benamrouz, S., Osman, M., Delaire, B., Creusy, C. et al. (2015). Identification of *Cryptosporidium* species in fish from lake Geneva (Lac Léman) in France. *PLOS ONE*, 10, e0133047.

Certad, G., Follet, J., Gantois, N., Hammouma-Ghelboun, O., Guyot, K., Benamrouz-Vanneste, S., Fréalle, E., Seesao, Y., Delaire, B., Creusy, C. et al. (2019). Prevalence, molecular identification, and risk factors for *Cryptosporidium* infection in edible marine fish: A survey across sea areas surrounding France. *Front. Microbiol.*, 1037.

Chai, J.Y. and Jung, B.K. (2017). Fishborne zoonotic heterophyid infections: An update. *Food Waterborne Parasitol.*, 8–9, 33–63.

Chai, J.Y. and Jung, B.K. (2020). Foodborne intestinal flukes: A brief review of epidemiology and geographical distribution. *Acta Trop.*, 201, 105210.

Chai, J.Y., Darwin Murrell, K., Lymbery, A.J. (2005). Fish-borne parasitic zoonoses: Status and issues. *Int. J. Parasitol.*, 35(11–12), 1233–1254.

Chi, T.T., Dalsgaard, A., Turnbull, J.F., Tuan, P.A., Darwin Murrell, K. (2008). Prevalence of zoonotic trematodes in fish from a Vietnamese fish-farming community. *J. Parasitol.*, 94(2), 423–428.

Choi, J.H., Kim, J.Y., Yi, M.H., Kim, M., Yong, T.S. (2021). *Anisakis pegreffii* extract induces airway inflammation with airway remodeling in a murine model system. *BioMed Res. Int.*, 2522305, 13.

Clavel, A., Bargues, M.D., Castillo, F.J., Rubio, M.D., Mas-Coma, S. (1997). Diplogonoporiasis presumably introduced into Spain: First confirmed case of human infection acquired outside the Far East. *Am. J. Trop. Med. Hyg.*, 57, 317–320.

Costello, M.J. (2009). The global economic cost of sea lice to the salmonid farming industry. *J. Fish. Dis.*, 32(1), 115–118.

Curry, A. (1999). Human microsporidial infection and possible animal sources. *Curr. Opin. Infect. Dis.*, 12, 473–480.

Daensvang, S. (1972). An experimental study on the life cycle of *Gnathostoma hispidum* (Fedchenko, 1872) in Thailand with special reference to the incidence and some significant morphological characters of the adult and larval stages. *Southeast Asian J. Trop. Med. Public Health*, 3, 376–389.

Daensvang, S. (1982). Gnathostomiasis. In *CRC Handbook Series in Zoonosis. Section C*, Schultz, M.G. (ed.). CRC Press, Boca Raton.

Dei-Cas, E. (1996). Larva migrans. *Encyclopédie Médico-Chirurgicale (Paris, France), Maladies Infectieuses*, 8-518-A-10, 16.

Dei-Cas, E., Rodriguez, N., Botto, C., Osimani, J.J. (1976). Larvas plerocercoides de *Spirometra* (Dibothriocephalidae) en el hombre y en animales silvestres de Uruguay. *Rev. Inst. Med. Trop. Sao Paulo*, 18, 165–172.

Didier, E.S., Stovall, M.E., Green, L.C., Brindley, P.J., Sestak, K., Didier, P.J. (2004). Epidemiology of microsporidiosis: Sources and modes of transmission. *Vet. Parasitol.*, 126, 145–166.

Dudley, J.P., Hoberg, E.P., Jenkins, E.J., Parkinson, A.J. (2015). Climate change in the North American Arctic: A one health perspective. *EcoHealth*, 12(4), 713–725.

Dupouy-Camet, J. and Peduzzi, R. (2004). Current situation of human diphyllobothriasis in Europe. *Eurosurveillance.*, 9(5), 31–35.

Dupouy-Camet, J., Haidar, M., Dei-Cas, E., Yera, H., Espinat, L., Benmostefa, A., Guillard, J., Aliouat-Denis, C.M. (2015). Prévalence de l'infestation par *Diphyllobothrium latum* de différents poissons des lacs Léman, du Bourget et d'Annecy et évaluation de l'incidence des cas humains auprès des laboratoires d'analyse médicale de la région (2011–2013). *Bulletin épidémiologique, santé animale et alimentation*, 67, 2–5.

Dupouy-Camet, J., Gay, M., Houin, R. (2020a). New eating habits, new parasitic risks: The example of fish. *Bull. Acad. Natl. Med.*, 204(9), 1010–1016.

Dupouy-Camet, J., Gay, M., Houin, R. (2020b). New eating habits, new parasitic risks: The example of fish. *Bulletin de l'académie naionale de médecine*, 204(9), 1010–1016.

Dykova, I. and Lom, J. (2004). Advances in the knowledge of amphizoic amoebae infecting fish. *Folia Parasitol. (Praha)*, 51, 81–97.

Eberhard, M.L., Hurwitz, H., Sun, A.M., Coletta, D. (1989). Intestinal perforation caused by larval Eustrongylides (Nematoda: Dioctophymatoidae) in New Jersey. *Am. J. Trop. Med. Hyg.*, 40, 648–650.

EFSA Panel on Biological Hazards (BIOHAZ) (2010). Scientific opinion on risk assessment of parasites in fishery products. *EFSA J.*, 8(4), 1543.

Espiñeira, M., Herrero, B., Vieites, J.M., Santaclara, F.J. (2010). Detection and identification of anisakids in seafood by fragment length polymorphism analysis and PCR-RFLP of ITS-1 region. *Food Control*, 21(7), 1051–1060.

Euzéby, J. (1964). *Les zoonoses helminthiques*. Vigot Frères Éditions, Paris.

Fæste, C.K., Plassen, C., Løvberg, K.E., Moen, A., Egaas, E. (2015). Detection of proteins from the fish parasite *Anisakis simplex* in Norwegian farmed salmon and processed fish products. *Food Anal. Methods*, 8, 1390–1402.

Fang, W., Xu, S., Zhang, S., Wang, Y., Chen, X., Luo, D. (2010). Multiple primer PCR for the identification of anisakid nematodes from Taiwan Strait. *Exp. Parasitol.*, 124(2), 197–201.

FAO (2020). *The State of World Fisheries and Aquaculture*. FAO, Rome.

Fiorenza, E.A., Wendt, C.A., Dobkowski, K.A., King, T.L., Pappaionou, M., Rabinowitz, P., Samhouri, J., Wood, C.L. (2020). It's a wormy world: Meta-analysis reveals several decades of change in the global abundance of the parasitic nematodes *Anisakis* spp. and *Pseudoterranova* spp. in marine fishes and invertebrates. *Glob. Chang. Biol.*, 26(5), 2854–2866.

Fitzsimmons, C., Falcone, F., Dunne, D. (2014). Helminth allergens, parasite-specific IgE, and its protective role in human immunity. *Front. Immunol.*, 5, 61.

Fournier, S., Liguory, O., Santillana-Hayat, M., Guillot, E., Sarfati, C., Dumoutier, N., Molina, J., Derouin, F. (2000). Detection of microsporidia in surface water: A one-year follow-up study. *FEMS Immunol. Med. Microbiol.*, 29, 95–100.

Franssen, F., Gerard, C., Cozma-Petruţ, A., Vieira-Pinto, M., Jambrak, A.R., Rowan, N., Paulsen, P., Rozycki, M., Tysnes, K., Rodriguez-Lazaro, D. et al. (2019). Inactivation of parasite transmission stages: Efficacy of treatments on food of animal origin. *Trends Food Sci. Technol.*, 83, 114–128.

Frenette, A.P., Benfey, T.J., Burt, M.D.B., Duffy, M.S. (2022). Experimental *Loma morhua* (Microsporidia) infections reveal early-onset production implications for Atlantic cod (*Gadus morhua*) aquaculture. *Aquaculture*, 549, 737663.

Furst, T., Keiser, J., Utzinger, J. (2012). Global burden of human food-borne trematodiasis: A systematic review and meta-analysis. *Lancet Infect. Dis.*, 12(3), 210–221.

Gantois, N., Lamot, A., Seesao, Y., Creusy, C., Li, L.L., Monchy, S., Benamrouz-Vanneste, S., Karpouzopoulos, J., Bourgain, J.L., Rault, C. et al. (2020). First report on the prevalence and subtype distribution of *Blastocystis* sp. in edible marine fish and marine mammals: A large scale-study conducted in Atlantic northeast and on the coasts of northern France. *Microorganisms*, 8(3), 460.

Garcia, L.S. and Bruckner, D.A. (1988). *Diagnostic Medical Parasitology*. Elsevier, New York.

Gay, M., Lailler, R., Bourgau, O., Rudloff, L., Guillotteau, S., Midelet, G. (2019). Niveaux d'infestation par les Anisakidae chez sept espèces de poisson prélevées au stade de la distribution en France. *FR Bulletin épidémiologique santé animale – alimentation – Anses*, 87(3).

Godinez-Gonzalez, C., Roca-Gerones, X., Cancino-Faure, B., Montoliu, I., Fisa, R. (2017). Quantitative SYBR Green qPCR technique for the detection of the nematode parasite Anisakis in commercial fish-derived food. *Int. J. Food Microbiol.*, 261, 89–94.

Gomes, T.L., Quiazon, K.M.A., Kotake, M., Itoh, N., Yoshinaga, T. (2020). *Anisakis* spp. in fishery products from Japanese waters: Updated insights on host prevalence and human infection risk factors. *Parasitol. Int.*, 78, 102137.

Gonzalez, L. (1998). Experimental infection of mice with *Hysterothylacium aduncum* (nematoda: Anisakidae) larvae from marine-farmed trout in Chile. *Arch. Med. Vet.*, 30, 139–142.

Graczyk, T.K., Conn, D.B., Lucy, F., Minchin, D., Tamang, L., Moura, L.N., DaSilva, A.J. (2004). Human waterborne parasites in zebra mussels (*Dreissena polymorpha*) from the Shannon River drainage area, Ireland. *Parasitol Res.*, 93, 385–391.

Greigert, V., Brunet, J., Pfaff, A.W., Lemoine, J.P., Candolfi, E., Abou-Bacar, A. (2020). Locally acquired infection with *Dibothriocephalus nihonkaiense* (=*diphyllobothrium nihonkaiense*) in France: The importance of molecular diagnosis. *Parasitol Res.*, 119(2), 513–518.

Guardone, L., Ricci, E., Susini, F., Polsinelli, E., Guglielmone, G., Armani, A. (2021). First detection of *Eustrongylides excisus* (Nematoda: Dioctophymatidae) in big-scale sand smelt (*Atherina boyeri*) from the lake Massaciuccoli (northwest Tuscany, Italy): Implications for public health and seafood quality. *Food Control*, 120, 107517.

Gustinelli, A., Menconi, V., Prearo, M., Caffara, M., Righetti, M., Scanzio, T., Raglio, A., Fioravanti, M.L. (2016). Prevalence of *Diphyllobothrium latum* (Cestoda: Diphyllobothriidae) plerocercoids in fish species from four Italian lakes and risk for the consumers. *Int. J. Food Microbiol.*, 235, 109–112.

Guyot, K., Follet-Dumoulin, A., Lelièvre, E., Sarfati, C., Rabodonirina, M., Nevez, G., Cailliez, J.C., Camus, D., Dei-Cas, E. (2001). Molecular characterization of *Cryptosporidium* obtained from humans in France. *J. Clin. Microbiol.*, 39, 3472–3480.

Guyot, K., Ngouanesavanh, T.M.E., Dei-Cas, E. (2005). Strategies for detecting pathogenic protists in water: The point on *Cryptosporidium*. *Eur. J. Water Qual.*, 36, 51–70.

Hira, P.R., Neafie, R., Peakash, B., Tammim, L., Behbehani, K. (1989). Human gnathostomiasis: Infection with an immature male *Gnathostoma spinigerum*. *Am. J. Trop. Med. Hyg.*, 41, 91–94.

Hoai, T.D., Nhinh, D.T., Giang, N.T.H., Senapin, S., Dong, H.T. (2022). Detection and characterization of *Kudoa thunni* from uncooked yellowfin tuna (*Thunnus albacares*) in Southeast Asia. *Parasitol Int.*, 87, 102536.

Hoevers, J.D. and Snowden, K.F. (2005). Analysis of the ITS region and partial ssu and lsu rRNA genes of *Blastocystis* and *Proteromonas lacertae*. *Parasitology*, 131, 187–196.

Huang, W. and Bussieras, J. (1988). Anisakidés et anisakidoses humaines – Première partie : données bibliographiques. *Ann. Parasitol. Hum. Comp.*, 63, 119–132.

Hyman, L.H. (1951). *The Invertebrates: Platyhelminthes and Rhynchocoela. The Acoelomate Bilateria*, volume II. McGraw-Hill, New York.

Jaroonvesama, N. (1988). Differential diagnosis of eosinophilic meningitis. *Parasitol. Today*, 4, 262–266.

Juhásová, L., Radačovská, A., Bazsalovicsová, E., Miklisová, D., Bindzárová-Gereľová, M., Králová-Hromadová, I. (2019). A study of the endohelminths of the European perch *Perca fluviatilis* L. from the central region of the Danube river basin in Slovakia. *Zookeys*, 2019(899), 47–58.

Junker, K., Wallace, K., Leslie, A.J., Boomker, J. (2006). Gastric nematodes of nile crocodiles, *Crocodylus niloticus* Laurenti, 1768, in the Okavango River, Botswana. *Onderstepoort J. Vet. Res.*, 73, 111–114.

Kent, M.L., Andree, K.B., Bartholomew, J.L., El-Matbouli, M., Desser, S.S., Devlin, R.H., Feist, S.W., Hedrick, R.P., Hoffmann, R.W., Khattra J. et al. (2001). Recent advances in our knowledge of the myxozoa. *J. Eukaryot. Microbiol.*, 48, 395–413.

Kijewska, A., Slominska, M., Wegrzyn, G., Rokicki, J. (2000). A PCR-RFLP assay for identification of *Anisakis simplex* from different geographical regions. *Mol. Cell Probes*, 14(6), 349–354.

Kim, J.J., Ryu, S., Lee, H. (2018). Foodborne illness outbreaks in Gyeonggi province, Korea, following seafood consumption potentially caused by *Kudoa septempunctata* between 2015 and 2016. *Osong Public Health Res. Perspect.*, 9(2), 66–72.

Klapper, R., Karl, H., Ostermeyer, U. (2021). Intrinsic properties of anisakid nematode larvae as a potential tool for the detection in fish products. *Int. J. Food Microbiol.*, 343, 109094.

Kobayashi, T., Kin, S., Funabiki, S., Sasahara, H., Kida, Y., Bunno, M., Kawaguchi, M., Ishii, M., Ishimoto, K. (2002). Case report of duodenal Bolbosoma incidentally found by screening endoscopy. *Gastroenterol. Endosc.*, 44, 2083–2086.

Kochanowski, M., Dąbrowska, J., Różycki, M., Karamon, J., Sroka, J., Cencek, T. (2020). Proteomic profiling reveals new insights into the allergomes of *Anisakis simplex*, *Pseudoterranova decipiens*, and *Contracaecum osculatum*. *J. Parasitol.*, 106(5), 572–588.

Koie, M. and Fagerholm, H.P. (1993). Third-stage larvae emerge from eggs of *Contracaecum osculatum* (Nematoda, Anisakidae). *J. Parasitol*, 79, 777–780.

Koo, H.S., Park, J.Y., Sung, G.H., Park, E.H., Ku, P.T., Lee, M.O. (2021). A study on Kudoa septempunctata infection from sashimi and sushi of olive flounder *Paralichthys olivaceus* in Busan, South Korea. *Fish. Aquat. Sci.*, 24(8), 277–283.

Le, T.H., Van De, N., Blair, D., Sithithaworn, P., McManus, D.P. (2006). *Clonorchis sinensis* and *Opisthorchis viverrini*: Development of a mitochondrial-based multiplex PCR for their identification and discrimination. *Exp. Parasitol.*, 112(2), 109–114.

Lecointre, G. and Le Guyader, H. (2001). *Classification phylogénétique du vivant*. Belin, Paris.

Leelayoova, S., Rangsin, R., Taamasri, P., Naaglor, T., Thathaisong, U., Mungthin, M. (2004). Evidence of waterborne transmission of *Blastocystis hominis*. *Am. J. Trop. Med. Hyg.*, 70, 658–662.

Leon-Regagnon, V., Osorio-Sarabia, D., Garcia-Prieto, L., Lamothe-Argumedo, R., Bertoni-Ruiz, F., Oceguera-Figueroa, A. (2005). New host records of the nematode *Gnathostoma* sp. in Mexico. *Parasitol. Int.*, 54, 51–53.

Levsen, A. and Maage, A. (2016). Absence of parasitic nematodes in farmed, harvest quality Atlantic salmon (*Salmo salar*) in Norway – Results from a large scale survey. *Food Control*, 68, 25–29.

Levsen, A., Lunestad, B.T., Berland, B. (2005). Low detection efficiency of candling as a commonly recommended inspection method for nematode larvae in the flesh of pelagic fish. *J. Food Prot.*, 68(4), 828–832.

Levsen, A., Svanevik, C.S., Cipriani, P., Mattiucci, S., Gay, M., Hastie, L.C., Bušelić, I., Mladineo, I., Karl, H., Ostermeyer, U. et al. (2018). A survey of zoonotic nematodes of commercial key fish species from major European fishing grounds – Introducing the FP7 PARASITE exposure assessment study. *Fish. Res.*, 202, 4–21.

Li, X., Guyot, K., Dei-Cas, E., Mallard, J.P., Ballet, J.J., Brasseur, P. (2006). *Cryptosporidium* oocysts in mussels (*Mytilus edulis*) from Normandy (France). *Int. J. Food Microbiol.*, 108, 321–325.

Ligon, B.L. (2005). Gnathostomiasis: A review of a previously localized zoonosis now crossing numerous geographical boundaries. *Semin Pediatr. Infect. Dis.*, 16, 137–143.

Liu, G.H., Sun, M.M., Elsheikha, H.M., Fu, Y.T., Sugiyama, H., Ando, K., Sohn, W.M., Zhu, X.Q., Yao, C. (2020). Human gnathostomiasis: A neglected food-borne zoonosis. *Parasit. Vectors*, 13(1), 616.

Lom, J. and Arthur, J.R. (1989). A guideline for the preparation of species descriptions in *Myxosporea. J. Fish Dis.*, 12(2), 151–156.

Long, X.L., Wang, S.P., Zhou, S.F., Li, Z.X., Zhuang, S.F., He, Z. (2021). Interspecies phylogenetic analysis of *Clonorchis sinensis* in high-incidence areas of Hunan Province, China. *Biomed. Environ. Sci.*, 34(11), 881–890.

Lopez, I. and Pardo, M.A. (2010). Evaluation of a real-time polymerase chain reaction (PCR) assay for detection of *Anisakis simplex* parasite as a food-borne allergen source in seafood products. *J. Agric. Food Chem.*, 58(3), 1469–1477.

Lu, L.H., Lin, M.R., Choi, W.M., Hwang, K.P., Hsu, Y.H., Bair, M.J., Liu, J.D., Wang, T.E., Liu, T.P., Chung, W.C. (2006). Human intestinal *Capillaria*sis (*Capillaria* philippinensis) in Taiwan. *Am. J. Trop. Med. Hyg.*, 74, 810–813.

Lun, Z.-R., Gasser, R.B., Lai, D.-H., Li, A.-X., Zhu, X.-Q., Yu, X.-B., Fang, Y.-Y. (2005). Clonorchiasis: A key foodborne zoonosis in China. *Lancet Infect. Dis.*, 5(1), 31–41.

Macpherson, C.N.L. (2005). Human behaviour and the epidemiology of parasitic zoonose. *Int. J. Parasitol.*, 35, 1319–1331.

Marcogliese, D.J. (2001). Implications of climate change for parasitism of animals in the aquatic environment. *Canadian Journal of Zoology*, 79(8), 1331–1352.

Matsuoka, H., Nakama, T., Kisanuki, H., Uno, H., Tachibana, N., Tsubouchi, H., Horii, Y., Nawa, Y. (1994). A case report of serologically diagnosed pulmonary anisakiasis with pleural effusion and multiple lesions. *Am. J. Trop. Med. Hyg.*, 51, 819–822.

Mattiucci, S. and Nascetti, G. (2008). Advances and trends in the molecular systematics of anisakid nematodes, with implications for their evolutionary ecology and host-parasite co-evolutionary processes. *Adv. Parasitol.*, 66, 47–148.

Ménard, A., Dos Santos, G., Dekumyoy, P., Ranque, S., Delmont, J., Danis, M., Bricaire, F., Caumes, E. (2003). Imported cutaneous gnathostomiasis: Report of five cases. *Trans. R. Soc. Trop. Med. Hyg.*, 97, 200–202.

Menconi, V., Manfrin, C., Pastorino, P., Mugetti, D., Cortinovis, L., Pizzul, E., Pallavicini, A., Prearo, M. (2020a). First report of *Clinostomum complanatum* (Trematoda: Digenea) in European Perch (*Perca fluviatilis*) from an Italian subalpine lake: A risk for public health? *Int. J. Environ. Res. Public Health*, 17(4), 1389.

Menconi, V., Pastorino, P., Momo, I., Mugetti, D., Bona, M.C., Levetti, S., Tomasoni, M., Pizzul, E., Ru, G., Dondo, A. et al. (2020b). Occurrence and spatial distribution of *Dibothriocephalus latus* (Cestoda: Diphyllobothriidea) in Lake Iseo (Northern Italy): An update. *Int. J. Environ. Res. Public Health*, 17(14), 5070.

Menconi, V., Riina, M.V., Pastorino, P., Mugetti, D., Canola, S., Pizzul, E., Bona, M.C., Dondo, A., Acutis, P.L., Prearo, M. (2020c). First occurrence of *Eustrongylides* spp. (nematoda: Dioctophymatidae) in a subalpine lake in Northwest Italy: New data on distribution and host range. *Int. J. Environ. Res. Public Health*, 17(11), 1–9.

Menconi, V., Zoppi, S., Pastorino, P., Di Blasio, A., Tedeschi, R., Pizzul, E., Mugetti, D., Tomasoni, M., Dondo, A., Prearo, M. (2021). Relationship between the prevalence of *Dibothriocephalus latus* (Cestoda: Diphyllobothriidea) and the load of *Escherichia coli*: New findings in a neglected fish-borne parasitic zoonosis. *Zoonoses Public Health*, 68(8), 965–972.

Mendoza, L., Vilela, R., Rosa, P.S., Fernandes Belone, A.F. (2005). *Lacazia loboi* and *Rhinosporidium seeberi*: A genomic perspective. *Rev. Iberoam. Micol.*, 22, 213–216.

Mercken, E., Van Damme, I., Serradell, A., Gabriel, S. (2020). Presence of Anisakidae in commercial fish species imported into the Belgian food markets: A systematic review and meta-analyses. *Int. J. Food Microbiol.*, 318, 108456.

Mercken, E., Van Damme, I., Soba, B., Vangeenberghe, S., Serradell, A., De Sterck, T., Lumain, J.P.L., Gabriel, S. (2022). Sensitivity of candling as routine method for the detection and recovery of ascaridoids in commercial fish fillets. *Sci. Rep.*, 12(1), 1358.

Mille, T., Soulier, L., Caill-Milly, N., Cresson, P., Morandeau, G., Monperrus, M. (2020). Differential micropollutants bioaccumulation in European hake and their parasites *Anisakis* sp. *Environ. Pollut.*, 265, 115021.

Mladineo, I., Segvic, T., Petric, M. (2011). Do captive conditions favor shedding of parasites in the reared Atlantic bluefin tuna (*Thunnus thynnus*)? *Parasitol. Int.*, 60(1), 25–33.

Mo, T.A., Gahr, A., Hansen, H., Hoel, E., Oaland, Ø., Poppe, T.T. (2014). Presence of *Anisakis simplex* (Rudolphi, 1809 det. Krabbe, 1878) and *Hysterothylacium aduncum* (Rudolphi, 1802) (Nematoda; Anisakidae) in runts of farmed Atlantic salmon, *Salmo salar* L. *J. Fish Dis.*, 37, 135–40.

Moneo, I., Carballeda-Sangiao, N., González-Muñoz, M. (2017). New perspectives on the diagnosis of allergy to *Anisakis* spp. *Curr. Allergy Asthma Rep.*, 17(5), 27.

Moreno-Ancillo, A., Caballero, M.T., Cabañas, R., Contreras, J., Martin-Barroso, J.A., Barranco, P., López-Serrano, M.C. (1997). Allergic reactions to *Anisakis simplex* parasitizing seafood. *Ann. Allergy Asthma Immunol.*, 79, 246–250.

Mudry, J., Lefebvre, P., Dei-Cas, E., Vernes, A., Poirriez, J., Debat, M., Marti, R., Binot, P., Cortot, A. (1986). Anisakiase humaine : 5 cas dans le Nord de la France. *Gastroenterol. Clin. Biol.*, 10, 83–87.

Narr, L.L., O'Donnell, J.G., Libster, B., Alessi, P., Abraham, D. (1996). Eustrongylidiasis – A parasitic infection acquired by eating live minnows. *J. Am. Osteopath. Assoc.*, 96, 400–402.

Nawa, Y. (1991). Historical review and current status of gnathostomiasis in Asia. *Southeast Asian J. Trop. Med. Public Health*, 22, 217–219.

Neva, F.A. and Brown, H.W. (1994). *Basic Clinical Parasitology*. Prentice Hall International Inc., London.

Noël, C., Dufernez, F., Gerbod, D., Edgcomb, V.P., Delgado-Viscogliosi, P., Ho, L.C., Singh, M., Wintjens, R., Sogin, M.L., Capron, M. et al. (2005). Molecular phylogenies of *Blastocystis* isolates from different hosts: Implications for genetic diversity, identification of species, and zoonosis. *J. Clin. Microbiol.*, 43, 348–355.

Paugam, A. and Yéra, H. (2021). A new French case of *Dibothriocephalus nihonkaiensis* (*Diphyllobothrium nihonkaiense*) Teniasis due to Pacific salmon consumption: An emerging and underdiagnosed parasitosis. *Infect. Dis. Now*, 51(6), 579–580.

Pengput, A. and Schwartz, D.G. (2020). Risk factors for *Opisthorchis viverrini* infection: A systematic review. *J. Infect. Public Health*, 13(9), 1265–1273.

Pippy, J.H.C. (1970). Use of ultraviolet light to find parasitic nematodes *in situ*. *J. Fish Res. Bd. Canada*, 27, 963–965.

Polimeno, L., Lisanti, M.T., Rossini, M., Giacovazzo, E., Polimeno, L., Debellis, L., Ballini, A., Topi, S., Santacroce, L. (2021). *Anisakis* allergy: Is aquacultured fish a safe and alternative food to wild-capture fisheries for *Anisakis simplex*-sensitized patients? *Biology*, 10, 106.

Pozio, E. (2020). How globalization and climate change could affect foodborne parasites. *Exp. Parasitol.*, 208, 107807.

Prasongwatana, J., Laummaunwai, P., Boonmars, T., Pinlaor, S. (2013). Viable metacercariae of *Opisthorchis viverrini* in northeastern Thai cyprinid fish dishes – As part of a rational program for control of *O. viverrini*-associated cholangiocarcinoma. *Parasitol Res.*, 112(3), 1323–1327.

Rahmati, A.R., Kiani, B., Afshari, A., Moghaddas, E., Williams, M., Shamsi, S. (2020). World-wide prevalence of *Anisakis* larvae in fish and its relationship to human allergic anisakiasis: A systematic review. *Parasitol. Res.*, 119(11), 3585–3594.

Robert-Gangneux, F., Autier, B., Gangneux, J.P. (2019). *Dibothriocephalus nihonkaiensis*: An emerging concern in western countries? *Expert Rev. Anti Infect. Ther.*, 17(9), 677–679.

Roberts, J.D., Silbergeld, E.K., Graczyk, T. (2007). A probabilistic risk assessment of *Cryptosporidium* exposure among Baltimore urban anglers. *J. Toxicol. Environ. Health A.*, 70, 1568–1576.

Saksirisampant, W. and Thanomsub, B.W. (2012). Positivity and intensity of *Gnathostoma spinigerum* infective larvae in farmed and wild-caught swamp eels in Thailand. *The Korean J. Parasitol*, 50(2), 113.

dos Santos, C.A.L. and Howgate, P. (2011). Fishborne zoonotic parasites and aquaculture: A review. *Aquaculture*, 318(3–4), 253–261.

Scaramozzino, P., Condoleo, R., Martini, E., Bossu, T., Aquilani, S., Spallucci, V., Aquilini, E., Marozzi, S. (2018). Behaviour and eating habits as determinants for human opisthorchiasis in the Bolsena Lake area, Italy. *Folia Parasitol. (Praha)*, 65.

Schmidt, G. (1971). Acanthocephalan infections of man, with two new records. *J. Parasitol.*, 57, 582–584.

Schmidt, L., Lepers, M., Brion, M., Abdellatifi, M.H., Dei-Cas, E., Vernes, A., Fourrier, A. (1982). Opisthorchiidoses chez les réfugiés de l'Asie du Sud-Est dans le Nord de la France. *Med. Mal. Infect.*, 12, 442–446.

Slifko, T.R., Smith, H.V., Rose, J.B. (2000). Emerging parasite zoonoses associated with water and food. *Int. J. Parasitol.*, 30(12–13), 1379–1393.

Smaldone, G., Abollo, E., Marrone, R., Bernardi, C.E.M., Chirollo, C., Anastasio, A., Del Hierro, S.P. (2020). Risk-based scoring and genetic identification for anisakids in frozen fish products from Atlantic FAO areas. *BMC Vet. Res.*, 16(1), 1–12.

Smith, J.W. and Wootten, R. (1978). *Anisakis* and Anisakiasis. *Adv. Parasitol.*, 16, 93–163.

Sokolov, S., Frolov, E., Novokreshchennykh, S., Atopkin, D. (2021). An opisthorchiid concept of the genus *Liliatrema* (Trematoda: Plagiorchiida: Opisthorchioidea): An unexpected systematic position. *Zool. J. Linn. Soc.*, 192(1), 24–42.

Sterling, C.R. (2006). Food-borne nematodes infections. In *Foodborne Parasites*, Ortega Y. (ed.). Springer, Cham.

Suzuki, J., Murata, R., Kodo, Y. (2021). Current status of anisakiasis and anisakis larvae in Tokyo, Japan. *Food Saf. (Tokyo)*, 9(4), 89–100.

Tada, I., Otsuji, Y., Kamiya, H., Mimori, T., Sakaguchi, Y., Makizumi, S. (1983). The first case of a human infected with an acanthocephalan parasite, *Bolbosoma* sp. *J. Parasitol.*, 69, 205–208.

Tang, Z.L., Huang, Y., Yu, X.B. (2016). Current status and perspectives of *Clonorchis sinensis* and clonorchiasis: Epidemiology, pathogenesis, omics, prevention and control. *Infect. Dis. Poverty*, 5(04), 14–25.

Timi, J.T., Paoletti, M., Cimmaruta, R., Lanfranchi, A.L., Alarcos, A.J., Garbin, L., Mattiucci, S. (2014). Molecular identification, morphological characterization and new insights into the ecology of larval *Pseudoterranova cattani* in fishes from the Argentine coast with its differentiation from the Antarctic species, *P. decipiens* sp. E (Nematoda: Anisakidae). *Vet. Parasitol.*, 199(1–2), 59–72.

Umehara, A., Kawakami, Y., Araki, J., Uchida, A. (2008). Multiplex PCR for the identification of Anisakis simplex sensu stricto, *Anisakis pegreffii* and the other anisakid nematodes. *Parasitol. Int.*, 57(1), 49–53.

Weiss, L.M. (2003). Microsporidia 2003: IWOP-8. *J. Eukaryot. Microbiol.*, 50, 566–568.

Werner, M.T., Fæste, C.K., Levsen, A., Egaas, E. (2011). A quantitative sandwich ELISA for the detection of *Anisakis simplex* protein in seafood. *Eur. Food Res. Technol.*, 232, 157–166.

WHO (2015). WHO estimates of the global burden of foodborne diseases: Foodborne disease burden epidemiology reference group 2007–2015. World Health Organization [Online]. Available at: https://apps.who.int/iris/handle/10665/199350.

WHO/IUIS Allergen Nomenclature Sub-Committee (2021). Database [Online]. Available at: http://allergen.org/search.php?allergensource=Anisakis+simplex [Accessed 27 October 2021].

Wunderink, H.F., Rozemeijer, W., Wever, P.C., Verweij, J.J., van Lieshout, L. (2014). Foodborne trematodiasis and *Opisthorchis felineus* acquired in Italy. *Emerg. Infect. Dis.*, 20(1), 154–155.

Yahata, Y., Sugita-Konishi, Y., Ohnishi, T., Toyokawa, T., Nakamura, N., Taniguchi, K., Okabe, N. (2015). *Kudoa septempunctata*-induced gastroenteritis in humans after flounder consumption in Japan: A case-controlled study. *Japan J. Infect. Dis.*, 68(2), 119–123.

Yera, H., Frealle, E., Dupouy-Camet, J. (2016). Molecular confirmation of *Anisakis pegreffii* as a causative agent of anisakidosis in France. *Dig. Liver Dis.*, 48(8), 970.

Yera, H., Frealle, E., Dutoit, E., Dupouy-Camet, J. (2018). A national retrospective survey of anisakidosis in France (2010–2014): Decreasing incidence, female predominance, and emerging allergic potential. *Parasite*, 25, 23.

Zahedi, A., Greay, T.L., Paparini, A., Linge, K.L., Joll, C.A., Ryan, U.M. (2019). Identification of eukaryotic microorganisms with 18S rRNA next-generation sequencing in wastewater treatment plants, with a more targeted NGS approach required for *Cryptosporidium* detection. *Water Res.*, 158, 301–312.

Zhong, Z.-H., Wu, H.-C., Li, Z.-C., Guo, Q.-K., Li, A.-X. (2022). Using galvanized materials to control *Cryptocaryon irritans* infestation. *Aquaculture*, 548, 737659.

Žilić, J. and Mladineo, I. (2006). Prevalence of the parasite *Anisakis simplex* (anisakidae, Nematoda) in the blue fin tuna (*Thunnus thynus*) from mariculture. *MESO: Prvi hrvatski časopis o mesu*, 8(4), 229–231.

7

Microplastics

Alexandre DEHAUT and Guillaume DUFLOS
ANSES, Laboratoire de Sécurité des Aliments, Boulogne-sur-Mer, France

Plastic is one of the most important inventions of the 20th century. It has been a part of our lives since the second half of this century. Practical, light and resistant, this material is used in many fields. Mass production and poor waste management has inevitably led to the presence of plastic waste in the environment, and in particular in the seas and oceans. This has been described since the early 1970s, as evidenced by observations made on the island of Sein in Brittany, France[1] in 1973, and also in one of the first publications on microplastics by Carpenter et al. (1972).

7.1. What are microplastics?

Microplastics are above all plastic composed of one or more polymers to which additives are added to give them special properties, and in some cases, a filler to stabilize the whole. Microplastics come mainly from the degradation of larger plastics, but can also be produced directly. Indeed, they can be found in the environment, when they are produced intentionally, it is then about primary sources as the microparticles added in cosmetic products or granules for the industry used to form the biggest plastics. Secondary sources include textile fibers or fragments from

For a color version of all figures in this chapter, see http://www.iste.co.uk/verrez-bagnis/aquaticproducts.zip.

1. See: https://www.ina.fr/video/CAF93027827 (accessed on 18 January, 2023).

the degradation of larger pieces of plastic, which are generated under the influence of exogenous factors such as ultraviolet light, oxidation or abrasion.

7.1.1. *Different chemical compositions*

Among the plastics, there are many molecules of different polymeric natures of which the most common are high and low density polyethylene (HDPE and LDPE), polypropylene (PP), polystyrene (PS), poly(ethylene terephthalate) (PET), poly(vinyl chloride) (PVC) and polyurethane (PUR). Figure 7.1 presents some manufactured products that have one of the polymers in their composition.

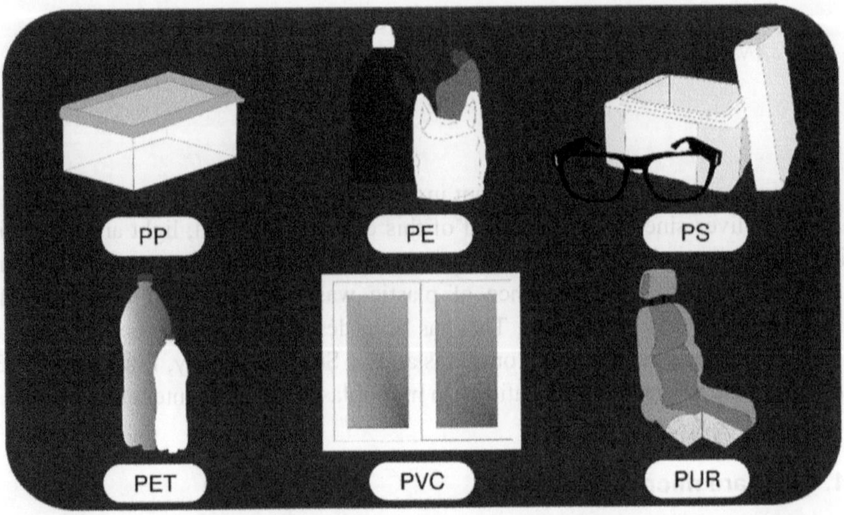

Figure 7.1. *Illustration of the main polymers produced and examples of manufactured products whose main composition corresponds to these polymers. PP (polypropylene), PE (polyethylene), PS (polystyrene), PET (poly(ethylene terephthalate)), PVC (poly(vinyl chloride)) and PUR (polyurethane)*

While it is relatively easy to determine the composition of macroplastics, it is important to establish the polymeric nature of MP particles in order to discriminate natural particles (sand grains, plant debris) from synthetic ones.

Another important point related to the nature of the polymer composing MPs is the greater or lesser presence of additives. Additives are intentionally added by manufacturers in their recipes in order to confer certain properties to plastics. For example, PVC can be composed of more than 50% of plasticizers. Among the targeted properties are flexibility conferred by plasticizers (phthalates, bisphenols, etc.), fire resistance made possible by flame retardants (brominated or phosphated), resistance to UV, oxidation, etc. The list of additives added to plastics is relatively long, and more than 400 compounds have recently been listed by the ECHA (2016).

In addition to intentionally added chemicals, plastics can also contain unintentionally added substances, known as non-intentionally added substances (NIAS). NIAS are the result of the presence of contaminants in the raw material or the generation of new molecules during the various manufacturing processes of plastics.

Finally, while there is no debate on many polymers, it should be kept in mind that there is a debate in the community on the inclusion of polymers derived from natural materials, such as rubber from rubber tree latex, or rayon and cellophane produced from cellulose (Hartmann et al. 2019). Again, the lack of a clear definition is problematic since, as an example, tire debris, based on rubber, could represent a significant portion of microplastics in the environment (Knight et al. 2020).

7.1.2. *Importance of the notion of size*

Microplastics are primarily a category of plastic particle size. As noted by Hartmann et al. (2019), while the upper limit of this category is relatively agreed upon and well defined, 5 mm for the largest particle size (see Figure 7.2), the lower limit is still subject to debate within the scientific community, between 100 nm and 1 μm (Gigault et al. 2018).

The reason for this discrepancy in the minimum size of microplastics originates from the difference between intentionally produced nanomaterials and nanoplastics resulting from the degradation of macroplastics or microplastics leading to the formation of heteroaggregates (Gigault et al. 2018). In fact, nanoplastic particles cannot be included in the definitions of nanomaterials formulated by the ISO (International Standard Organization).

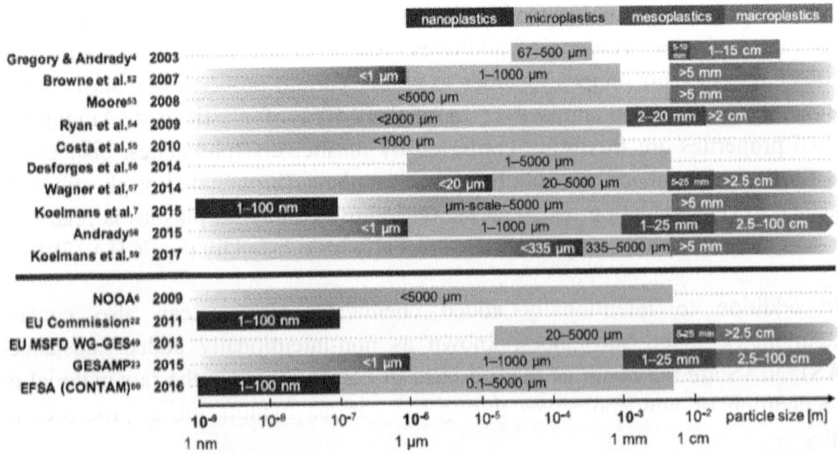

Figure 7.2. *Examples of categorization of plastic waste size classes, nanoplastics, microplastics, mesoplastics and microplastics based on different literature references. Reprinted with permission from Hartmann et al. (2019, pp. 1039–1047). (Copyright © 2019, American Chemical Society)*

Beyond the debate, the lack of consensus on size limits remains problematic when an aggregation of results must be performed by comparing data from various studies. Indeed, the estimate of the number of particles is not the same according to the established limits, and it thus becomes complex to compare the exposure levels according to the environments or matrices compared.

7.1.3. *A diversity of forms within microplastics*

Beyond size and chemical composition, a third important aspect to consider is the shape of the particle: spheres, granules, fibers, foams, films, spherules and fragments. These are all types and terms that can characterize MPs. Some classifications have been proposed (Lusher et al. 2017; Hartmann et al. 2019) in order to use common terms to standardize the communication of results.

While the shape of MPs matters for data transmission, it is of primary importance for characterizing the effects observed following exposure to MPs. Indeed, as the relationship between structures and effects is quite complex to study, it should be noted that, for the same size, the volumes and surfaces of the particles can be drastically different. Thus, the difference in volume of a 1-mm fiber for 20 µm thickness and a 1-mm sphere is 1–1,600 and their difference in surface is

1–50. Moreover, one of the limitations of the first toxicology studies conducted on microplastics was the use of microbeads that are not representative of what is encountered in the environment. Similarly, the fibers that correspond to the microparticles of anthropic origin are mostly found in the environment (Dris et al. 2018; Hermabessiere et al. 2019), and are, because of their spatial conformations, difficult to identify. Nevertheless, some studies have reported that although present in significant quantities, they are not all composed of synthetic polymers (Dris et al. 2016; Athey et al. 2020; Athey and Erdle 2021).

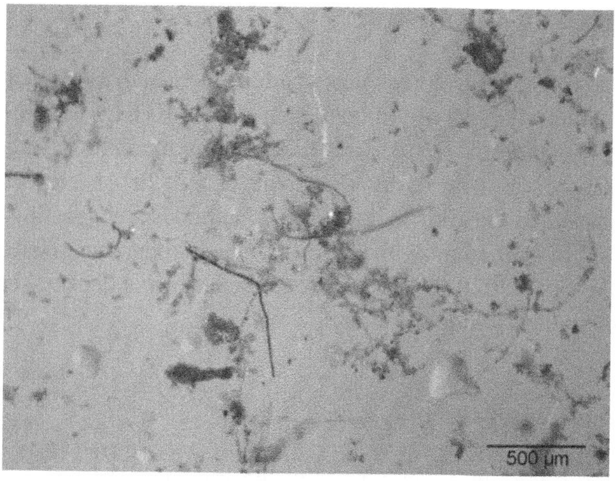

Figure 7.3. *Picture of various particles, including microplastics observed in dust. The scale bar corresponds to a 500-µm measurement*

7.1.4. *Carrying of chemical and biological contaminants by microplastics*

Microplastics have the capacity to vectorize chemical contaminants by the adsorption of polluting substances present in the environment. It has been shown that persistent organic pollutants and heavy metals can adsorb to plastic particles, and also marine biotoxins. Nevertheless, microplastics are not the major source of these pollutants in the environment.

From a biological perspective, microplastics are also potential vectors for bacteria and Archaea (Amaral-Zettler et al. 2020). Through this bacterial carriage, microplastics can vectorize both pathogens (Kirstein et al. 2016; Virsek et al. 2017; Wu et al. 2019) and genetic material involved in antibiotic resistance (Arias-Andres et al. 2018; Yang et al. 2019; Zhang et al. 2020c).

7.2. Analysis of microplastics in seafood

Although research on microplastics is relatively recent, numerous studies have been carried out around the world, reporting contamination in many marine species (Appendix, section 7.7). However, these studies have been conducted with different protocols, with information reported that is not equivalent from one publication to another.

7.2.1. *Isolation of microplastics*

Two main approaches have been described for the isolation of microplastics: one by hydrolysis under alkaline conditions using a 10% KOH solution, adapted from the work of Foekema et al. (2013), and the other by oxidation using a solution of H_2O_2 30%, adapted from the work of Nuelle et al. (2014). Nevertheless, recent studies have reinforced the knowledge of these methods by suggesting modifications of the protocol, in particular with regard to the temperatures used to treat the samples (Treilles et al. 2020). Some authors also perform a density separation using zinc chloride or sodium iodide solutions.

7.2.2. *Identification of the polymeric nature*

The search for MPs in a sample cannot be limited to a simple particle count. Identification by recognized technologies, whether spectroscopic, vibrational or spectrometric is now the rule thanks to micro Raman, micro FT-IR or thermoanalytical techniques coupled with mass spectrometry. Nevertheless, there is no consensus on the number of particles to be studied, whole versus subsample. A second problem is the detection of small MNPs for which the analysis is not necessarily easy and difficult to automate. More globally, working with common databases is a first step towards the harmonization of identification data processing. In this respect, important work has been achieved in this field, notably with the Openspecy platform[2] (Cowger et al. 2021), which allows the consultation of a common database using the same search algorithm, or the free software siMPle (Primpke et al. 2020), which allows practices to be standardized in the field of spectral imaging analysis.

2. See: www.openanalysis.org/openspecy (accessed on 18 January, 2023).

7.2.3. *Contamination management*

Management of contamination due to the work environment is a primary control focus for the study of microplastics. Working under a high laminar flow or other means of controlling the laboratory environment has helped to control the risk of contamination (Wiesheu et al. 2016).

However, as the size, shape and color of the particles isolated in the samples and controls are very diverse, there is currently no common method of accounting for contamination in the processing of results.

Unfortunately, there are still many works that did not perform adequate controls or report contamination data properly.

7.2.4. *A problem that is beginning to emerge*

All of the points raised during this section are areas of improvement that should be considered regarding the practices of the broader fishery products community. Methods of isolation and identification of particles and contamination management are only a few highlights of a broader issue. Method harmonization is beginning to be the subject of specific work such as that of Hermsen et al. (2018), Dehaut et al. (2019) and Cowger et al. (2020). In the context of the study of seafood products, a set of points to be verified for the realization of manipulations in correct conditions has been proposed, one of them, the MIMS, proposes essential and desired points to be verified by scientists publishing data on the subject.

7.3. Contamination status in consumed marine species

In order to write this section, aggregated data from several studies have been used based on work initiated by Hermabessiere (2018). These were complemented by a set of publications released between 2019 and the first quarter of 2021. The set of names of consumed species, for which contamination by microplastics has been documented, is shown in the Appendix (see section 7.7). For fish, the consultation of the FishBase database[3] allowed us to ascertain the consumption of the species cited.

---

3. See: https://www.fishbase.se/search.php (accessed September 2021).

REMARK 7.1.– For a given article, several species or locations may have been analyzed, and the data have therefore been recorded on separate lines. Thus, an article may have generated many study cases. This notion of cases will therefore be repeated hereafter.

7.3.1. *Shellfish*

The presence of microplastics has been documented for 41 species of shellfish in 45 scientific articles.

One-third of the cases studied show correct sampling in accordance with the recommendations of the Marine Strategy Directive Framework (MSDF) technical group (Hanke et al. 2013), i.e. the study of a minimum of 50 individuals per species and age group (see Figure 7.4).

Concerning the percentages of ingestion, it should be noted that the information is not available in half of the cases. On the contrary, when it is mentioned, it is noted that the frequency of ingestion is relatively significant since it is between 80 and 100% (see Figure 7.4). This significant ingestion can be explained by the filtering character of many of the species studied.

The number of microplastics per individual shows a certain dichotomy with cases where the quantities are relatively low, between 0 and 4 MP per individual (42% of cases), and others where the levels of contamination are very high, i.e. greater than 10 MP per individual (20%) (see Figure 7.4).

Finally, concerning the lower limit of the particle sizes sought, two-thirds of the studies (see Figure 7.4) targeted particles with a lower size either between 20 and 50 μm or less than 20 μm.

7.3.2. *Crustaceans*

The crustacean category is the third most documented category, with 18 published papers and 20 species studied.

The distribution of the ranges of individuals studied is similar to that of shellfish (see section 7.3.1), with nearly 30% of the studies complying with MSDF recommendations (see Figure 7.4).

There is no characteristic percentage of ingestion for crustaceans, as shown by the particularly heterogeneous distribution of the different ranges (see Figure 7.4). It should be noted that for 25% of the cases the percentage of ingestion is higher than 80%, and that in almost a third of the cases it is lower than 60%. Here again, the number of cases for which data are missing is significant, since it corresponds to 25% of the cases.

Like the ingestion percentage, the number of microplastics per individual is extremely variable between the different cases (see Figure 7.4). The most significant class of ingestion level remains the class 0–2 particles per individual, which corresponds to nearly 40% of cases.

Regarding the minimum size sought by the authors, in 34% of cases particles smaller than 20 μm were found (see Figure 7.4). However, in comparison with the other classes of seafood, the studies that targeted only particles of 100 μm or more was more significant, representing more than a third of cases.

7.3.3. *Fish*

Fish is the category for which the most data can be found regarding contamination levels of fish products. More than 90 articles were reviewed reporting contamination levels for over 300 species.

For this category of products, it is mainly small samples that were taken in the various cases, i.e. less than 20 individuals in nearly two-thirds of cases (see Figure 7.4), and only 16% of the cases meet the criterion of 50 individuals defined above. These small samples can be explained by the opportunistic catches reported in several studies.

Regarding ingestion rates, for half of the cases less than 40% of the individuals tested ingested microplastics (see Figure 7.4). It should be noted here that some studies only focused on certain parts of the animals, especially those that were only slightly consumed, such as the gills or the digestive system. The intercomparison of data is therefore particularly difficult in the context of the study of fish.

The number of plastics found per individual was relatively low compared to the number of plastics found per individual.

This is consistent with the other species, since in 62% of the cases, only 0–2 particles are found per individual (see Figure 7.4). Nevertheless, these data should be put into perspective with respect to the size data provided below.

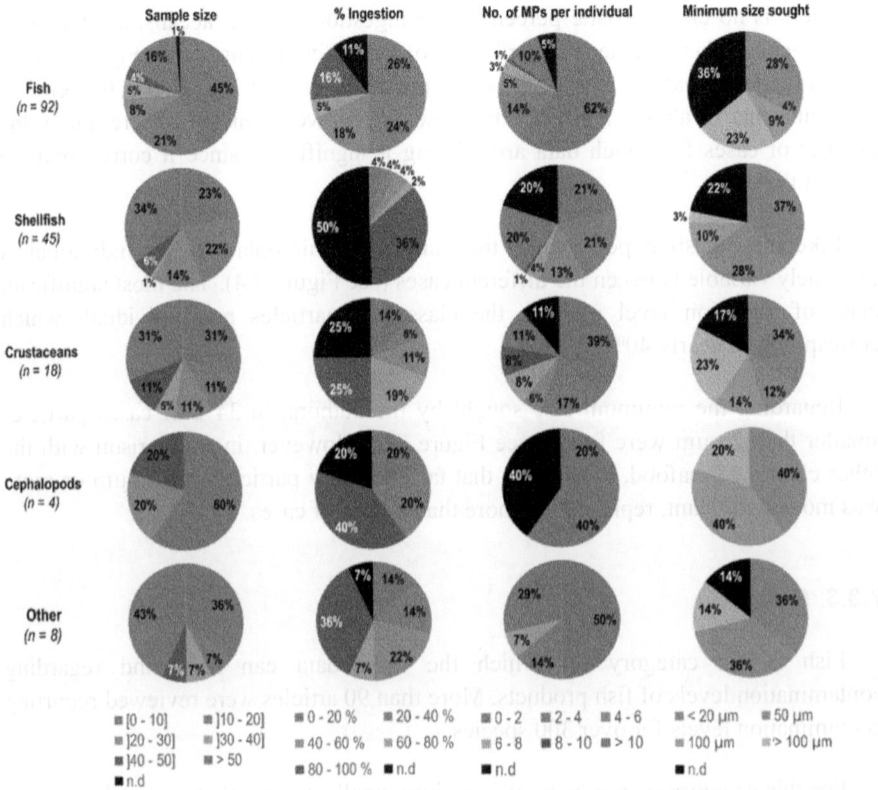

Figure 7.4. *Aggregated data on sample size, percentage of intake, average number per individual and minimum size sought in different studies (n=150) involving different types of fish products (fish, shellfish, crustaceans, cephalopods and other species). For each type of fish product, the number of studies used is shown in brackets. The complete list of species studied is shown in the Appendix (see section 7.7). n.d.: not determined*

Indeed, first, in 36% of the cases, the minimum size targeted is not indicated. Finally, when this is the case, only a quarter of the cases target very small MPs, i.e. less than 50 µm, which is quite different from other categories such as shellfish seen previously (see section 4.3.1). Thus, it is less likely to find significant amounts of MPs when the target sizes are large.

7.3.4. *Cephalopods*

Only four cases were studied among three species of cephalopods: the most studied was cuttlefish (Abidli et al. 2019; Oliveira et al. 2020). This is the category for which the least information is available.

The samples tested on these few cases fall short of the recommended criteria, since in two-thirds of the cases, the sampling is less than 10 individuals (see Figure 7.4).

The percentage of ingestion by cephalopods is again dichotomous since in 40% of the cases, the ingestion level described is between 80 and 100%, and for the rest, the levels are lower than 40% of ingestion.

When contamination levels are reported, i.e. for 60% of cases, they are mostly significant since 40% of cases report contamination levels greater than 10 MP per individual (see Figure 7.4).

Finally, studies on cephalopods have focused on large microplastic particles since only 40% of cases offer data on smaller particles between 50 and 100 μm.

7.3.5. *Other species consumed*

The final category includes several types of seafood products without a specific category such as elasmobranchs, sea cucumbers or sea urchins. Eight papers have listed data on nine distinct species.

In terms of sampling, for 50% of the cases grouped in this category, the number of individuals selected is almost acceptable, i.e. greater than 40 for each species analyzed (see Figure 7.4). However, for almost a third of the cases, the number remains too low, ranging from 1 to 10 individuals. These low numbers of samples can still be explained, as in the case of the study of elasmobranchs where the collection of 50 individuals can be more complex in comparison with other species.

The ingestion rates recorded in this category are among the highest of those documented in all previous categories, with more than 40% of cases reporting an ingestion rate greater than 60% of individuals analyzed (see Figure 7.4).

The number of particles ingested by the different species in this category is quite low since the most represented class, in 50% of the cases, is the one ranging from 0 to 2 PM per individual. Nevertheless, for some cases, 29%, more than 10 particles were found per individual.

In contrast to the other categories, for the species analyzed here, the lower limit of targeted particle size was relatively low as 72% of the cases targeted particles with a size of 50 µm or less.

IMPORTANT.– The lack of harmonization of methods means that the practices of different authors can vary greatly from one article to another. It should therefore be noted that in the aggregation work carried out here, the results obtained can only be compared with each other with difficulty and the conclusions drawn must be taken with caution.

7.4. Contamination status in processed products containing fish

Although the presence of MPs has been described in many consumed marine species (see section 7.3), there have been a few studies conducted on processed seafood products, as noted by Karami et al. (2018). The articles presented below were retrieved from literature databases. Before 2021, all published studies focused on canned fish (Karami et al. 2018; Akhbarizadeh et al. 2020). In 2021, a Korean study (Kim et al. 2021) looked at new matrices such as salted, cooked products and fish or fish product-based sauces such as oyster sauces.

Echoing sections 7.2 and 7.3 here again, it is difficult to compare the results, given that different methodologies were used by the authors.

The study by Karami et al. (2018), which pioneered the field, found relatively low levels of contamination, as only nine microplastics[4] were found in 21 cans (see Table 7.1). This low level of contamination can be explained by the filter porosity used by the authors (149 µm), which led to counts of large microplastics, i.e. larger than 149 µm. This high porosity is explained by the difficulty the authors faced in processing complex samples, notably because of the presence of oils.

4. The authors reported isolating six microplastics and three particles of phthalocyanine, which is a blue pigment found in many microplastics.

	Species	Scientific names	Number	Concentrations	References
Canned goods	Sardine Sprat	– –	12 cans 9 cans	9 MPs for 21 cans 0.05 ± 0.01 to 0.13 ± 0.02 MPs/g	[1]
	Yellowfin tuna Longtail tuna Spanish mackerel	*T. albacares* *T. tonggol* *S. commerson*	25 cans 20 cans 5 cans	0.09 ± 0.01 to 0.16 ± 0.01 MPs/g 0.15 ± 0.03 MPs/g	[2]
	Tuna, cockles, sea snails	–	19 cans	0.48 ± 0.45 MPs/g	[3]
Salted products	Salted shrimps, clams in sauce, intestines in sauce	–	11	1.73 ± 1.73 MPs/g	[3]
Cooked products	Mussels, shrimp meat, clam meat	–	8	0.47 ± 0.36 MPs/g	[3]
Sauces	Oyster sauce, Tuna extract	–	8	0.32 ± 0.21 MPs/g	[3]
Other sauces	Anchovy sauce, Sand eel extract	–	4	0.26 ± 0.19 MPs/g	[3]

[1] Karami et al. (2018), [2] Akhbarizadeh et al. (2020) and [3] Kim et al. (2021).

Table 7.1. *Levels of microplastic contamination identified in the literature dealing with processed fishery products*

Akhbarizadeh et al. (2020) were able to perform counts on smaller particles, thanks to the use of filters with lower porosity (5 µm). However, the study actually required the use of multiple filters per sample due to clogging issues for the same reasons described by Karami et al. (2018). Of the 50 cans analyzed, 128 microplastics were isolated, with an average frequency of 80% of cans containing at least one particle.

Although it covered a wide range of processed products, the study by Kim et al. (2021) gives little information on the results obtained. Also, no data on the frequency of contamination or the total number of particles isolated are available.

To conclude this presentation of the different published studies, in a more anecdotal way, a recent study conducted on canned fish (salmon and sardines) was able to identify plastic polymers, without any counting having been carried out or any contamination controls having been set up (Hussien et al. 2021).

Overall, the contamination levels range from 0.05 ± 0.01 to 1.73 ± 1.73 MP/g depending on the studies and matrices. This large disparity is likely explained by the origin of the products, the different production processes and the difference in the methods used by the authors, as the size limits studied were roughly equivalent.

With respect to the isolated particle sizes, no clear trend emerged. In their study, Akhbarizadeh et al. (2020) observed a majority of MPs in the size range 1–5 mm, while Kim et al. (2021) noted that particles in the 100–500 μm range are the most numerous. These data are in contradiction with the environmental data, which overwhelmingly conclude a preponderance of MP particles in the smaller size classes.

MPs are found in different forms: they are mainly fibers, fragments and films (Karami et al. 2018; Akhbarizadeh et al. 2020). Concerning the polymeric compositions of the isolated particles, again no general trend appears. While PP is described in the three studies as the polymer composing several particles, some results are diametrically opposed. Thus, in the study by Akhbarizadeh et al. (2020), as in the Karami et al.'s (2018) study, PET was the most common polymer found (36.6% of identifications), while this same polymer was only found in 2.9% of the samples from Kim et al. (2021). Similarly, the study by Kim et al. (2021) accounts for almost half of the MPs as being PE, whereas this polymer only made up 3.2% of the particles isolated by Akhbarizadeh et al. (2020).

Among the sources of product contamination, a first pathway is translocation of MP from the gastrointestinal tract to animal tissues (Karami et al. 2018; Akhbarizadeh et al. 2020). While this explanation is plausible for the small particle sizes evoked by the second study, i.e. the class lower than 50 μm, it is not conceivable for the classes higher than 149 μm or the class [1–5 mm]. A more likely possibility, mentioned by the same authors, are accidents that occur during the evisceration phase, which leads to the contamination of edible tissue. More likely still, and as described in a relatively exhaustive manner in the study by Akhbarizadeh et al. (2020), contamination occurs during the transformation process, which involves many steps and as many risks of contamination. Finally, Akhbarizadeh et al. (2020) note that the use of ingredients in addition to fish products, such as oils, water, salt or flavorings can also be a source of contamination. Kim et al. (2021) observe that the more processing the product has undergone and the lower its fish product content, the lower the level of contamination. These observations are qualified, knowing that the samples taken are small and are probably not representative of the real contamination levels.

From these studies on processed products, it is clear that this type of product is not free of microplastics. Nevertheless, there is still too little data to obtain a clear overview of these contamination levels and therefore of the exposure levels via these products. Different sources of contamination are mentioned by the authors, such as the consumption of seafood during its life in the marine environment or during processing in the industry. At present, the relative share of contamination levels brought about by the environment and by the processing of the products remains little known.

7.5. Importance of risk assessment in the context of microplastic contamination

7.5.1. *Human exposure*

There are three types of exposure: breathing, eating and skin contact (see Figure 7.5).

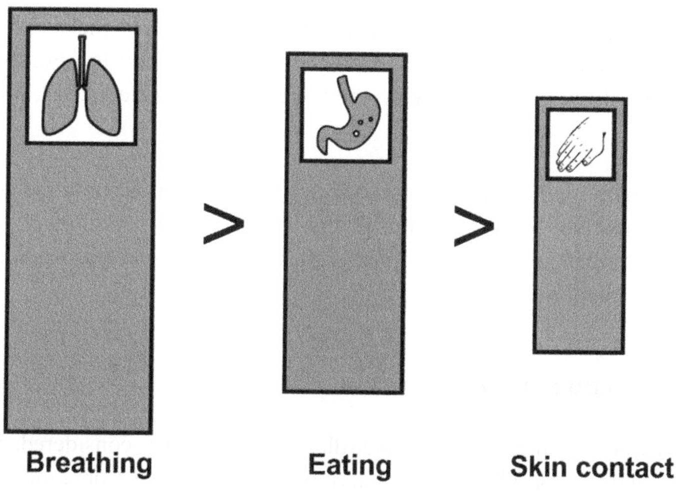

Figure 7.5. *Schematic representation of the three major pathways of exposure to microplastics and their relative importance*

Fishery and aquaculture products were the first foods that attracted attention regarding dietary exposure, particularly filter-feeding organisms. A complete list is provided in Table 7.1 and the Appendix (see section 7.7).

Other foods that have been studied and shown to contain microplastics include salt, sugar, honey, beer and water (Barboza et al. 2020).

As discussed in a previous chapter, the lack of method harmonization or consensus methods makes it difficult to use the different data to perform this exposure part of the risk assessment. Nevertheless, the dietary intake of microplastics has been confirmed by their presence in human feces (Schwabl et al. 2019; Zhang et al. 2021a).

7.5.2. *Toxicological impact*

The toxicological impact via the intake of microplastics in food has been clearly demonstrated for some organisms such as fish, oysters, marine worms, mice, clams and crabs by actions on their development, metabolism or reproduction (Lu et al. 2016; Sussarellu et al. 2016; Qiao et al. 2019; Barboza et al. 2020). Concerning humans, the toxicological impact via food is still to be demonstrated, although the first studies on cellular models tend to show impacts. On the contrary, chronic exposure by inhalation of fibers on workers of the textile industry reveals pulmonary inflammatory diseases that can be severe in certain cases (Prata 2018; Rahman et al. 2021).

More generally, the main impacts of microplastics will occur at the cellular level through oxidative stress, immune disruption and neurological alterations. Translocation phenomena through membranes are also suspected, but this is only for very small microplastics.

7.6. Concluding remarks

The search for microplastics, regardless of the matrix considered, is relatively recent, leading to the absence of a harmonized method for obtaining comparable data.

Despite this lack of common methodology, efforts have been made and common practices are beginning to emerge. In recent years, the number of publications related to the investigation of microplastics in fish products has increased significantly (see the Appendix in section 7.7), but these studies often remain focused on the digestive systems of the individuals analyzed.

However, the fish product matrix as food, a fortiori when it is processed, is not taken into account enough at the moment, as evidenced by the few studies published on the subject. In a more general way, foods are not targeted because of the difficulty encountered in analyzing them properly in the context of the search for microplastics. It is not by chance that the food matrix for which the work is the most successful is drinking water.

Although a standardized method for the study of fish products is not yet available, more data on this type of food should be available.

In addition, it is now necessary to go beyond raw materials, as fish products are increasingly consumed after having undergone processing, whether summary (salting, drying) or complex (sauces, preserves). The production of data on the levels of contamination of these products is therefore necessary.

7.7. Acknowledgements

The authors would like to thank Dorian Lescure for his help in developing Figure 7.4.

7.8. Appendix: List of consumed species for which microplastic contamination has been documented

	List of species
Various species	*Apostichopus japonicus* (Mohsen et al. 2019), *Hemicentrotus pulcherrimus* (Feng et al. 2020), *Paracentrotus lividus* (Avio et al. 2020), *Raja asterias* (Neves et al. 2015), *Scyliorhinus canicula* (Neves et al. 2015; Bellas et al. 2016; Mancia et al. 2020; Morgan et al. 2021), *Squalus acanthias* (Avio et al. 2015), *Squalus blainville* (Anastasopoulou et al. 2013), *Strongylocentrotus intermedius* (Feng et al. 2020), *Temnopleurus hardwickii* (Feng et al. 2020), *Temnopleurus reesevii* (Feng et al. 2020)
Bivalves	*Alectryonella plicatula* (Li et al. 2015), *Aulacomya atra* (Sparks 2020), *Bolinus brandaris* (Abidli et al. 2019), *Cerastoderma edule* (Hermabessiere et al. 2019), *Cerastoderma lamarcki* (Bagheri et al. 2020), *Choromytilus meridionalis* (Sparks 2020), *Corbicula fluminea* (Klasios et al. 2021), *Crassostrea gasar* (Vieira et al. 2021), *Crassostrea gigas* (Van Cauwenberghe and Janssen 2014; Rochman et al. 2015; Phuong et al. 2018; Abidli et al. 2019;

	List of species
Bivalves (continued)	Baechler et al. 2019; Cho et al. 2019; 2021; Teng et al. 2019; Ding et al. 2020; Martinelli et al. 2020), *Cyclina sinensis* (Li et al. 2015), *Donax cuneatus* (Narmatha Sathish et al. 2020), *Gari solida* (Vinay Kumar et al. 2021), *Hexaplex trunculus* (Abidli et al. 2019), *Magallana bilineata* (Patterson et al. 2019), *Meretrix lusoria* (Li et al. 2015), *Meretrix meretrix* (Dowarah et al. 2020), *Mytilaster lineatus* (Bagheri et al. 2020), *Mytilus californianus/Mytilus* spp. (Klasios et al. 2021), *Mytilus chilensis* (Perez et al. 2020), *Mytilus edulis* (De Witte et al. 2014; Mathalon and Hill 2014; Van Cauwenberghe and Janssen 2014; Van Cauwenberghe et al. 2015; Vandermeersch et al. 2015; Li et al. 2016, 2018b; Courtene-Jones et al. 2017; Phuong et al. 2018; Cho et al. 2019; Hermabessiere et al. 2019; Scott et al. 2019; Perez et al. 2020; Cho et al. 2021; Vinay Kumar et al. 2021), *Mytilus galloprovincialis* (Li et al. 2015; Vandermeersch et al. 2015; Digka et al. 2018; Abidli et al. 2019; Gomiero et al. 2019; Avio et al. 2020; Ding et al. 2020; Gedik and Eryasar 2020; Sparks 2020; Wakkaf et al. 2020; Klasios et al. 2021; Mercogliano et al. 2021), *Mytilus meridionalis* (Sparks et al. 2021), *Mytilus* spp. (Marques et al. 2021), *Mytilus trossulus* (Railo et al. 2018), *Ostrea denselamellosa* (Wu et al. 2020), *Ostrea edulis* (Avio et al. 2020), *Paphia undulata* (Vinay Kumar et al. 2021), *Patinopecten yessoensis* (Li et al. 2015; Cho et al. 2019), *Pecten maximus* (Akoueson et al. 2020), *Perna canaliculus* (Webb et al. 2019; Vinay Kumar et al. 2021), *Perna perna* (Santana et al. 2016), *Perna vieridis* (Ding et al. 2020; Dowarah et al. 2020), *Ruditapes decussatus* (Abidli et al. 2019), *Ruditapes philippinarum* (Ding et al. 2020; Cho et al. 2021), *Saccostrea cucullata* (Li et al. 2018a; Wang et al. 2021a), *Scapharca subcrenata* (Li et al. 2015; Ding et al. 2020), *Siliqua patula* (Baechler et al. 2019), *Sinonovacula constricta* (Li et al. 2015; Ding et al. 2020; Wu et al. 2020), *Tapes philippinarum* (Cho et al. 2019), *Tegillarca granosa* (Li et al. 2015), *Venerupis philippinarum* (Li et al. 2015; Davidson and Dudas 2016), *Zygochlamys patagonica* (Akoueson et al. 2020)
Cephalopods	*Dosidicus gigas* (Braid et al. 2011), *Sepia officinalis* (Abidli et al. 2019; Oliveira et al. 2020), *Uroteuthis* (*Photololigo*) *duvaucelii* (Daniel et al. 2021)
Crustaceans	*Aristeus antennatus* (Cau et al. 2019; Carreras-Colom et al. 2020), *Charybdis japonica* (Zhang et al. 2021b), *Crangon Crangon* (Devriese et al. 2015), *Eriocheir sinensis* (Wojcik-Fudalewska et al. 2016), *Fenneropenaeus indicus* (Curren et al. 2020; Daniel et al. 2020, 2021), *Litopenaeus vannamei* (Bagheri et al. 2020; Curren et al. 2020), *Maja squinado* (Welden et al. 2018), *Matuta planipes* (Zhang et al. 2021b), *Metapenaeus dobsoni* (Daniel et al. 2021), *Metapenaeus monoceros* (Hossain et al. 2020), *Nephrops norvegicus* (Murray and Cowie 2011; Welden and Cowie 2016; Cau et al. 2019; Avio et al. 2020; Hara et al. 2020), *Palaemon* sp. (Avio et al. 2020), *Parapenaeopsis hardwickii* (Wu et al. 2020),

	List of species
Crustaceans (continued)	*Penaeus indicus* (Goswami et al. 2020), *Penaeus kerathurus* (Avio et al. 2020), *Penaeus monodon* (Hossain et al. 2020), *Pleoticus muelleri* (Curren et al. 2020; Fernandez Severini et al. 2020), *Portunus pelagicus* (Daniel et al. 2020; Goswami et al. 2020), *Portunus trituberculatus* (Zhang et al. 2021b), *Squilla mantis* (Avio et al. 2020)
Fish	*Abudefduf sordidus* (Chen et al. 2021), *Acanthocybium solandri* (Chen et al. 2021), *Acanthogobius flavimanus* (Lin et al. 2020), *Acanthopagrus latus* (Al-Salem et al. 2020), *Acanthurus dussumieri* (Chen et al. 2021), *Acanthurus lineatus* (Nie et al. 2019), *Acanthurus mata* (Chen et al. 2021), *Acanthurus pyroferus* (Nie et al. 2019), *Acanthurus triostegus* (Chen et al. 2021), *Acropoma japonicum* (Koongolla et al. 2020), *Alepes djedaba* (Goswami et al. 2020; Klangnurak and Chunniyom, 2020; Zhang et al. 2020a), *Alosa fallax* (Neves et al. 2015), *Amblychaeturichthys hexanema* (Feng et al. 2019), *Ammodytes personatus* (Sun et al. 2019; Zhao et al. 2019), *Anchoa mitchili* (Parker et al. 2020), *Anchoviella commersonii* (Sun et al. 2019; Zhao et al. 2019), *Anodontostoma chacunda* (Daniel et al. 2020), *Antigonia capros* (Zhu et al. 2019), *Apogon semilineatus* (Koongolla et al. 2020), *Arctozenus risso* (Lusher et al. 2016), *Argentina silus* (Murphy et al. 2017), *Argyrosomus argentatus* (Wang et al. 2021b), *Argyrosomus regius* (Neves et al. 2015), *Argyrozona argyrozona* (Sparks and Immelman 2020), *Aspitrigla cuculus* (Lusher et al. 2013), *Astronesthes indopacifica* (Boerger et al. 2010), *Atherinopsis californiensis* (Rochman et al. 2015), *Atule mate* Klangnurak (Klangnurak and Chunniyom, 2020), *Balistapus undulatus* (Nie et al. 2019), *Balistes capistratus* (Nie et al. 2019), *Benthosema glaciale* (Lusher et al. 2016), *Boleophthalmus pectinirostris* (Su et al. 2019), *Boops boops* (Neves et al. 2015; Nadal et al. 2016) (Rios-Fuster et al. 2019; Tsangaris et al. 2020), *Boreogadus saida* (Kuhn et al. 2018), *Bostrychus sinensis* (Lin et al. 2020), *Brama brama* (Neves et al. 2015), *Branchiostegus argentatus* (Koongolla et al. 2020), *Branchiostegus auratus* (Lin et al. 2020), *Branchiostegus japonicus* (Lin et al. 2020), *Brevoortia tyrannus* (Parker et al. 2020), *Buglossisium luteum* (Lusher et al. 2013), *Caesio cuning* (Klangnurak and Chunniyom 2020), *Callionymus lyra* (Lusher et al. 2013; Filgueiras et al. 2020), *Carangoides armatus* (James et al. 2020), *Carangoides chrysophrys* (Koongolla et al. 2020), *Carangoides malabaricus* (Goswami et al. 2020; Baalkhuyur et al. 2020), *Caranx pectoralis* (Koongolla et al. 2020), *Cathorops agassizii* (Possatto et al. 2011), *Cathorops spixii* (Possatto et al. 2011), *Centroberyx lineatus* (Zhu et al. 2019), *Cephalopholis argus* (Chen et al. 2021), *Cephalopholis urodeta* (Nie et al. 2019; Chen et al. 2021), *Cepola macrophthalma* (Lusher et al. 2013), *Chaeturichthys stigmatias* (Feng et al. 2019; Wang et al. 2021b), *Chanos* spp. (Ferreira et al. 2020), *Cheilinus chlororus* (Chen et al. 2021),

	List of species
Fish (continued)	*Cheilopogon simus* (Garnier et al. 2019), *Chelidonichthys capensis* (Sparks and Immelman, 2020), *Chelidonichthys kumu* (Sun et al. 2019; Zhao et al. 2019; Zhang et al. 2020b), *Chelidonichthys lucerna* (Avio et al. 2020), *Chelidonichtys lucernus* (Avio et al. 2015), *Chelon aurata* (Zakeri et al. 2020), *Chelon saliens* (Gündoğdu et al. 2020), *Chirocentrus dorab* (Sathish et al. 2020), *Chlorophthalmus agassizi* (Zhu et al. 2019), *Chlorophthalmus albatrossis* (Zhu et al. 2019), *Ciliata mustela* (Kartar et al. 1976), *Cirrhitus pinnulatus* (Chen et al. 2021), *Citharichthys sordidus* (Rochman et al. 2015) *Cleisthenes herzensteini* (Sun et al. 2019; Wang et al. 2021b; Zhao et al. 2019), *Clupea harengus* (Carpenter et al. 1972; Foekema et al. 2013; Collard et al. 2017; Hermsen et al. 2017; Ogonowski et al. 2019), *Coilia dussumieri* (Karuppasamy et al. 2020), *Coilia ectenes* (Su et al. 2019), *Coilia mystus* (Su et al. 2019; Lin et al. 2020), *Collichthys lucidus* (Feng et al. 2019; Su et al. 2019; Lin et al. 2020; Zhang et al. 2020b), *Cololabis saira* (Boerger et al. 2010), *Conge myriaster* (Wang et al. 2021b), *Coryphaena hippurus* (Choy and Drazen 2013; Chen et al. 2021), *Cyanoglossus macrostomus* (James et al. 2020), *Cynoglossus joyneri* (Wang et al. 2021b), *Cynoglossus lida* (Goswami et al. 2020), *Cynoglossus robustus* (Su et al. 2019), *Cynoglossus semilaevis* (Feng et al. 2019; Wang et al. 2021b), *Cynoglossus trigrammus* (Lin et al. 2020), *Cynoscion jamaicensis* (Neto et al. 2020), *Cynoscion nebulosus* (Parker et al. 2020), *Dascyllus trimaculatus* (Chen et al. 2021), *Dasyatis kuhlii* (Nie et al. 2019), *Decapterus macarellus* (Karuppasamy et al. 2020), *Decapterus macrosoma* (Rochman et al. 2015), *Decapterus maruadsi* (Koongolla et al. 2020), *Decapterus maruadsi* (Sun et al. 2019), *Decapterus maruadsi* (Zhao et al. 2019), *Decapterus muroadsi* (Ory et al. 2017), *Decapterus russelli* (Karuppasamy et al. 2020), *Dentex angolensis* (Adika et al. 2020), *Dentex macrophthalmus* (Neves et al. 2015), *Diaphus andernesi* (Davison and Asch 2011), *Diaphus fulgens* (Davison and Asch 2011), *Diaphus phillipsi* (Davison and Asch 2011), *Diapterus rhombeus* (Ramos et al. 2012), *Dicentrarchus labrax* (Bessa et al. 2018; Akoueson et al. 2020; Barboza et al. 2020), *Diplodus vulgaris* (Bessa et al. 2018; Avio et al. 2020), *Dussumieria acuta* (Daniel et al. 2020), *Elates ransonnettii* (Klangnurak and Chunniyom 2020), *Eleginops maclovinus* (Pozo et al. 2019), *Eleutheronema tetradactylum* (Al-Salem et al. 2020; Karuppasamy et al. 2020), *Eleutheronema tridactylum* (Karbalaei et al. 2019), *Engraulis encrasicolus* (Collard et al. 2017; Filgueiras et al. 2020); (Renzi et al. 2019; Rios-Fuster et al. 2019; Pennino et al. 2020; Savoca et al. 2020), *Engraulis japonicus* (Sun et al. 2019; Zhao et al. 2019), *Engraulis mordax* (Rochman et al. 2015), *Ephinephelus merra* (Garnier et al. 2019), *Epinephelus coioides* (Karbalaei et al. 2019; Baalkhuyur et al. 2020), *Epinephelus diacanthus* (James et al. 2020),

	List of species
Fish (continued)	*Epinephelus hexagonatus* (Chen et al. 2021), *Epinephelus merra* (Chen et al. 2021), *Escualosa thoracata* (Karuppasamy et al. 2020), *Etelis* spp. (Bakir et al. 2020), *Etrumeus whiteheadi* (Sparks and Immelman 2020), *Eubleekeria splendens* (Klangnurak and Chunniyom 2020), *Eucinostomus melanopterus* (Ramos et al. 2012), *Eugeres brasilianus* (Ramos et al. 2012), *Eugraulis japonicus* (Tanaka and Takada 2016), *Eupleurogrammus muticus* (Sun et al. 2019; Zhao et al. 2019; Wang et al. 2021b), *Euthynnus affinis* (Karbalaei et al. 2019; Baalkhuyur et al. 2020), *Gadus macrocephalus* (Sun et al. 2019; Zhao et al. 2019), *Gadus morhua* (Foekema et al. 2013; Brate et al. 2016; Liboiron et al. 2016; Rummel et al. 2016), *Gempylus serpens* (Choy and Drazen 2013), *Genidens genidens* (Dantas et al. 2019), *Gerres filamentous* (Goswami et al. 2020), *Glossogobius giuris* (Lin et al. 2020), *Halichoeres hortulanus* (Chen et al. 2021), *Harpadon nehereus* (Hossain et al. 2019; Lin et al. 2020), *Harpadon translucens* (Hossain et al. 2019), *Harpodon nehereus* (Lin et al. 2020), *Harpodon nehereus* (Sathish et al. 2020), *Helicolenus dactylopterus* (Pereira et al. 2020), *Hemibarbus maculatus* (Su et al. 2019), *Hexagrammos otakii* (Sun et al. 2019), *Hexagrammos otakii* (Wang et al. 2021b); (Zhao et al. 2019), *Hygophum reinhardtii* (Boerger et al. 2010), *Idiacanthus antrostomus* (Davison and Asch 2011), *Inegocia japonicus* (Lin et al. 2020), *Istiophorus platypterus* (Sathish et al. 2020), *Johnius belengerii* (Lin et al. 2020; Wang et al. 2021b), *Katsuwonus pelamis* (Neto et al. 2020; Pereira et al. 2020; Sathish et al. 2020), *Konosirus punctatus* (Wu et al. 2019; Lin et al. 2020; Zhang et al. 2020a, Wang et al. 2021b), *Kyphosus cinerascens* (Chen et al. 2021), *Lactarius lactarius* (James et al. 2020), *Larimichthys crocea* (Wu et al. 2019; Zhang et al. 2020b), *Larimichthys polyactis* (Sun et al. 2019; Zhao et al. 2019), *Lateolabrax maculatus* (Wang et al. 2021b), *Leiognathus brevirostris* (Zhang et al. 2020a), *Leiognathus ruconius* (*Lin* et al. 2020), *Leiognathus splendens* (*James* et al. 2020), *Lelostomus xanthurus* (Parker et al. 2020), *Lepidorhombus boscii* (Neves et al. 2015), *Lepidorhombus whiffiagonis* (Murphy et al. 2017), *Lepidotrigla alata* (Koongolla et al. 2020), *Lepidotrigla guentheri* (Zhu et al. 2019), *Lethrinus nebulosus* (Baalkhuyur et al. 2020), *Lethrinus* spp. (Ferreira et al. 2020), *Limanda limanda* (Rummel et al. 2016; Hermsen et al. 2017; Murphy et al. 2017), *Liparis liparis* (Kartar et al. 1976), *Liparis tanakae* (Sun et al. 2019; Zhao et al. 2019), *Lithognathu mormyrus* (Avio et al. 2020; Gündoğdu et al. 2020), *Liza haematocheila* (Su et al. 2019; Wang et al. 2021b), *Liza parsia* (Baalkhuyur et al. 2020), *Liza saliens* (Bagheri et al. 2020), *Lobianchia gemellarii* (Davison and Asch 2011), *Lophius litulon* (Sun et al. 2019; Zhao et al. 2019), *Lophius piscatorius* (Neves et al. 2015), *Loweina interrupta* (Boerger et al. 2010), *Lutjanus quinquelineatus* (Al-Salem et al. 2020), *Lutjanus* spp.

	List of species
Fish (continued)	(Ferreira et al. 2020), *Maurolicus muelleri* (Lusher et al. 2016), *Megalaspis cordyla* (Karbalaei et al. 2019; Daniel et al. 2020), *Melanogrammus aeglefinus* (Foekema et al. 2013; Akoueson et al. 2020), *Melichthys vidua* (Nie et al. 2019), *Menidia menidia* (Carpenter et al. 1972), *Merlangius merlangus* (Foekema et al. 2013; Lusher et al. 2013; Hermsen et al. 2017), *Merluccius capensis* (Sparks and Immelman 2020), *Merluccius gayi* (Pozo et al. 2019), *Merluccius merluccius* (Avio et al. 2015; Neves et al. 2015; Bellas et al. 2016; Giani et al. 2019; Mancuso et al. 2019; Avio et al. 2020), *Merluccius paradoxus* (Sparks and Immelman 2020), *Microchirus variegates* (Lusher et al. 2013), *Micromesistius poutassou* (Lusher et al. 2013), *Micropogonias furnieri* (Arias et al. 2019), *Micropogonias furnieri* (Neto et al. 2020), *Morone americana* (Carpenter et al. 1972), *Morone saxatilis* (Rochman et al. 2015), *Mugil cephalus* (Lin et al. 2020; Parker et al. 2020; Zhang et al. 2020a), *Mugil* spp. (Ferreira et al. 2020; Naidoo et al. 2020), *Mullus barbatus* (Avio et al. 2015; Bellas et al. 2016; Digka et al. 2018; Giani et al. 2019; Avio et al. 2020; Gündoğdu et al. 2020), *Mullus surmuletus* (Neves et al. 2015; Filgueiras et al. 2020; Gündoğdu et al. 2020), *Muraenesox cinereus* (Lin et al. 2020; Zhang et al. 2020b), *Myctophum aurolanternatum* (Boerger et al. 2010), *Myctophum nitidulum* (Davison and Asch 2011), *Myoxocephalus aeneus* (Carpenter et al. 1972), *Naso brevirostris* (Nie et al. 2019), *Naso lituratus* (Nie et al. 2019), *Nemichthys scolopaceus* (Lusher et al. 2016), *Nemipterus bipunctatus* (Karbalaei et al. 2019), *Nemipterus hexodon* (Klangnurak and Chunniyom 2020), *Nemipterus japonicus* (Karuppasamy et al. 2020), *Nemipterus peronii* (Goswami et al. 2020), *Nemipterus randalli* (James et al. 2020), *Nemipterus virgatus* (Koongolla et al. 2020; Lin et al. 2020), *Neogobius melanostomus* (Bagheri et al. 2020), *Nibea maculata* (James et al. 2020), *Notoscopelus kroyeri* (Lusher et al. 2016), *Odontamblyopus rubicundus* (Feng et al. 2019), *Odonus niger* (Nie et al. 2019), *Oncorhynchus tshawytscha* (Rochman et al. 2015; Collicutt et al. 2019), *Ophiodon elongatus* (Rochman et al. 2015), *Oplegnathus fasciatus* (Zhang et al. 2020b), *Osmerus eperlanus* (McGoran et al. 2017), *Ostorhinchus aureus* (Chen et al. 2021), *Pagellus acarne* (Neves et al. 2015), *Pagellus bogaraveo* (Anastasopoulou et al. 2013; Pereira et al. 2020), *Pagellus erythrinus* (Digka et al. 2018; Avio et al. 2020), *Pampus argenteus* (Sun et al. 2019; Zhao et al. 2019; Lin et al. 2020; Wang et al. 2021b), *Pampus cinereus* (Su et al. 2019), *Paracirrhites forsteri* (Chen et al. 2021), *Paralichthys olivaceus* (Wang et al. 2021b), *Parupeneus multifasciatus* (Chen et al. 2021), *Pennahia macrocephalus* (Koongolla et al. 2020), *Pentaprion longimanus* (Karuppasamy et al. 2020), *Platichtyes flesus* (Kartar et al. 1973; Kartar et al. 1976; Rummel et al. 2016; McGoran et al. 2017; Murphy et al. 2017; Bessa et al. 2018),

	List of species
Fish (continued)	*Platycephalus indicus* (Lin et al. 2020; Wang et al. 2021b), *Plectorhinchus cinctus* (Lin et al. 2020), *Pleuronectes platessa* (Murphy et al. 2017; Welden et al. 2018; Akoueson et al. 2020), *Pneumatophorus japonicus* (Wang et al. 2021b), *Pollachius virens* (Carpenter et al. 1972), *Polydactylus sexfilis* (Lin et al. 2020), *Pomacanthus imperator* (Nie et al. 2019), *Pomatomus saltatrix* (Neto et al. 2020), *Pomatoschistus minutus* (*Kartar* et al. 1976), *Priacanthus macracanthus* (Koongolla et al. 2020), *Prionotus evolans* (Carpenter et al. 1972), *Prionotus punctatus* (Neto et al. 2020), *Pristipomoides* spp. (Bakir et al. 2020), *Psenopsis anomala* (Sun et al. 2019; Zhao et al. 2019; Koongolla et al. 2020), *Pseudobalistes fuscus* (Nie et al. 2019), *Pseudopleuronectes americanus* (Carpenter et al. 1972), *Pseudopleuronectes yokohamae* (Wang et al. 2021b), *Pseudorhombus oligodon* (Koongolla et al. 2020), *Pseudosciaena crocea* (Lin et al. 2020), *Rastrelliger faughni* (Karuppasamy et al. 2020), *Rastrelliger kanagurta* (Rochman et al. 2015; Karbalaei et al. 2019; Baalkhuyur et al. 2020; Daniel et al. 2020; James et al. 2020; Klangnurak and Chunniyom 2020; Sathish et al. 2020), *Rexea solandri* (Zhu et al. 2019), *Rutilus frisii kutum* (Taghizadeh Rahmat Abadi et al. 2021), *Rutilus kutum* (Zakeri et al. 2020), *Sardina pilchardus* (Avio et al. 2015; Collard et al. 2017; Digka et al. 2018; Renzi et al. 2019; Rios-Fuster et al. 2019; Avio et al. 2020; Filgueiras et al. 2020; Maaghloud et al. 2020; Pennino et al. 2020; Savoca et al. 2020), *Sardinella albella* (Sathish et al. 2020), *Sardinella aurita* (Adika et al. 2020; Avio et al. 2020), *Sardinella gibbosa* (Hossain et al. 2019; Daniel et al. 2020; James et al. 2020; Klangnurak and Chunniyom 2020), *Sardinella longiceps* (Daniel et al. 2020; James et al. 2020; Karuppasamy et al. 2020), *Sardinella maderensis* (Adika et al. 2020), *Sardinella zunasi* (Wang et al. 2021b), *Saurida elongata* (Wang et al. 2021b), *Saurida tumbil* (Goswami et al. 2020; James et al. 2020; Koongolla et al. 2020), *Saurida undosquamis* (Klangnurak and Chunniyom 2020), *Scarus frenatus* (Chen et al. 2021), *Sciades herzbergii* (Possatto et al. 2011), *Scolopsis lineata* (Chen et al. 2021), *Scolopsis taenioptera* (Klangnurak and Chunniyom 2020), *Scomber colias* (Herrera et al. 2019; Barboza et al. 2020; Pereira et al. 2020), *Scomber japonicus* (Neves et al. 2015; Su et al. 2019; Sparks and Immelman 2020), *Scomber scombrus* (Neves et al. 2015; Rummel et al. 2016; Akoueson et al. 2020; Avio et al. 2020; Maaghloud et al. 2020), *Scomberoides lysan* (Chen et al. 2021), *Scomberomorous guttatus* (Karuppasamy et al. 2020), *Scomberomorus commerson* (Baalkhuyur et al. 2020; Chen et al. 2021), *Scomberomorus niphonius* (Wang et al. 2021b), *Scorpaena hatizyoensis* (Koongolla et al. 2020), *Sebastes flavidus* (Rochman et al. 2015), *Sebastes mystinus* (Rochman et al. 2015), *Sebastiscus marmoratus* (Zhang et al. 2020b), *Sebastods schlegelii* (Wang et al. 2021b), *Selar boops* (Karbalaei et al. 2019), *Selar*

	List of species
Fish (continued)	*crumenophthalmus* (Klangnurak and Chunniyom 2020), *Selaroides leptolepis* (Klangnurak and Chunniyom 2020), *Seriola aureovittata* (Wang et al. 2021b), *Seriola lalandi* (Gassel et al. 2013), *Serranus scriba* (Zitouni et al. 2020), *Setipinna taty* (Sun et al. 2019; Zhang et al. 2020b), *Setipinna taty* (Zhao et al. 2019), *Siganus argenteus* (Rochman et al. 2015), *Siganus canaliculatus* (Rochman et al. 2015; Baalkhuyur et al. 2020; Klangnurak and Chunniyom 2020; Koongolla et al. 2020), *Siganus fuscescens* (Bucol et al. 2020; Lin et al. 2020; Zhang et al. 2020a), *Siganus* spp. (Garnier et al. 2019; Ferreira et al. 2020), *Sillago japonica* (Zhang et al. 2020a), *Sillago sihama* (Lin et al. 2020; Wang et al. 2021b), *Sirembo imberbis* (Koongolla et al. 2020), *Solea solea* (Pellini et al. 2018; Avio et al. 2020), *Sphyraena obtusata* (Daniel et al. 2020), *Spondilosoma cantharus* (Avio et al. 2020), *Spratelloides gracilis* (Rochman et al. 2015), *Sprattus sprattus* (Hermsen et al. 2017), *Stellifer brasilensis* (Dantas et al. 2012), *Stellifer stellifer* (Dantas et al. 2012), *Sternoptyx diaphna* (Davison and Asch 2011), *Sternoptyx pseudobscura* (Davison and Asch 2011), *Stolephorus commersonii* (Kripa et al. 2014), *Stolephorus indicus* (Daniel et al. 2020; James et al. 2020; Karuppasamy et al. 2020), *Stolephorus* spp. (Ningrum and Patria 2019), *Stomias boa boa* (Lusher et al. 2016), *Strangomera bentincki* (Pozo et al. 2019), *Sufflamen chrysopterum* (Chen et al. 2021), *Symbolophorus californiensis* (Boerger et al. 2010), *Synagrops japonicus* (Zhu et al. 2019), *Synechogobius hasta* (Wang et al. 2021b), *Terapon jarbua* (Naidoo et al. 2020), *Terapon puta* (James et al. 2020; Karuppasamy et al. 2020), *Terapon theraps* (Klangnurak and Chunniyom 2020), *Thalassoma hardwicke* (Chen et al. 2021), *Thalassoma lutescens* (Chen et al. 2021), *Thryssa dussumieri* (Daniel et al. 2020), *Thryssa kammalensis* (Feng et al. 2019), *Thryssa mystax* (James et al. 2020; Wang et al. 2021b), *Thunnus alalunga* (Romeo et al. 2015), *Thunnus albacares* (Bakir et al. 2020), *Thunnus obesus* (Choy and Drazen 2013), *Thunnus thynnus* (Romeo et al. 2015), *Thunnus tonggol* (Karbalaei et al. 2019), *Trachinotus ovatus* (Lin et al. 2020), *Trachiocephalus myops* (Koongolla et al. 2020), *Trachurus capensis* (Sparks and Immelman 2020), *Trachurus japonicus* (Koongolla et al. 2020), *Trachurus mediterraneus* (Gündoğdu et al. 2020), *Trachurus mediterraneus* (Rios-Fuster et al. 2019), *Trachurus murphyi* (Pozo et al. 2019), *Trachurus picturatus* (Neves et al. 2015; Pereira et al. 2020), *Trachurus trachurus* (Foekema et al. 2013; Lusher et al. 2013; Neves et al. 2015; Avio et al. 2020; Barboza et al. 2020; Maaghloud et al. 2020), *Tracinus draco* (Avio et al. 2020), *Trichiurus haumela* (Koongolla et al. 2020), *Trichiurus lepturus* (Chen et al. 2021), *Trigla lyra* (Neves et al. 2015), *Trisopterus minutus* (Lusher et al. 2013), *Umbrina canosai* (Neto et al. 2020), *Upeneus bensasi* (Koongolla et al. 2020), *Upeneus*

	List of species
Fish (continued)	*sulphureus* (Koongolla et al. 2020), *Upeneus vittatus* (Goswami et al. 2020; Klangnurak and Chunniyom 2020), *Xanthichthys caeruleolineatus* (Nie et al. 2019), *Xenodermichthys copei* (Lusher et al. 2016), *Xiphias gladius* (Choy and Drazen 2013; Romeo et al. 2015), *Zebrasoma veliferum* (Nie et al. 2019), *Zeus faber* (Lusher et al. 2013; Neves et al. 2015).

7.9. References

Abidli, S., Lahbib, Y., Trigui El Menif, N. (2019). Microplastics in commercial molluscs from the lagoon of Bizerte (Northern Tunisia). *Mar Pollut Bull*, 142, 243–252. doi: 10.1016/j.marpolbul.2019.03.048.

Adika, S.A., Mahu, E., Crane, R., Marchant, R., Montford, J., Folorunsho, R., Gordon, C. (2020). Microplastic ingestion by pelagic and demersal fish species from the Eastern Central Atlantic Ocean, off the Coast of Ghana. *Mar Pollut Bull*, 153, 110998. doi: 10.1016/j.marpolbul.2020.110998.

Akhbarizadeh, R., Dobaradaran, S., Nabipour, I., Tajbakhsh, S., Darabi, A.H., Spitz, J. (2020). Abundance, composition, and potential intake of microplastics in canned fish. *Mar Pollut Bull*, 160, 111633. doi: 10.1016/j.marpolbul.2020.111633.

Akoueson, F., Sheldon, L.M., Danopoulos, E., Morris, S., Hotten, J., Chapman, E., Li, J., Rotchell, J.M. (2020). A preliminary analysis of microplastics in edible versus non-edible tissues from seafood samples. *Environ Pollut*, 263, 114452. doi: 10.1016/j.envpol.2020.114452.

Al-Salem, S.M., Uddin, S., Lyons, B. (2020). Evidence of microplastics (MP) in gut content of major consumed marine fish species in the State of Kuwait (of the Arabian/Persian Gulf). *Mar Pollut Bull*, 154, 111052. doi: 10.1016/j.marpolbul.2020.111052.

Amaral-Zettler, L.A., Zettler, E.R., Mincer, T.J. (2020). Ecology of the plastisphere. *Nat Rev Microbiol*, 18, 139–151. doi: 10.1038/s41579-019-0308-0.

Anastasopoulou, A., Mytilineou, C., Smith, C.J., Papadopoulou, K.N. (2013). Plastic debris ingested by deep-water fish of the Ionian Sea (Eastern Mediterranean). *Deep Sea Research Part I: Oceanographic Research Papers*, 74, 11–13. doi: 10.1016/j.dsr.2012.12.008.

Arias, A.H., Ronda, A.C., Oliva, A.L., Marcovecchio, J.E. (2019). Evidence of microplastic ingestion by fish from the Bahia Blanca Estuary in Argentina, South America. *Bull Environ Contam Toxicol*, 102, 750–756. doi: 10.1007/s00128-019-02604-2.

Arias-Andres, M., Klumper, U., Rojas-Jimenez, K., Grossart, H.P. (2018). Microplastic pollution increases gene exchange in aquatic ecosystems. *Environ Pollut*, 237, 253–261. doi: 10.1016/j.envpol.2018.02.058.

Athey, S.N. and Erdle, L.M. (2021). Are we underestimating anthropogenic microfiber pollution? A critical review of occurrence, methods and reporting. *Environ Toxicol Chem*. doi: 10.1002/etc.5173.

Athey, S.N., Adams, J.K., Erdle, L.M., Jantunen, L.M., Helm, P.A., Finkelstein, S.A., Diamond, M.L. (2020). The widespread environmental footprint of indigo denim microfibers from blue jeans. *Environ Sci Technol Lett*, 7, 840–847. doi: 10.1021/acs.estlett.0c00498.

Avio, C.G., Gorbi, S., Regoli, F. (2015). Experimental development of a new protocol for extraction and characterization of microplastics in fish tissues: First observations in commercial species from Adriatic Sea. *Mar Environ Res*, 111, 18–26. doi: 10.1016/j.marenvres.2015.06.014.

Avio, C.G., Pittura, L., d'Errico, G., Abel, S., Amorello, S., Marino, G., Gorbi, S., Regoli, F. (2020). Distribution and characterization of microplastic particles and textile microfibers in Adriatic food webs: General insights for biomonitoring strategies. *Environ Pollut*, 258, 113766. doi:10.1016/j.envpol.2019.113766.

Baalkhuyur, F.M., Qurban, M.A., Panickan, P., Duarte, C.M. (2020). Microplastics in fishes of commercial and ecological importance from the Western Arabian Gulf. *Mar Pollut Bull*, 152, 110920. doi: 10.1016/j.marpolbul.2020.110920.

Baechler, B.R., Granek, E.F., Hunter, M.V., Conn, K.E. (2019). Microplastic concentrations in two Oregon bivalve species: Spatial, temporal, and species variability. *Limnol Oceanogr Lett*, 5, 54–65. doi: 10.1002/lol2.10124.

Bagheri, T., Gholizadeh, M., Abarghouei, S., Zakeri, M., Hedayati, A., Rabaniha, M., Aghaeimoghadam, A., Hafezieh, M. (2020). Microplastics distribution, abundance and composition in sediment, fishes and benthic organisms of the Gorgan Bay, Caspian sea. *Chemosphere*, 257, 127201. doi: 10.1016/j.chemosphere.2020.127201.

Bakir, A., Desender, M., Wilkinson, T., Van Hoytema, N., Amos, R., Airahui, S., Graham, J., Maes, T. (2020). Occurrence and abundance of meso and microplastics in sediment, surface waters, and marine biota from the South Pacific region. *Mar Pollut Bull*, 160, 111572. doi: 10.1016/j.marpolbul.2020.111572.

Barboza, L.G.A., Lopes, C., Oliveira, P., Bessa, F., Otero, V., Henriques, B., Raimundo, J., Caetano, M., Vale, C., Guilhermino, L. (2020). Microplastics in wild fish from North East Atlantic Ocean and its potential for causing neurotoxic effects, lipid oxidative damage, and human health risks associated with ingestion exposure. *Sci Total Environ*, 717, 134625. doi: 10.1016/j.scitotenv.2019.134625.

Bellas, J., Martinez-Armental, J., Martinez-Camara, A., Besada, V., Martinez-Gomez, C. (2016). Ingestion of microplastics by demersal fish from the Spanish Atlantic and Mediterranean coasts. *Mar Pollut Bull*, 109, 55–60. doi: 10.1016/j.marpolbul.2016.06.026.

Bessa, F., Barria, P., Neto, J.M., Frias, J., Otero, V., Sobral, P., Marques, J.C. (2018). Occurrence of microplastics in commercial fish from a natural estuarine environment. *Mar Pollut Bull*, 128, 575–584. doi: 10.1016/j.marpolbul.2018.01.044.

Boerger, C.M., Lattin, G.L., Moore, S.L., Moore, C.J. (2010). Plastic ingestion by planktivorous fishes in the North Pacific Central Gyre. *Mar Pollut Bull*, 60, 2275–2278. doi: 10.1016/j.marpolbul.2010.08.007.

Braid, H.E., Deeds, J., Degrasse, S.L., Wilson, J.J., Osborne, J., Hanner, R.H. (2011). Preying on commercial fisheries and accumulating paralytic shellfish toxins: A dietary analysis of invasive *Dosidicus gigas* (Cephalopoda Ommastrephidae) stranded in Pacific Canada. *Mar Biol*, 159, 25–31. doi: 10.1007/s00227-011-1786-4.

Brate, I.L.N., Eidsvoll, D.P., Steindal, C.C., Thomas, K.V. (2016). Plastic ingestion by Atlantic cod (*Gadus morhua*) from the Norwegian coast. *Mar Pollut Bull*, 112, 105–110. doi: 10.1016/j.marpolbul.2016.08.034.

Bucol, L.A., Romano, E.F., Cabcaban, S.M., Siplon, L.M.D., Madrid, G.C., Bucol, A.A., Polidoro, B. (2020). Microplastics in marine sediments and rabbitfish (*Siganus fuscescens*) from selected coastal areas of Negros Oriental, Philippines. *Mar Pollut Bull*, 150, 110685. doi: 10.1016/j.marpolbul.2019.110685.

Carpenter, E.J., Anderson, S.J., Harvey, G.R., Miklas, H.P., Peck, B.B. (1972). Polystyrene spherules in coastal waters. *Science*, 178, 749–750. doi: 10.1126/science.178.4062.749.

Carreras-Colom, E., Constenla, M., Soler-Membrives, A., Cartes, J.E., Baeza, M., Carrasson, M. (2020). A closer look at anthropogenic fiber ingestion in *Aristeus antennatus* in the NW Mediterranean Sea: Differences among years and locations and impact on health condition. *Environ Pollut*, 263, 114567. doi: 10.1016/j.envpol.2020.114567.

Cau, A., Avio, C.G., Dessi, C., Follesa, M.C., Moccia, D., Regoli, F., Pusceddu, A. (2019). Microplastics in the crustaceans *Nephrops norvegicus* and *Aristeus antennatus*: Flagship species for deep-sea environments? *Environ Pollut*, 255, 113107. doi: 10.1016/j.envpol.2019.113107.

Chen, K.J., Chen, M.C., Chen, T.H. (2021). Plastic ingestion by fish in the coastal waters of the Hengchun Peninsula, Taiwan: Associated with human activity but no evidence of biomagnification. *Ecotoxicol Environ Saf*, 213, 112056. doi: 10.1016/j.ecoenv.2021.112056.

Cho, Y., Shim, W.J., Jang, M., Han, G.M., Hong, S.H. (2019). Abundance and characteristics of microplastics in market bivalves from South Korea. *Environ Pollut*, 245, 1107–1116. doi: 10.1016/j.envpol.2018.11.091.

Cho, Y., Shim, W.J., Jang, M., Han, G.M., Hong, S.H. (2021). Nationwide monitoring of microplastics in bivalves from the coastal environment of Korea. *Environ Pollut*, 270, 116175. doi: 10.1016/j.envpol.2020.116175.

Choy, C.A. and Drazen, J.C. (2013). Plastic for dinner? Observations of frequent debris ingestion by pelagic predatory fishes from the central North Pacific. *Mar Ecol Progr Ser*, 485, 155–163. doi: 10.3354/meps10342.

Collard, F., Gilbert, B., Eppe, G., Roos, L., Compere, P., Das, K., Parmentier, E. (2017). Morphology of the filtration apparatus of three planktivorous fishes and relation with ingested anthropogenic particles. *Mar Pollut Bull*, 116, 182–191. doi: 10.1016/j.marpolbul.2016.12.067.

Collicutt, B., Juanes, F., Dudas, S.E. (2019). Microplastics in juvenile Chinook salmon and their nearshore environments on the east coast of Vancouver Island. *Environ Pollut*, 244, 135–142. doi: 10.1016/j.envpol.2018.09.137.

Courtene-Jones, W., Quinn, B., Murphy, F., Gary, S.F., Narayanaswamy, B.E. (2017). Optimisation of enzymatic digestion and validation of specimen preservation methods for the analysis of ingested microplastics. *Analytical Methods*, 9, 1437–1445. doi: 10.1039/C6AY02343F.

Cowger, W., Booth, A.M., Hamilton, B.M., Thaysen, C., Primpke, S., Munno, K., Lusher, A.L., Dehaut, A., Vaz, V.P., Liboiron, M. et al. (2020). Reporting guidelines to increase the reproducibility and comparability of research on microplastics. *Appl Spectrosc*, 74, 1066–1077. doi: 10.1177/0003702820930292.

Cowger, W., Steinmetz, Z., Gray, A., Munno, K., Lynch, J., Hapich, H., Primpke, S., De Frond, H., Rochman, C., Herodotou, O. (2021). Microplastic spectral classification needs an open source community: Open specy to the rescue! *Anal Chem*, 93, 7543–7548. doi: 10.1021/acs.analchem.1c00123.

Curren, E., Leaw, C.P., Lim, P.T., Leong, S.C.Y. (2020). Evidence of marine microplastics in commercially harvested seafood. *Front Bioeng Biotechnol*, 8, 562760. doi: 10.3389/fbioe.2020.562760.

Daniel, D.B., Ashraf, P.M., Thomas, S.N. (2020). Abundance, characteristics and seasonal variation of microplastics in Indian white shrimps (*Fenneropenaeus indicus*) from coastal waters off Cochin, Kerala, India. *Sci Total Environ*, 737, 139839. doi: 10.1016/j.scitotenv.2020.139839.

Daniel, D.B., Ashraf, P.M., Thomas, S.N., Thomson, K.T. (2021). Microplastics in the edible tissues of shellfishes sold for human consumption. *Chemosphere*, 264, 128554. doi: 10.1016/j.chemosphere.2020.128554.

Dantas, D.V., Barletta, M., Da Costa, M.F. (2012). The seasonal and spatial patterns of ingestion of polyfilament nylon fragments by estuarine drums (Sciaenidae). *Environ Sci Pollut Res Int*, 19, 600–606. doi: 10.1007/s11356-011-0579-0.

Dantas, D.V., Ribeiro, C.I.R., Frischknecht, C.C.A., Machado, R., Farias, E.G.G. (2019). Ingestion of plastic fragments by the Guri sea catfish *Genidens genidens* (Cuvier, 1829) in a subtropical coastal estuarine system. *Environ Sci Pollut Res Int*, 26, 8344–8351. doi: 10.1007/s11356-019-04244-9.

Davison, P. and Asch, R.G. (2011). Plastic ingestion by mesopelagic fishes in the North Pacific Subtropical Gyre. *Mar Ecol Progr Ser*, 432, 173–180. doi: 10.3354/meps09142.

Davidson, K. and Dudas, S.E. (2016). Microplastic ingestion by wild and cultured Manila clams (*Venerupis philippinarum*) from Baynes Sound, British Columbia. *Arch Environ Contam Toxicol*, 71, 147–56. doi: 10.1007/s00244-016-0286-4.

De Witte, B., Devriese, L., Bekaert, K., Hoffman, S., Vandermeersch, G., Cooreman, K., Robbens, J. (2014). Quality assessment of the blue mussel (*Mytilus edulis*): Comparison between commercial and wild types. *Mar Pollut Bull*, 85, 146–155. doi: 10.1016/j.marpolbul.2014.06.006.

Dehaut, A., Hermabessiere, L., Duflos, G. (2019). Current frontiers and recommendations for the study of microplastics in seafood. *TrAC – Trends Anal Chem*, 116, 346–359. doi: 10.1016/j.trac.2018.11.011.

Devriese, L.I., Van Der Meulen, M.D., Maes, T., Bekaert, K., Paul-Pont, I., Frere, L., Robbens, J., Vethaak, A.D. (2015). Microplastic contamination in brown shrimp (*Crangon crangon*, Linnaeus 1758) from coastal waters of the Southern North Sea and Channel area. *Mar Pollut Bull*, 98, 179–187. doi: 10.1016/j.marpolbul.2015.06.051.

Digka, N., Tsangaris, C., Torre, M., Anastasopoulou, A., Zeri, C. (2018). Microplastics in mussels and fish from the Northern Ionian Sea. *Mar Pollut Bull*, 135, 30–40. doi: 10.1016/j.marpolbul.2018.06.063.

Ding, J., Li, J., Sun, C., Jiang, F., He, C., Zhang, M., Ju, P., Ding, N.X. (2020). An examination of the occurrence and potential risks of microplastics across various shellfish. *Sci Total Environ*, 739, 139887. doi: 10.1016/j.scitotenv.2020.139887.

Dowarah, K., Patchaiyappan, A., Thirunavukkarasu, C., Jayakumar, S., Devipriya, S.P. (2020). Quantification of microplastics using Nile Red in two bivalve species *Perna viridis* and *Meretrix meretrix* from three estuaries in Pondicherry, India and microplastic uptake by local communities through bivalve diet. *Mar Pollut Bull*, 153, 110982. doi: 10.1016/j.marpolbul.2020.110982.

Dris, R., Gaspéri, J., Saad, M., Mirande, C., Tassin, B. (2016). Synthetic fibers in atmospheric fallout: A source of microplastics in the environment? *Mar Pollut Bull*, 104, 290–293. doi: 10.1016/j.marpolbul.2016.01.006.

Dris, R., Gaspéri, J., Tassin, B. (2018). Sources and fate of microplastics in urban areas: A focus on Paris Megacity. In *Freshwater Microplastics: Emerging Environmental Contaminants?* Wagner, M. and Lambert, S. (eds). Springer International Publishing, Cham.

Echa (2016). Plastic Additives Initiative [Online]. Available at: https://echa.europa.eu/fr/plastic-additives-initiative [Accessed 18 August 2021].

Feng, Z., Zhang, T., Li, Y., He, X., Wang, R., Xu, J., Gao, G. (2019). The accumulation of microplastics in fish from an important fish farm and mariculture area, Haizhou Bay, China. *Sci Total Environ*, 696, 133948. doi: 10.1016/j.scitotenv.2019.133948.

Feng, Z., Wang, R., Zhang, T., Wang, J., Huang, W., Li, J., Xu, J., Gao, G. (2020). Microplastics in specific tissues of wild sea urchins along the coastal areas of northern China. *Sci Total Environ*, 728, 138660. doi: 10.1016/j.scitotenv.2020.138660.

Fernandez Severini, M.D., Buzzi, N.S., Forero Lopez, A.D., Colombo, C.V., Chatelain Sartor, G.L., Rimondino, G.N., Truchet, D.M. (2020). Chemical composition and abundance of microplastics in the muscle of commercial shrimp *Pleoticus muelleri* at an impacted coastal environment (Southwestern Atlantic). *Mar Pollut Bull*, 161, 111700. doi: 10.1016/j.marpolbul.2020.111700.

Ferreira, M., Thompson, J., Paris, A., Rohindra, D., Rico, C. (2020). Presence of microplastics in water, sediments and fish species in an urban coastal environment of Fiji, a Pacific small island developing state. *Mar Pollut Bull*, 153, 110991. doi: 10.1016/j.marpolbul.2020.110991.

Filgueiras, A.V., Preciado, I., Carton, A., Gago, J. (2020). Microplastic ingestion by pelagic and benthic fish and diet composition: A case study in the NW Iberian Shelf. *Mar Pollut Bull*, 160, 111623. doi: 10.1016/j.marpolbul.2020.111623.

Foekema, E.M., De Gruijter, C., Mergia, M.T., Van Franeker, J.A., Murk, A.J., Koelmans, A.A. (2013). Plastic in north sea fish. *Environ Sci Technol*, 47, 8818–8824. doi: 10.1021/es400931b.

Garnier, Y., Jacob, H., Guerra, A.S., Bertucci, F., Lecchini, D. (2019). Evaluation of microplastic ingestion by tropical fish from Moorea Island, French Polynesia. *Mar Pollut Bull*, 140, 165–170. doi: 10.1016/j.marpolbul.2019.01.038.

Gassel, M., Harwani, S., Park, J.S., Jahn, A. (2013). Detection of nonylphenol and persistent organic pollutants in fish from the North Pacific Central Gyre. *Mar Pollut Bull*, 73, 231–42. doi: 10.1016/j.marpolbul.2013.05.014.

Gedik, K. and Eryasar, A.R. (2020). Microplastic pollution profile of Mediterranean mussels (*Mytilus galloprovincialis*) collected along the Turkish coasts. *Chemosphere*, 260, 127570. doi: 10.1016/j.chemosphere.2020.127570.

Giani, D., Baini, M., Galli, M., Casini, S., Fossi, M.C. (2019). Microplastics occurrence in edible fish species (*Mullus barbatus* and *Merluccius merluccius*) collected in three different geographical sub-areas of the Mediterranean Sea. *Mar Pollut Bull*, 140, 129–137. doi: 10.1016/j.marpolbul.2019.01.005.

Gigault, J., Halle, A.T., Baudrimont, M., Pascal, P.Y., Gauffre, F., Phi, T.L., El Hadri, H., Grassl, B., Reynaud, S. (2018). Current opinion: What is a nanoplastic? *Environ Pollut*, 235, 1030–1034. doi: 10.1016/j.envpol.2018.01.024.

Gomiero, A., Strafella, P., Oysaed, K.B., Fabi, G. (2019). First occurrence and composition assessment of microplastics in native mussels collected from coastal and offshore areas of the northern and central Adriatic Sea. *Environ Sci Pollut Res Int*, 26, 24407–24416. doi: 10.1007/s11356-019-05693-y.

Goswami, P., Vinithkumar, N.V., Dharani, G. (2020). First evidence of microplastics bioaccumulation by marine organisms in the Port Blair Bay, Andaman Islands. *Mar Pollut Bull*, 155, 111163. doi: 10.1016/j.marpolbul.2020.111163.

Gündoğdu, S., Çevik, C., Temiz Ataş, N. (2020). Occurrence of microplastics in the gastrointestinal tracts of some edible fish species along the Turkish coast. *Turkish Journal of Zoology*, 44, 312–323. doi: 10.3906/zoo-2003-49.

Hanke, G., Galgani, F., Werner, S., Oosterbaan, L., Nilsson, P., Fleet, D., Kinsey, S., Thompson, R.C., Palatinus, A., Van Franeker, J.A. et al. (2013). Guidance on monitoring of marine litter in European seas. *Office of the European Union*, JRC83985. doi: 10.2788/05469.

Hara, J., Frias, J., Nash, R. (2020). Quantification of microplastic ingestion by the decapod crustacean *Nephrops norvegicus* from Irish waters. *Mar Pollut Bull*, 152, 110905. doi: 10.1016/j.marpolbul.2020.110905.

Hartmann, N.B., Hüffer, T., Thompson, R.C., Hassellöv, M., Verschoor, A., Daugaard, A.E., Rist, S., Karlsson, T., Brennholt, N., Cole, M. et al. (2019). Are we speaking the same language? Recommendations for a definition and categorization framework for plastic debris. *Environ Sci Technol*, 53, 1039–1047. doi: 10.1021/acs.est.8b05297.

Hermabessiere, L. (2018). Les microplastiques et leurs additifs dans les produits de la pêche : développements méthodologiques et prévalence. PhD, Université du Littoral Côte d'Opale [Online]. Available at: https://tel.archives-ouvertes.fr/tel-02055801.

Hermabessiere, L., Paul-Pont, I., Cassone, A.L., Himber, C., Receveur, J., Jezequel, R., El Rakwe, M., Rinnert, E., Riviere, G., Lambert, C. et al. (2019). Microplastic contamination and pollutant levels in mussels and cockles collected along the channel coasts. *Environ Pollut*, 250, 807–819. doi: 10.1016/j.envpol.2019.04.051.

Hermsen, E., Pompe, R., Besseling, E., Koelmans, A.A. (2017). Detection of low numbers of microplastics in North Sea fish using strict quality assurance criteria. *Mar Pollut Bull*, 122, 253–258. doi: 10.1016/j.marpolbul.2017.06.051.

Hermsen, E., Mintenig, S.M., Besseling, E., Koelmans, A.A. (2018). Quality criteria for the analysis of microplastic in biota samples: A critical review. *Environ Sci Technol*, 52, 10230–10240. doi: 10.1021/acs.est.8b01611.

Herrera, A., Stindlova, A., Martinez, I., Rapp, J., Romero-Kutzner, V., Samper, M.D., Montoto, T., Aguiar-Gonzalez, B., Packard, T., Gomez, M. (2019). Microplastic ingestion by Atlantic chub mackerel (*Scomber colias*) in the Canary Islands coast. *Mar Pollut Bull*, 139, 127–135. doi: 10.1016/j.marpolbul.2018.12.022.

Hossain, M.S., Sobhan, F., Uddin, M.N., Sharifuzzaman, S.M., Chowdhury, S.R., Sarker, S., Chowdhury, M.S.N. (2019). Microplastics in fishes from the Northern Bay of Bengal. *Sci Total Environ*, 690, 821–830. doi: 10.1016/j.scitotenv.2019.07.065.

Hossain, M.S., Rahman, M.S., Uddin, M.N., Sharifuzzaman, S.M., Chowdhury, S.R., Sarker, S., Nawaz Chowdhury, M.S. (2020). Microplastic contamination in penaeid shrimp from the Northern Bay of Bengal. *Chemosphere*, 238, 124688. doi: 10.1016/j.chemosphere. 2019.124688.

Hussien, N.A., Mohammadein, A., Tantawy, E.M., Khattab, Y., Al Malki, J.S. (2021). Investigating microplastics and potentially toxic elements contamination in canned Tuna, Salmon, and Sardine fishes from Taif markets, KSA. *Open Life Sciences*, 16, 827–837. doi: 10.1515/biol-2021-0086.

James, K., Vasant, K., Padua, S., Gopinath, V., Abilash, K.S., Jeyabaskaran, R., Babu, A., John, S. (2020). An assessment of microplastics in the ecosystem and selected commercially important fishes off Kochi, south eastern Arabian Sea, India. *Mar Pollut Bull*, 154, 111027. doi: 10.1016/j.marpolbul.2020.111027.

Karami, A., Golieskardi, A., Choo, C.K., Larat, V., Karbalaei, S., Salamatinia, B. (2018). Microplastic and mesoplastic contamination in canned sardines and sprats. *Sci Total Environ*, 612, 1380–1386. doi: 10.1016/j.scitotenv.2017.09.005.

Karbalaei, S., Golieskardi, A., Hamzah, H.B., Abdulwahid, S., Hanachi, P., Walker, T.R., Karami, A. (2019). Abundance and characteristics of microplastics in commercial marine fish from Malaysia. *Mar Pollut Bull*, 148, 5–15. doi: 10.1016/j.marpolbul.2019.07.072.

Kartar, S., Milne, R.A., Sainsbury, M. (1973). Polystyrene waste in the Severn estuary. *Mar Pollut Bull*, 4. doi: 10.1016/0025-326X(73)90010-6.

Kartar, S., Abou-Seedo, F., Sainsbury, M. (1976). Polystyrene spherules in the Severn estuary – A progress report. *Mar Pollut Bull*, 7. doi: 10.1016/0025-326X(76)90092-8.

Karuppasamy, P.K., Ravi, A., Vasudevan, L., Elangovan, M.P., Dyana Mary, P., Vincent, S.G.T., Palanisami, T. (2020). Baseline survey of micro and mesoplastics in the gastro-intestinal tract of commercial fish from Southeast coast of the Bay of Bengal. *Mar Pollut Bull*, 153, 110974. doi: 10.1016/j.marpolbul.2020.110974.

Kim, J.S., Whang, K.H., Jang, S.E., Song, S.M., Im, Y.S., Han, Y.S., Heo, M.J. (2021). Microplastics pollution of seafoods and processed seafood products distributed in Incheon area. *Korean J Food Preserv*, 28, 161–168. doi: 10.11002/kjfp.2021.28.2.161.

Kirstein, I.V., Kirmizi, S., Wichels, A., Garin-Fernandez, A., Erler, R., Loder, M., Gerdts, G. (2016). Dangerous hitchhikers? Evidence for potentially pathogenic *Vibrio* spp. on microplastic particles. *Mar Environ Res*, 120, 1–8. doi: 10.1016/j.marenvres.2016.07.004.

Klangnurak, W. and Chunniyom, S. (2020). Screening for microplastics in marine fish of Thailand: The accumulation of microplastics in the gastrointestinal tract of different foraging preferences. *Environ Sci Pollut Res Int*, 27, 27161–27168. doi: 10.1007/s11356-020-09147-8.

Klasios, N., De Frond, H., Miller, E., Sedlak, M., Rochman, C.M. (2021). Microplastics and other anthropogenic particles are prevalent in mussels from San Francisco Bay, and show no correlation with PAHs. *Environ Pollut*, 271, 116260. doi: 10.1016/j.envpol.2020.116260.

Knight, L.J., Parker-Jurd, F.N.F., Al-Sid-Cheikh, M., Thompson, R.C. (2020). Tyre wear particles: An abundant yet widely unreported microplastic? *Environ Sci Pollut Res Int*, 27, 18345–18354. doi: 10.1007/s11356-020-08187-4.

Koongolla, J.B., Lin, L., Pan, Y.F., Yang, C.P., Sun, D.R., Liu, S., Xu, X.R., Maharana, D., Huang, J.S., Li, H.X. (2020). Occurrence of microplastics in gastrointestinal tracts and gills of fish from Beibu Gulf, South China Sea. *Environ Pollut*, 258, 113734. doi: 10.1016/j.envpol.2019.113734.

Kripa, V., Preetha, G.N., Dhanya, A.M., Pravitha, V.P., Abhilash, K.S., Abbas, A.M., Dhannesh, V., Vishnu, P.G., Gishnu, M., Anil Kumar, P.S. et al. (2014). Microplastics in the gut of anchovies caught from the mud bank area of Alappuzha, Kerala. *Mar Fish Infor Serv, T, & E Ser*, 219, 27–28.

Kuhn, S., Schaafsma, F.L., Van Werven, B., Flores, H., Bergmann, M., Egelkraut-Holtus, M., Tekman, M.B., Van Franeker, J.A. (2018). Plastic ingestion by juvenile polar cod (*Boreogadus saida*) in the Arctic Ocean. *Polar Biol*, 41, 1269–1278. doi: 10.1007/s00300-018-2283-8.

Li, J., Yang, D., Li, L., Jabeen, K., Shi, H. (2015). Microplastics in commercial bivalves from China. *Environ Pollut*, 207, 190–195. doi: 10.1016/j.envpol.2015.09.018.

Li, J., Qu, X., Su, L., Zhang, W., Yang, D., Kolandhasamy, P., Li, D., Shi, H. (2016). Microplastics in mussels along the coastal waters of China. *Environ Pollut*, 214, 177–184. doi: 10.1016/j.envpol.2016.04.012.

Li, H.X., Ma, L.S., Lin, L., Ni, Z.X., Xu, X.R., Shi, H.H., Yan, Y., Zheng, G.M., Rittschof, D. (2018a). Microplastics in oysters *Saccostrea cucullata* along the Pearl River Estuary, China. *Environ Pollut*, 236, 619–625. doi: 10.1016/j.envpol.2018.01.083.

Li, J., Green, C., Reynolds, A., Shi, H., Rotchell, J.M. (2018b). Microplastics in mussels sampled from coastal waters and supermarkets in the United Kingdom. *Environ Pollut*, 241, 35–44. doi: 10.1016/j.envpol.2018.05.038.

Liboiron, M., Liboiron, F., Wells, E., Richard, N., Zahara, A., Mather, C., Bradshaw, H., Murichi, J. (2016). Low plastic ingestion rate in Atlantic cod (*Gadus morhua*) from Newfoundland destined for human consumption collected through citizen science methods. *Mar Pollut Bull*, 113, 428–437. doi: 10.1016/j.marpolbul.2016.10.043.

Lin, L., Ma, L.S., Li, H.X., Pan, Y.F., Liu, S., Zhang, L., Peng, J.P., Fok, L., Xu, X.R., He, W.H. (2020). Low level of microplastic contamination in wild fish from an urban estuary. *Mar Pollut Bull*, 160, 111650. doi: 10.1016/j.marpolbul.2020.111650.

Lu, Y., Zhang, Y., Deng, Y., Jiang, W., Zhao, Y., Geng, J., Ding, L., Ren, H. (2016). Uptake and accumulation of polystyrene microplastics in zebrafish (*Danio rerio*) and toxic effects in liver. *Environ Sci Technol*, 50, 4054–4060. doi: 10.1021/acs.est.6b00183.

Lusher, A.L., McHugh, M., Thompson, R.C. (2013). Occurrence of microplastics in the gastrointestinal tract of pelagic and demersal fish from the English Channel. *Mar Pollut Bull*, 67, 94–99. doi: 10.1016/j.marpolbul.2012.11.028.

Lusher, A.L., O'Donnell, C., Officer, R., O'Connor, I. (2016). Microplastic interactions with North Atlantic mesopelagic fish. *ICES J Mar Sci*, 73, 1214–1225. doi: 10.1093/icesjms/fsv241.

Lusher, A.L., Welden, N.A., Sobral, P., Cole, M. (2017). Sampling, isolating and identifying microplastics ingested by fish and invertebrates. *Anal Methods*, 9, 1346–1360. doi: 10.1039/C6AY02415G.

Maaghloud, H., Houssa, R., Ouansafi, S., Bellali, F., El Bouqdaoui, K., Charouki, N., Fahde, A. (2020). Ingestion of microplastics by pelagic fish from the Moroccan Central Atlantic Coast. *Environ Pollut*, 261, 114194. doi: 10.1016/j.envpol.2020.114194.

Mancia, A., Chenet, T., Bono, G., Geraci, M.L., Vaccaro, C., Munari, C., Mistri, M., Cavazzini, A., Pasti, L. (2020). Adverse effects of plastic ingestion on the Mediterranean small-spotted catshark (*Scyliorhinus canicula*). *Mar Environ Res*, 155, 104876. doi: 10.1016/j.marenvres.2020.104876.

Mancuso, M., Savoca, S., Bottari, T. (2019). First record of microplastics ingestion by European hake *Merluccius merluccius* from the Tyrrhenian Sicilian coast (Central Mediterranean Sea). *J Fish Biol*, 94, 517–519. doi: 10.1111/jfb.13920.

Marques, F., Vale, C., Rudnitskaya, A., Moreirinha, C., Costa, S.T., Botelho, M.J. (2021). Major characteristics of microplastics in mussels from the Portuguese coast. *Environ Res*, 197, 110993. doi: 10.1016/j.envres.2021.110993.

Martinelli, J.C., Phan, S., Luscombe, C.K., Padilla-Gamino, J.L. (2020). Low incidence of microplastic contaminants in Pacific oysters (*Crassostrea gigas* Thunberg) from the Salish Sea, USA. *Sci Total Environ*, 715, 136826. doi: 10.1016/j.scitotenv.2020.136826.

Mathalon, A. and Hill, P. (2014). Microplastic fibers in the intertidal ecosystem surrounding Halifax Harbor, Nova Scotia. *Mar Pollut Bull*, 81, 69–79. doi: 10.1016/j.marpolbul.2014.02.018.

McGoran, A.R., Clark, P.F., Morritt, D. (2017). Presence of microplastic in the digestive tracts of European flounder, *Platichthys Flesus*, and European smelt, *Osmerus eperlanus*, from the River Thames. *Environ Pollut*, 220, 744–751. doi: 10.1016/j.envpol.2016.09.078.

Mercogliano, R., Santonicola, S., Raimo, G., Gasperi, M., Colavita, G. (2021). Extraction and identification of microplastics from mussels: Method development and preliminary results. *Ital J Food Saf*, 10, 9264. doi: 10.4081/ijfs.2021.9264.

Mohsen, M., Wang, Q., Zhang, L., Sun, L., Lin, C., Yang, H. (2019). Microplastic ingestion by the farmed sea cucumber *Apostichopus japonicus* in China. *Environ Pollut*, 245, 1071–1078. doi: 10.1016/j.envpol.2018.11.083.

Morgan, E., Hutchinson, D., Gaion, A. (2021). Plastic ingestion by the small-spotted catshark (*Scyliorhinus canicula*) from the South West coast of the United Kingdom. *Bull Environ Contam Toxicol*, 106, 910–915. doi: 10.1007/s00128-021-03129-3.

Murphy, F., Russell, M., Ewins, C., Quinn, B. (2017). The uptake of macroplastic & microplastic by demersal & pelagic fish in the Northeast Atlantic around Scotland. *Mar Pollut Bull*, 122, 353–359. doi: 10.1016/j.marpolbul.2017.06.073.

Murray, F. and Cowie, P.R. (2011). Plastic contamination in the decapod crustacean *Nephrops norvegicus* (Linnaeus, 1758). *Mar Pollut Bull*, 62, 1207–1217. doi: 10.1016/j.marpolbul.2011.03.032.

Nadal, M.A., Alomar, C., Deudero, S. (2016). High levels of microplastic ingestion by the semipelagic fish bogue *Boops boops* (L.) around the Balearic Islands. *Environ Pollut*, 214, 517–523. doi: 10.1016/j.envpol.2016.04.054.

Naidoo, T., Sershen, Thompson, R.C., Rajkaran, A. (2020). Quantification and characterisation of microplastics ingested by selected juvenile fish species associated with mangroves in KwaZulu-Natal, South Africa. *Environ Pollut*, 257, 113635. doi: 10.1016/j.envpol.2019.113635.

Narmatha Sathish, M., Immaculate Jeyasanta, K., Patterson, J. (2020). Monitoring of microplastics in the clam *Donax cuneatus* and its habitat in Tuticorin coast of Gulf of Mannar (GoM), India. *Environ Pollut*, 266, 115219. doi: 10.1016/j.envpol.2020.115219.

Neto, J.G.B., Rodrigues, F.L., Ortega, I., Rodrigues, L.D.S., Lacerda, A., Coletto, J.L., Kessler, F., Cardoso, L.G., Madureira, L., Proietti, M.C. (2020). Ingestion of plastic debris by commercially important marine fish in southeast-south Brazil. *Environ Pollut*, 267, 115508. doi: 10.1016/j.envpol.2020.115508.

Neves, D., Sobral, P., Ferreira, J.L., Pereira, T. (2015). Ingestion of microplastics by commercial fish off the Portuguese coast. *Mar Pollut Bull*, 101, 119–126. doi: 10.1016/j.marpolbul.2015.11.008.

Nie, H., Wang, J., Xu, K., Huang, Y., Yan, M. (2019). Microplastic pollution in water and fish samples around Nanxun Reef in Nansha Islands, South China Sea. *Sci Total Environ*, 696, 134022. doi: 10.1016/j.scitotenv.2019.134022.

Ningrum, E.W. and Patria, M.P. (2019). Ingestion of microplastics by anchovies from east Lombok Harbour, Lombok Island, Indonesia. *AIP Conference Proceedings*. doi: 10.1063/1.5115640.

Nuelle, M.T., Dekiff, J.H., Remy, D., Fries, E. (2014). A new analytical approach for monitoring microplastics in marine sediments. *Environ Pollut*, 184, 161–169. doi: 10.1016/j.envpol.2013.07.027.

Ogonowski, M., Wenman, V., Barth, A., Hamacher-Barth, E., Danielsson, S., Gorokhova, E. (2019). Microplastic intake, its biotic drivers, and hydrophobic organic contaminant levels in the Baltic herring. *Front Environ Sci*, 7. doi: 10.3389/fenvs.2019.00134.

Oliveira, A.R., Sardinha-Silva, A., Andrews, P.L.R., Green, D., Cooke, G.M., Hall, S., Blackburn, K., Sykes, A.V. (2020). Microplastics presence in cultured and wild-caught cuttlefish, *Sepia officinalis*. *Mar Pollut Bull*, 160, 111553. doi: 10.1016/j.marpolbul.2020.111553.

Ory, N.C., Sobral, P., Ferreira, J.L., Thiel, M. (2017). Amberstripe scad *Decapterus muroadsi* (Carangidae) fish ingest blue microplastics resembling their copepod prey along the coast of Rapa Nui (Easter Island) in the South Pacific subtropical gyre. *Sci Total Environ*, 586, 430–437. doi: 10.1016/j.scitotenv.2017.01.175.

Parker, B.W., Beckingham, B.A., Ingram, B.C., Ballenger, J.C., Weinstein, J.E., Sancho, G. (2020). Microplastic and tire wear particle occurrence in fishes from an urban estuary: Influence of feeding characteristics on exposure risk. *Mar Pollut Bull*, 160, 111539. doi: 10.1016/j.marpolbul.2020.111539.

Patterson, J., Jeyasanta, K.I., Sathish, N., Booth, A.M., Edward, J.K.P. (2019). Profiling microplastics in the Indian edible oyster, *Magallana bilineata* collected from the Tuticorin coast, Gulf of Mannar, Southeastern India. *Sci Total Environ*, 691, 727–735. doi: 10.1016/j.scitotenv.2019.07.063.

Pellini, G., Gomiero, A., Fortibuoni, T., Ferra, C., Grati, F., Tassetti, A.N., Polidori, P., Fabi, G., Scarcella, G. (2018). Characterization of microplastic litter in the gastrointestinal tract of *Solea solea* from the Adriatic Sea. *Environ Pollut*, 234, 943–952. doi: 10.1016/ j.envpol.2017.12.038.

Pennino, M.G., Bachiller, E., Lloret-Lloret, E., Albo-Puigserver, M., Esteban, A., Jadaud, A., Bellido, J.M., Coll, M. (2020). Ingestion of microplastics and occurrence of parasite association in Mediterranean anchovy and sardine. *Mar Pollut Bull*, 158, 111399. doi: 10.1016/j.marpolbul.2020.111399.

Pereira, J.M., Rodriguez, Y., Blasco-Monleon, S., Porter, A., Lewis, C., Pham, C.K. (2020). Microplastic in the stomachs of open-ocean and deep-sea fishes of the North-East Atlantic. *Environ Pollut*, 265, 115060. doi: 10.1016/j.envpol.2020.115060.

Perez, A.F., Ojeda, M., Rimondino, G.N., Chiesa, I.L., Di Mauro, R., Boy, C.C., Calcagno, J.A. (2020). First report of microplastics presence in the mussel *Mytilus chilensis* from Ushuaia Bay (Beagle Channel, Tierra del Fuego, Argentina). *Mar Pollut Bull*, 161, 111753. doi: 10.1016/j.marpolbul.2020.111753.

Phuong, N.N., Poirier, L., Pham, Q.T., Lagarde, F., Zalouk-Vergnoux, A. (2018). Factors influencing the microplastic contamination of bivalves from the French Atlantic coast: Location, season and/or mode of life? *Mar Pollut Bull*, 129, 664–674. doi: 10.1016/ j.marpolbul.2017.10.054.

Possatto, F.E., Barletta, M., Costa, M.F., Do Sul, J.A., Dantas, D.V. (2011). Plastic debris ingestion by marine catfish: An unexpected fisheries impact. *Mar Pollut Bull*, 62, 1098–102. doi: 10.1016/j.marpolbul.2011.01.036.

Pozo, K., Gomez, V., Torres, M., Vera, L., Nunez, D., Oyarzun, P., Mendoza, G., Clarke, B., Fossi, M.C., Baini, M. et al. (2019). Presence and characterization of microplastics in fish of commercial importance from the Biobio region in central Chile. *Mar Pollut Bull*, 140, 315–319. doi: 10.1016/j.marpolbul.2019.01.025.

Prata, J.C. (2018). Airborne microplastics: Consequences to human health? *Environ Pollut*, 234, 115–126. doi: 10.1016/j.envpol.2017.11.043.

Primpke, S., Cross, R.K., Mintenig, S.M., Simon, M., Vianello, A., Gerdts, G., Vollertsen, J. (2020). Toward the systematic identification of microplastics in the environment: Evaluation of a new independent software tool (siMPle) for spectroscopic analysis. *Appl Spectrosc*, 74, 1127–1138. doi: 10.1177/0003702820917760.

Qiao, R., Sheng, C., Lu, Y., Zhang, Y., Ren, H., Lemos, B. (2019). Microplastics induce intestinal inflammation, oxidative stress, and disorders of metabolome and microbiome in zebrafish. *Sci Total Environ*, 662, 246–253. doi: 10.1016/j.scitotenv.2019.01.245.

Rahman, A., Sarkar, A., Yadav, O.P., Achari, G., Slobodnik, J. (2021). Potential human health risks due to environmental exposure to nano- and microplastics and knowledge gaps: A scoping review. *Sci Total Environ*, 757, 143872. doi: 10.1016/j.scitotenv.2020.143872.

Railo, S., Talvitie, J., Setala, O., Koistinen, A., Lehtiniemi, M. (2018). Application of an enzyme digestion method reveals microlitter in *Mytilus trossulus* at a wastewater discharge area. *Mar Pollut Bull*, 130, 206–214. doi: 10.1016/j.marpolbul.2018.03.022.

Ramos, J.A.A., Barletta, M., Costa, M.F. (2012). Ingestion of nylon threads by Gerreidae while using a tropical estuary as foraging grounds. *Aquat Biol*, 17, 29–34. doi: 10.3354/ab00461.

Renzi, M., Specchiulli, A., Blaskovic, A., Manzo, C., Mancinelli, G., Cilenti, L. (2019). Marine litter in stomach content of small pelagic fishes from the Adriatic Sea: Sardines (*Sardina pilchardus*) and anchovies (*Engraulis encrasicolus*). *Environ Sci Pollut Res Int*, 26, 2771–2781. doi: 10.1007/s11356-018-3762-8.

Rios-Fuster, B., Alomar, C., Compa, M., Guijarro, B., Deudero, S. (2019). Anthropogenic particles ingestion in fish species from two areas of the western Mediterranean Sea. *Mar Pollut Bull*, 144, 325–333. doi: 10.1016/j.marpolbul.2019.04.064.

Rochman, C.M., Tahir, A., Williams, S.L., Baxa, D.V., Lam, R., Miller, J.T., Teh, F.C., Werorilangi, S., Teh, S.J. (2015). Anthropogenic debris in seafood: Plastic debris and fibers from textiles in fish and bivalves sold for human consumption. *Sci Rep*, 5, 14340. doi: 10.1038/srep14340.

Romeo, T., Pietro, B., Peda, C., Consoli, P., Andaloro, F., Fossi, M.C. (2015). First evidence of presence of plastic debris in stomach of large pelagic fish in the Mediterranean Sea. *Mar Pollut Bull*, 95, 358–361. doi: 10.1016/j.marpolbul.2015.04.048.

Rummel, C.D., Loder, M.G., Fricke, N.F., Lang, T., Griebeler, E.M., Janke, M., Gerdts, G. (2016). Plastic ingestion by pelagic and demersal fish from the North Sea and Baltic Sea. *Mar Pollut Bull*, 102, 134–141. doi: 10.1016/j.marpolbul.2015.11.043.

Santana, M.F., Ascer, L.G., Custodio, M.R., Moreira, F.T., Turra, A. (2016). Microplastic contamination in natural mussel beds from a Brazilian urbanized coastal region: Rapid evaluation through bioassessment. *Mar Pollut Bull*, 106, 183–189. doi: 10.1016/j.marpolbul.2016.02.074.

Sathish, M.N., Jeyasanta, I., Patterson, J. (2020). Occurrence of microplastics in epipelagic and mesopelagic fishes from Tuticorin, Southeast coast of India. *Sci Total Environ*, 720, 137614. doi: 10.1016/j.scitotenv.2020.137614.

Savoca, S., Bottari, T., Fazio, E., Bonsignore, M., Mancuso, M., Luna, G.M., Romeo, T., D'urso, L., Capillo, G., Panarello, G. et al. (2020). Plastics occurrence in juveniles of *Engraulis encrasicolus* and *Sardina pilchardus* in the Southern Tyrrhenian Sea. *Sci Total Environ*, 718, 137457. doi: 10.1016/j.scitotenv.2020.137457.

Schwabl, P., Koppel, S., Konigshofer, P., Bucsics, T., Trauner, M., Reiberger, T., Liebmann, B. (2019). Detection of various microplastics in human stool: A prospective case series. *Ann Intern Med*, 171, 453–457. doi: 10.7326/M19-0618.

Scott, N., Porter, A., Santillo, D., Simpson, H., Lloyd-Williams, S., Lewis, C. (2019). Particle characteristics of microplastics contaminating the mussel *Mytilus edulis* and their surrounding environments. *Mar Pollut Bull*, 146, 125–133. doi: 10.1016/j.marpolbul.2019.05.041.

Sparks, C. (2020). Microplastics in mussels along the coast of Cape Town, South Africa. *Bull Environ Contam Toxicol*, 104, 423–431. doi: 10.1007/s00128-020-02809-w.

Sparks, C. and Immelman, S. (2020). Microplastics in offshore fish from the Agulhas Bank, South Africa. *Mar Pollut Bull*, 156, 111216. doi: 10.1016/j.marpolbul.2020.111216.

Sparks, C., Awe, A., Maneveld, J. (2021). Abundance and characteristics of microplastics in retail mussels from Cape Town, South Africa. *Mar Pollut Bull*, 166, 112186. doi: 10.1016/j.marpolbul.2021.112186.

Su, L., Deng, H., Li, B., Chen, Q., Pettigrove, V., Wu, C., Shi, H. (2019). The occurrence of microplastic in specific organs in commercially caught fishes from coast and estuary area of east China. *J Hazard Mater*, 365, 716–724. doi: 10.1016/j.jhazmat.2018.11.024.

Sun, X., Li, Q., Shi, Y., Zhao, Y., Zheng, S., Liang, J., Liu, T., Tian, Z. (2019). Characteristics and retention of microplastics in the digestive tracts of fish from the Yellow Sea. *Environ Pollut*, 249, 878–885. doi: 10.1016/j.envpol.2019.01.110.

Sussarellu, R., Suquet, M., Thomas, Y., Lambert, C., Fabioux, C., Pernet, M.E., Le Goic, N., Quillien, V., Mingant, C., Epelboin, Y. et al. (2016). Oyster reproduction is affected by exposure to polystyrene microplastics. *Proc Natl Acad Sci USA*, 113, 2430–2435. doi: 10.1073/pnas.1519019113.

Taghizadeh Rahmat Abadi, Z., Abtahi, B., Grossart, H.P., Khodabandeh, S. (2021). Microplastic content of Kutum fish, *Rutilus frisii kutum* in the southern Caspian Sea. *Sci Total Environ*, 752, 141542. doi: 10.1016/j.scitotenv.2020.141542.

Tanaka, K. and Takada, H. (2016). Microplastic fragments and microbeads in digestive tracts of planktivorous fish from urban coastal waters. *Sci Rep*, 6, 34351. doi: 10.1038/srep34351.

Teng, J., Wang, Q., Ran, W., Wu, D., Liu, Y., Sun, S., Liu, H., Cao, R., Zhao, J. (2019). Microplastic in cultured oysters from different coastal areas of China. *Sci Total Environ*, 653, 1282–1292. doi: 10.1016/j.scitotenv.2018.11.057.

Treilles, R., Cayla, A., Gasperi, J., Strich, B., Ausset, P., Tassin, B. (2020). Impacts of organic matter digestion protocols on synthetic, artificial and natural raw fibers. *Sci Total Environ*, 748, 141230. doi: 10.1016/j.scitotenv.2020.141230.

Tsangaris, C., Digka, N., Valente, T., Aguilar, A., Borrell, A., De Lucia, G.A., Gambaiani, D., Garcia-Garin, O., Kaberi, H., Martin, J. et al. (2020). Using *Boops boops* (osteichthyes) to assess microplastic ingestion in the Mediterranean Sea. *Mar Pollut Bull*, 158, 111397. doi: 10.1016/j.marpolbul.2020.111397.

Van Cauwenberghe, L. and Janssen, C.R. (2014). Microplastics in bivalves cultured for human consumption. *Environ Pollut*, 193, 65–70. doi: 10.1016/j.envpol.2014.06.010.

Van Cauwenberghe, L., Claessens, M., Vandegehuchte, M.B., Janssen, C.R. (2015). Microplastics are taken up by mussels (*Mytilus edulis*) and lugworms (*Arenicola marina*) living in natural habitats. *Environ Pollut*, 199, 10–17. doi: 10.1016/j.envpol.2015.01.008.

Vandermeersch, G., Van Cauwenberghe, L., Janssen, C.R., Marques, A., Granby, K., Fait, G., Kotterman, M.J., Diogene, J., Bekaert, K., Robbens, J. et al. (2015). A critical view on microplastic quantification in aquatic organisms. *Environ Res*, 143, 46–55. doi: 10.1016/j.envres.2015.07.016.

Vieira, K.S., Baptista Neto, J.A., Crapez, M.A.C., Gaylarde, C., Pierri, B.D.S., Saldana-Serrano, M., Bainy, A.C.D., Nogueira, D.J., Fonseca, E.M. (2021). Occurrence of microplastics and heavy metals accumulation in native oysters *Crassostrea gasar* in the Paranagua estuarine system, Brazil. *Mar Pollut Bull*, 166, 112225. doi: 10.1016/ j.marpolbul.2021.112225.

Vinay Kumar, B.N., Loschel, L.A., Imhof, H.K., Loder, M.G.J., Laforsch, C. (2021). Analysis of microplastics of a broad size range in commercially important mussels by combining FTIR and Raman spectroscopy approaches. *Environ Pollut*, 269, 116147. doi: 10.1016/j.envpol.2020.116147.

Virsek, M.K., Lovsin, M.N., Koren, S., Krzan, A., Peterlin, M. (2017). Microplastics as a vector for the transport of the bacterial fish pathogen species *Aeromonas salmonicida. Mar Pollut Bull*, 125, 301–309. doi: 10.1016/j.marpolbul.2017.08.024.

Wakkaf, T., El Zrelli, R., Kedzierski, M., Balti, R., Shaiek, M., Mansour, L., Tlig-Zouari, S., Bruzaud, S., Rabaoui, L. (2020). Microplastics in edible mussels from a southern Mediterranean lagoon: Preliminary results on seawater-mussel transfer and implications for environmental protection and seafood safety. *Mar Pollut Bull*, 158, 111355. doi: 10.1016/j.marpolbul.2020.111355.

Wang, D., Su, L., Ruan, H.D., Chen, J., Lu, J., Lee, C.H., Jiang, S.Y. (2021a). Quantitative and qualitative determination of microplastics in oyster, seawater and sediment from the coastal areas in Zhuhai, China. *Mar Pollut Bull*, 164, 112000. doi: 10.1016/j.marpolbul. 2021.112000.

Wang, Q., Zhu, X., Hou, C., Wu, Y., Teng, J., Zhang, C., Tan, H., Shan, E., Zhang, W., Zhao, J. (2021b). Microplastic uptake in commercial fishes from the Bohai Sea, China. *Chemosphere*, 263, 127962. doi: 10.1016/j.chemosphere.2020.127962.

Webb, S., Ruffell, H., Marsden, I., Pantos, O., Gaw, S. (2019). Microplastics in the New Zealand green lipped mussel *Perna canaliculus. Mar Pollut Bull*, 149. doi: 10.1016/j.marpolbul.2019.110641.

Welden, N.A.C. and Cowie, P.R. (2016). Environment and gut morphology influence microplastic retention in langoustine, *Nephrops norvegicus. Environ Pollut*, 214, 859–865. doi: 10.1016/j.envpol.2016.03.067.

Welden, N.A.C., Abylkhani, B., Howarth, L.M. (2018). The effects of trophic transfer and environmental factors on microplastic uptake by plaice, *Pleuronectes plastessa*, and spider crab, *Maja squinado. Environ Pollut*, 239, 351–358. doi: 10.1016/j.envpol. 2018.03.110.

Wiesheu, A.C., Anger, P.M., Baumann, T., Niessner, R., Ivleva, N.P. (2016). Raman microspectroscopic analysis of fibers in beverages. *Anal Methods*, 8, 5722–5725. doi: 10.1039/c6ay01184e.

Wojcik-Fudalewska, D., Normant-Saremba, M., Anastacio, P. (2016). Occurrence of plastic debris in the stomach of the invasive crab *Eriocheir sinensis*. *Mar Pollut Bull*, 113, 306–311. doi: 10.1016/j.marpolbul.2016.09.059.

Wu, X., Pan, J., Li, M., Li, Y., Bartlam, M., Wang, Y. (2019). Selective enrichment of bacterial pathogens by microplastic biofilm. *Water Res*, 165, 114979. doi: 10.1016/j.watres.2019.114979.

Wu, F., Wang, Y., Leung, J.Y.S., Huang, W., Zeng, J., Tang, Y., Chen, J., Shi, A., Yu, X., Xu, X. et al. (2020). Accumulation of microplastics in typical commercial aquatic species: A case study at a productive aquaculture site in China. *Sci Total Environ*, 708, 135432. doi: 10.1016/j.scitotenv.2019.135432.

Yang, Y., Liu, G., Song, W., Ye, C., Lin, H., Li, Z., Liu, W. (2019). Plastics in the marine environment are reservoirs for antibiotic and metal resistance genes. *Environ Int*, 123, 79–86. doi: 10.1016/j.envint.2018.11.061.

Zakeri, M., Naji, A., Akbarzadeh, A., Uddin, S. (2020). Microplastic ingestion in important commercial fish in the southern Caspian Sea. *Mar Pollut Bull*, 160, 111598. doi: 10.1016/j.marpolbul.2020.111598.

Zhang, C., Wang, S., Pan, Z., Sun, D., Xie, S., Zhou, A., Wang, J., Zou, J. (2020a). Occurrence and distribution of microplastics in commercial fishes from estuarine areas of Guangdong, South China. *Chemosphere*, 260, 127656. doi: 10.1016/j.chemosphere.2020.127656.

Zhang, D., Cui, Y., Zhou, H., Jin, C., Yu, X., Xu, Y., Li, Y., Zhang, C. (2020b). Microplastic pollution in water, sediment, and fish from artificial reefs around the Ma'an Archipelago, Shengsi, China. *Sci Total Environ*, 703, 134768. doi: 10.1016/j.scitotenv.2019.134768.

Zhang, Y., Lu, J., Wu, J., Wang, J., Luo, Y. (2020c). Potential risks of microplastics combined with superbugs: Enrichment of antibiotic resistant bacteria on the surface of microplastics in mariculture system. *Ecotoxicol Environ Saf*, 187, 109852. doi: 10.1016/j.ecoenv.2019.109852.

Zhang, N., Li, Y.B., He, H.R., Zhang, J.F., Ma, G.S. (2021a). You are what you eat: Microplastics in the feces of young men living in Beijing. *Sci Total Environ*, 767, 144345. doi: 10.1016/j.scitotenv.2020.144345.

Zhang, S., Sun, Y., Liu, B., Li, R. (2021b). Full size microplastics in crab and fish collected from the mangrove wetland of Beibu gulf: Evidences from Raman tweezers (1–20 mum) and spectroscopy (20–5000 mum). *Sci Total Environ*, 759, 143504. doi: 10.1016/j.scitotenv.2020.143504.

Zhao, Y., Sun, X., Li, Q., Shi, Y., Zheng, S., Liang, J., Liu, T., Tian, Z. (2019). Data on microplastics in the digestive tracts of 19 fish species from the Yellow Sea, China. *Data Brief*, 25, 103989. doi: 10.1016/j.dib.2019.103989.

Zhu, L., Wang, H., Chen, B., Sun, X., Qu, K., Xia, B. (2019). Microplastic ingestion in deep-sea fish from the South China Sea. *Sci Total Environ*, 677, 493–501. doi: 10.1016/j.scitotenv.2019.04.380.

Zitouni, N., Bousserrhine, N., Belbekhouche, S., Missawi, O., Alphonse, V., Boughatass, I., Banni, M. (2020). First report on the presence of small microplastics (</= 3 µm) in tissue of the commercial fish *Serranus scriba* (Linnaeus. 1758) from Tunisian coasts and associated cellular alterations. *Environ Pollut*, 263, 114576. doi: 10.1016/j.envpol.2020.114576.

8

Smoking: A Flavoring and Preservation Technique

Régis BARON
Ifremer, Nantes, France

8.1. Introduction

Smoking is a significant way of enhancing seafood products, particularly fish. It also concerns many delicatessen products. The process probably dates back to prehistoric times and is one of the oldest known methods of preserving meat products, along with drying and salting. The method remains in the spirit, not far from that seen in antiquity, except the scale of production due to industrialization and the extreme diversity of the proposed products. The method consists, generally after a salting and a light preliminary drying at least in the countries where the cold chain is very developed, of subjecting the product to be smoked to the action of the smoke's compounds resulting from the slow and incomplete combustion of wood to ensure flavoring and to improve the shelf life of the products. In practice, smoking is mainly used for organoleptic purposes (color, smell and flavor) rather than as a means of preservation, at least in industrialized countries. In some tropical countries where fish is a staple food, smoking is still a preservation method and the products obtained have little to do with their cousins in industrialized countries. The species concerned are clearly differentiated. In France, the tradition of cold smoking of seafood products, mainly salmon, herring and trout, is strong. Smoking has a very special place in France. For a long time in France, only herring was smoked

For a color version of all figures in this chapter, see http://www.iste.co.uk/verrez-bagnis/aquaticproducts.zip.

Current Challenges for the Aquatic Products Processing Industry,
coordinated by Véronique VERREZ-BAGNIS.

industrially, especially in the Boulonnais region. Production reached record annual tonnages of around 15,000 tons in the mid-1980s. The smokehouses were large chimneys with natural draught, called “coresses”. The hearth was located at the base, under the floor where the smoke was produced by self-combustion of sawdust or wood shavings. The coresses were generally equipped with two floors of carts where the fish was dried on the lower floor near the hearth and then brought up to the second floor where the smoking was completed. This historical industrial smoking method is nowadays very little used. It has been replaced by indirect smoking in an air-conditioned chamber, which is currently the most commonly used process throughout France. Since the end of the 1970s, the sector has undergone progressive change. The smoking of clupeids (herring, sardines) gradually gave way to the smoking of salmonids (salmon and trout). Today, smoked fish represents more than 30,000 tons per year, making France the second European producer after Poland. National consumption was 28,400 t in 2018 according to France Agrimer 2019 (corresponding to more than 58,000 t of whole fish). Smoked salmon is the first smoked product consumed at national level since it represents, alone, more than 16,700 t, which makes France one of the largest consumers in the world (the first in Europe). At the same time, its average selling price has been divided by two in about 20 years from the 1980s before stabilizing and then rising slightly. This production is spread over the whole territory with both artisanal workshops and industrial production plants. The sector counts about 30 companies and about 3,000 employees. However, the bulk of the production is held by a small number of companies. Thus, eight companies have production capacities higher than 1,000 t per year (and sometimes largely higher). The largest companies export part of their production, notably to Europe, but more than 80% of this production is destined for national consumption. Smoked salmon has recently experienced a slight decrease in consumption, while smoked trout has experienced a still interesting growth which compensates for this decrease. The consumption of smoked trout was just under 6,000 t in 2018. While the salmon (fresh whole fish and fresh fillets) processed in French plants are mostly from farms, an increasingly significant number (although still modest) is also from fishing. Thus, among farmed salmon: 60% now come from Norway, 24% from Scotland and 4% from Ireland. Approximately 9% of Smoked Salmon in France is of wild origin (vs. 6% in 2016). Most of these are salmon caught in Alaska, recently joined by salmon from the Russian Far East (Siberia) (source: Adepale information). This tendency to increase the share of wild salmon, which remains however, as mentioned above, a very small minority, has been very significant over the last 10 years. Note the growing importance given by French consumers to the main official signs such as the “Label Rouge” (several constraints in terms of farming and processing and specifications on the fat content of between 10% and 17% depending on the size) or the AB logo (farming conditions that respect the consumer, the environment and the animal, with a diet aimed at preserving marine resources). The average price of smoked salmon is about 35€ per kilo. However, it is possible to find smoked salmon for less than 50% of the average price as well as products whose price per kilo is multiplied by 4 or 5 compared to the average price.

During 2016, the Canadian authorities authorized the introduction and marketing of genetically modified salmon on their territory in order to accelerate its growth. More than 5 tons of GMO-type salmon were already distributed by Canadian supermarkets. In fact, the transgenic salmon produced by AquaBounty is faster to produce (16–18 months compared to 30 months for a "classic" salmon) and consumes up to 75% less food. These attractive characteristics reduce the price of the salmon as well as its ecological footprint, but these characteristics seem far from contributing to a good acceptability for a large proportion of the population.

The content of this chapter aims to review the manufacturing and preservation practices, from the raw material to the consumer, and to address their impacts on the main characteristics of the product and their evolution over time. A second part reviews the technologies related to the three main unit operations of smoking, i.e. salting, drying and smoking processes. A non-exhaustive review of certain other practices that are sometimes anecdotal in this sector is also proposed. Finally, a brief review of certain aspects of fish quality and safety is provided.

8.2. Role of the main manufacturing steps

Initially, a luxury product consumed during festive periods, smoked salmon has become an everyday foodstuff that can be found in all seasons on the shelves of supermarkets and hypermarkets. However, a large proportion of sales are still made in December and January, which sometimes forces manufacturers to work with frozen raw materials or to freeze/thaw the smoked product from a fresh raw material. The risk associated with bacterial development then potentially increased which requires a good control of the freezing/thawing operations. The product sold is then generally simply labeled with the mention "do not refreeze".

Atlantic salmon (*Salmo salar*) is the main species used and is most often farmed in Norway, Scotland, Ireland or Chile. The fish are slaughtered, bled and eviscerated on or near the farm. Then, the fish are put on ice and transported to the smoking plant.

Cold smoking is the most classical form. There are also hot smoking processes (smoking at temperatures of 70–80°C most often and generally also associated with a prior salting stage) which will lead to a cooking of the fish flesh and thus considerably modify its sensory characteristics. There are also many variants and many names (regional or more global, but practices may vary according to the region for the same name, for example, for "boucanés" fish).

Cold smoking, which is the main focus of this chapter, consists essentially of filleting, salting, drying, smoking, packaging and storage in a refrigerated state until it reaches the consumer's plate (Knockaert 1990). Cold-smoked fish is therefore a microbiologically fragile product that is also subject to numerous manipulations. The salting phase first gives the product a specific flavor. Its bactericidal power is weak, but its preservative action is also associated with the reduction of water content. The alterations of the flesh are due to chemical, enzymatic and bacterial reactions for which the available water plays a predominant role. Preliminary drying allows the surface of the fish to be placed in the right hygrometric state for good smoking. It should be noted that the drying, i.e. desiccation by evaporation, also continues at a lower speed during the smoking operation. These two operations (salting and drying) allow a reduction in the activity of water (aw) leading to a reduction in the reaction rates mentioned above and particularly a more significant reduction in the growth rate of bacteria present on the surface of the product. These two operations also play a role in the organoleptic characteristics of the product (color, texture, surface effect and flavor obviously). The smoke compounds deposited during smoking provide a specific surface coloring (mainly carbonyls) and a significant modification of the flavor (via phenolic compounds, carbonyls and acids). Some compounds, especially phenols, are also responsible for a slight antioxidant and antibacterial action (Maga 1988).

8.3. Factors influencing conservation

The product, during cold smoking, never undergoes a temperature higher than 30°C during all these stages, and thus remains a "raw fish". The manufacturing practices, as well as the control of the cold chain, will thus play a major role in the conservation of the smoked product and its sanitary quality. Smoked seafood is therefore a perishable product. The salting, drying and smoking stages are "barrier" technologies, i.e. they slow down/hinder the development of micro-organisms naturally present or resulting from external contamination during the different phases, but without ever stopping it completely. They also fall into the category of semi-preserved products, i.e. the products that have undergone a very light transformation treatment resulting in a final pH higher than 5 and a salt content lower than 6% in the aqueous phase. The shelf life of these smoked products, which is mainly associated with the development of psychotropic germs, is one of the main aspects of product quality. The use-by date is the responsibility of the manufacturer;

it is primarily determined in relation to health risks, including the risk of listeriosis, and to a lesser degree in relation to the development of non-pathogenic bacterial flora responsible for the alteration of sensory characteristics. It is clear, as we will describe later, that the temperature history of the product between its manufacture and consumption, a factor that the manufacturer can no longer control after leaving the factory, is also a critical point of quality and risk.

8.3.1. *Factors influencing alteration*

The shelf life of vacuum-packed sliced smoked salmon stored at +4°C generally varies from three to six weeks for smoked products according to the most common French practices. However, in some cases and under certain conditions, a loss of taste qualities, even the appearance of bad odors or flavors and a pasty texture, can be observed from the end of the second week of storage. This alteration is mainly linked to the activity of micro-organisms because autolytic or oxidative reactions play a more limited role in the sensory perception.

No step in the manufacturing process allows a total elimination of the germs contaminating the product, salting, drying and smoking only leading to a partial reduction of the microbial load. Furthermore, the finished product can be stored for several weeks at low temperatures and then consumed without cooking. Psychrotrophic germs can therefore grow (Leroi 2002).

In terms of raw material, the skin, the mucus, the gills and the gastrointestinal tract contain an initial bacterial flora represented by the genera *Pseudomonas*, *Aeromonas*, *Photobacterium*, *Vibrio*, *Clostridium*, etc. During the preparation of the fillets (gutting, heading, filleting, trimming and possibly skinning), these micro-organisms can contaminate the fish flesh. On the contrary, during the processing phases and especially during slicing and packaging, contamination by germs present in the environment can also occur. The genera encountered are much more variable (*Photobacterium*, *Shewanella*, *Vibrio*, *Serratia*, *Brochothrix*, *Carnobacterium*, *Lactobacilli*, etc.) and depend, among other things, on the hygiene procedures in the company (Leroi 2002).

The initial contamination of the finished product depends on the initial contamination of the fillets, the intensity of the salting, drying and smoking treatments and the packaging method. The bacterial flora is complex and varies according to the origin of the fillets and the manufacturing sites. Within this flora, there are competitions between bacterial groups that do not necessarily develop at the same speed depending on the environmental conditions. The "biodiversity" of the bacterial flora of vacuum-packed smoked salmon is large, complex to understand

and evolves over time at potentially different rates depending on the species and conditions (characteristics of the smoked fish, temperatures, etc.) and with many possible interactions between the different bacteria. In work on the identification of bacterial species present on smoked salmon, more than 15 different bacterial species are commonly found. The variability on the species and the relative concentrations observed is strong according to the batches even for an identical manufacturing site. Alteration odors are mainly due to some of the bacteria present within this biodiversity (Leroi et al. 2000, 2001; Stohr et al. 2001). Some of the main alteration bacteria in smoked salmon are *Brochothrix thermosphacta*, *Lactobacillus sakei*, *Serratia liquefaciens*, *Photobacterium phosphoreum* and *Shewanella putrefaciens*. Note that the presence of certain altering bacteria can attenuate the perception of certain alteration odors. This is the case when *Lactobacillus sakei*, a highly altering bacterium, is in the presence of *Serratia liquefaciens*, leading to an antagonistic effect on this perception. Note also that the association of certain bacteria will cause alteration odors of the product while when these bacteria are not associated they do not present any characteristic of alteration even at significant concentrations. This is the case of the association of certain genera *Carnobacterium* + *Vibrio*.

The prediction of bacterial development therefore remains difficult, even though certain predictive microbiology models applied to smoked salmon already allow a first approach as a first "roughing-in". Predicting the sensory alteration of products is, for all the reasons mentioned, an even greater challenge than predicting bacterial development.

Bacterial growth is significantly slowed when products are stored at refrigerated temperatures. The relative rate of spoilage R, defined as the ratio of shelf life at 0°C to shelf life at a constant temperature T, is very sensitive to positive temperature changes (Dalggard et al. 2004). For example, according to the models discussed above, a smoked salmon that would last 42 days at 0°C would only last about 25 days at 4°C (almost twice as fast) and less than 15 days at 12°C (almost three times as fast).

The effect of phenolic compounds on the growth of spoilage flora remains weak for the concentrations deposited under the current manufacturing conditions, the range of concentration of total phenols generally encountered ranging from 0.2 to 2 mg/100 g of flesh. The salt concentration, which varies from 2% to 4% of the wet weight, has a more marked effect. After three weeks, some models predict that the total flora can be reduced linearly by about 0.7 log per percentage of added salt (Leroi et al. 2000, 2001). On the contrary, the salty taste is strongly marked for contents of 3.5% and more. Moreover, it is commonly accepted that high dietary salt intake increases the risk of cardiovascular problems. A compromise between limiting the development of flora and a too high salt content is therefore to be determined. In general, it is recommended to have a level of at least 0.5 mg/100 g of total phenols (colorimetric assay) and a salt content of at least 2.8% on the finished product, in order to guarantee an acceptable level for at least four weeks after smoking (associated with vacuum packaging or modified atmosphere packaging).

Various research studies have been conducted to determine the role of different phenolic compounds (guaiacol, syringol, eugenol) and other families of compounds on the perception of the smoked flavor (Cardinal et al. 1997, 2006; Sérot et al. 2003; Varlet et al. 2006).

Some works also aim to understand and control the mechanisms of generation and deposition of these compounds in order to propose ways to manage the quantities and ratios of compounds adsorbed by the fish. Thus, the temperature of pyrolysis, the quantity of oxygen within the incomplete combustion, the species of wood used, its granulometry will influence the quantity and the nature of the generated compounds. It is advisable not to generate a ratio between undesirable compounds (PAHs, see below, or too pungent or acidic compounds, for example) and the desired compounds (certain phenols, for example) that are too high. If, in order to respect this ratio, the quantity of smoke can be lower, the smoking time will be lengthened.

The antibacterial action of the different compounds, particularly with regard to the risk associated with the growth of *Listeria*, has also been the subject of some specific work (without really taking into account, however, the possible interactions between compounds).

8.3.2. *Factors influencing sanitary quality*

Initial contamination of the fish by parasites, such as *Anisakis*, for example, is possible. This parasite, associated with food, is frequently observed in wild salmon. Therefore, the legislation imposes for the latter a freezing step to kill these parasites.

Similarly, as already mentioned above, some compounds of the PAH family (Polycyclic Aromatic Hydrocarbons) resulting from the combustion of wood can present a health risk (carcinogenic, genotoxic, teratogenic, endocrine disruptor) when present in large quantities and for regular consumption (Maga 1988; Varlet et al. 2007b). These compounds appear in greater quantities at high temperatures of wood thermodegradation and oxidation. Most industrial smoke generators allow a pyrolysis temperature below 500°C which generates only a small amount of PAHs. The temperature of the smoking chamber influences the proportion of PAHs deposited. A higher amount of PAH is deposited for hot smoking (above 60°C) than for cold smoking. The amount of benzo[a]pyrene, the marker of PAHs, is regulated as well as the sum of four PAHs namely Benzo[a]pyrene + Chrysene + Benz[a]anthracene + Benzo[b]fluoranthene. The corresponding regulation is the regulation (EC) No. 835/2011 which sets, for meat and smoked fish, at 2 µg/kg the threshold of benzo[a]pyrene and at 12 µg/kg the threshold on the sum of the four PAHs. However, values for smoked salmon under properly controlled conditions rarely exceed this threshold value. A number of the PAHs in smoked fish are of "environmental pollution" origin and found directly in the raw material to be smoked and sometimes constitutes a significant part of the final contamination of the product (this is particularly true in the case of a final quantity below or slightly above the regulatory thresholds mentioned before). Exceeding the threshold is therefore essentially associated with poor practice during smoke production and smoking conditions. The French General Directorate of Competition, Consumption and Fraud Control, in a recent survey, indicates that the consideration of the PAH risk and the implementation of self-controls on this criterion is very variable according to the companies: the largest industrial sites and companies with large production volumes are aware of the risk and carry out frequent analyses; small- and medium-sized companies carry out spot checks, at the start of their activity or when they obtain approval, or even with insufficient frequency; at the artisanal stage, the PAH risk in smoked products is often ignored.

Pathogenic flora can also appear (*Listeria monocytogenes*, *Escherichia coli*, *Staphylococcus*, *Salmonella*) during processing. However, the only pathogenic flora really able to develop for salt contents higher than 2% of the total wet weight and slightly positive temperatures is *Listeria*. *Listeria monocytogenes* is therefore the main risk associated with the consumption of smoked salmon which can lead to listeriosis. This bacterium is not responsible for the organoleptic degradation. Therefore, a smoked salmon without the presence of a sour or bitter odor does not mean that there is no risk associated with *Listeria*. Manufacturers therefore attach particular importance to the development of this flora in determining the shelf life of their products. One of the protocols often applied by the industry to validate the shelf life of products during storage in a refrigerated state is based on storage at 4°C for 1/3 of the time, followed by storage at 8°C, then a simulation of a break in the cold chain at 20°C for 2 hours, and finally a return to storage at 8°C for the rest of the time. Predictive models (Jorgensen 2000; Gimenez and Dalgaard 2004) taking into account the presence of lactic acid bacteria confirm the importance of temperature in the development of *Listeria monocytogenes*. At 5°C, lactic acid bacteria compete with *Listeria*, preventing them from exceeding the threshold of 10^3 germs/g. Above 10°C, *Listeria* grows as if they were almost alone in the product and reach levels of 10^8 g/g. The salt content only affects the growth of *Listeria* when the phenol content is at least of the order of 1 mg/100 g, at least in model liquid media (Thurette et al. 1998). High phenol concentrations (around 2 mg/100 g) seem to have a significant inhibitory effect on the growth of *Listeria*, contrary to low concentrations especially those below 1 mg/100.

Listeria contamination is common in smoked salmon but generally at extremely low levels. However, in view of this frequent presence, the risk of the development of these bacteria over time during conservation, even in a refrigerated state, cannot be completely ruled out. The main industrial producers of smoked salmon effectively complement the controls carried out by the authorities by carrying out numerous self-checks on this point. The regulatory health threshold of 100 *Listeria monocytogenes* per gram of product is therefore rarely exceeded. Batch recalls are therefore rare. However, in November 2021, a major French smoked salmon brand experienced a massive recall of batches of smoked salmon due to the risk of *Listeria*. Vigilance is therefore still required around smoked fish. Indeed, in case of human infection, *Listeria* can be responsible for serious complications, especially in fragile populations. After an incubation period ranging from a few days to several weeks (an average of eight days depending on the clinical form), the disease usually results in a more or less high fever, accompanied by headaches and, sometimes, digestive problems (nausea, diarrhea, vomiting, etc.). In these more severe cases, listeriosis can manifest itself either by affecting the central nervous system, causing meningitis or meningoencephalitis, or by septicemia (presence of the bacteria in the blood). Listeriosis mainly affects immunocompromised people (people with cancer or untreated HIV, transplant patients, hemodialysis patients). Pregnant women, newborns and elderly people can also be affected. In pregnant women, it can cause fetal loss or premature delivery but can also be responsible for a neonatal infection (infection at birth of a newborn).

The diagnosis of *Listeria* infection can be made through medical testing. The disease is confirmed after isolation of *Listeria monocytogenes* from blood, placenta, cerebrospinal fluid, etc. The knowledge of virulence factors of *Listeria monocytogenes* (strain dependent) is now a major challenge in the fight against the disease, both to improve treatments and to develop new methods of detection of pathogenic strains.

Finally, human listeriosis has been a reportable disease since 1998. This means that the physician or medical analysis laboratory must inform the health authorities when a case of listeriosis is detected. Each year, approximately 300 cases of listeriosis are recorded in France. These cases are isolated from each other (sporadic), and no epidemic has been identified in France since 2003. In 2006, the number of recorded cases increased, without any identified reason, and then stabilized in 2008. In 2010, there were 4.9 cases of listeriosis per million inhabitants and 5.2 cases per 100,000 births. To date, about 70 epidemics have been reported worldwide, including nine in France (on seven different products). The foods implicated were pork tongue in jelly (in 1992 and 2000), rillettes (1993 and 2000), brie cheese (1995), pont l'évêque cheese (1997), epoisses cheese (1999), tartinettes (2002) and mortadella (2003). In 2011, an epidemic in the United States occurred due to the contamination of cantaloupes (fruits of the melon family). This is, to my knowledge, the last known outbreak of listeriosis. Smoked fish has rarely been incriminated as a cause of listeriosis.

8.3.3. *Other manufacturing steps that may affect preservation*

Many steps exist around the salting, drying and smoking steps, as shown in the attached smoked salmon process flow diagram (for farmed smoked salmon without a freezing/storage/thawing step). Some of these steps can play an important role in the quality and preservation of the smoked product.

Filleting is generally made up of the following phases: heading, i.e. eliminating the head by cutting behind the gills, filleting by letting the blade "slide" along the central edge, and then trimming to eliminate the areas with too much fat. During the whole filleting process, the fish is rinsed abundantly to eliminate most of the bacterial contamination associated with these phases.

1) We distinguish between farmed salmon and wild salmon. The latter, because of the risk of parasites, must have undergone or be undergoing a deep freezing stage.

2) Some smoked salmon may be sold as unsliced fillets with skin.

3) This step is not a systematic step.

4) Slicing can, in some cases depending on the slicing technique, be preceded by a stiffening step, sometimes called crusting. The aim is to quickly obtain a core temperature between −6°C and −10°C before slicing. The slicing is generally followed by a reconstitution phase of the sliced fillets. The notion of weight grading can lead, in some cases, to a redistribution of the slices of different fillets.

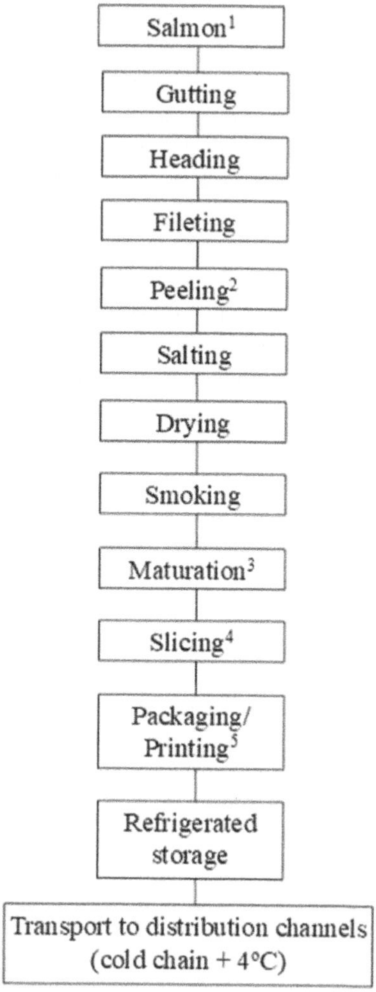

Figure 8.1. *Manufacturing diagram*

It is also possible to proceed to a chilling step, i.e. the salmon is placed at a slight minus temperature between 0 and −2°C. This last temperature corresponds to the freezing point of the product. Chilling does not lead to partial freezing. It allows for a more important storage of smoked salmon. The packaging is generally completed under vacuum or modified atmosphere to increase its conservation. The printing on the package is the last phase included in the manufacturing process. The chilling can be considered as part of the manufacturing process.

A migration phase, often called "maturation", at a slightly positive temperature after smoking and before slicing, can be practiced to allow for the diffusion and progressive homogenization of the compounds between the surface and the heart of the product. Indeed, the slicing slows down the diffusion/migration of the compounds. However, the practices of this migratory phase are very variable, ranging from 0 to 48 hours, and are poorly known. It also acts on the texture of the final product.

Most of the smoked salmon production is marketed in strips or slices, sometimes separated by spacers, vacuum packed to avoid oxidation reactions and limit microbial growth. The brown muscle part of the salmon is more sensitive to spoilage reactions than the rest of the fillet, so it can be interesting to remove this part during slicing. For ease of slicing and packaging, it is often preferable to remove all fish bones. This operation is most often performed after the filleting and trimming phase. It is important to ensure that the vacuum in the package is maintained correctly during storage as the shelf life of the product is estimated in relation to standard packaging conditions. Modified atmosphere packaging is also used, although it is less common than vacuum packaging.

Various additives such as organic acids, nitrites, enzymes, bacteriocins or essential oils, deposited after smoking, have been tested. They give little results in smoked salmon or often modify the flavor of the product too perceptibly. These additives are not authorized in France in the designation "smoked salmon". The reasoned use of strains of lactic acid bacteria isolated from smoked salmon, selected for their inhibiting capacities, neutral with regard to the characteristics of the product, and reimplanted at a high level in the product (Brillet et al. 2005), makes it possible, however, to increase the conservation without modifying the organoleptic characteristics. Biopreservation of seafood in general is described in more detail in another chapter of this book.

Chilling is a process that consists of subjecting smoked salmon to a temperature between 0°C and the start of freezing (around –2°C) for one to several weeks before distribution. This approach can be an important step in controlling the logistics of smoked salmon production and distribution as it affects the shelf life of the product. It is therefore an option within the industrial practices of the sector.

The next section of this chapter focuses on the three central stages of the manufacturing process, namely, salting, drying and smoking, and indicates the different technical solutions associated with them.

8.4. The different technologies of salting – drying – smoking

8.4.1. *Salting*

Figure 8.2. *Manual dry salt salting*

Different salting practices and equipment exist. The simplest and oldest salting method is dry salting. A layer of salt is applied to the surface of the fillet. The fillet itself can be placed on a layer of salt, especially when the fillet has been peeled, in order to increase the amount of salt penetrating per unit of time. The osmotic effect causes the water to migrate to the surface of the product. A brine is then formed at the fish–salt interface. Fick's law describes the diffusive exchanges within the product. Two flows are generally opposed and interact: the penetration of salt and the migration to the surface of water. These flows are dependent on the salt concentration in the product, the temperature conditions, the lipid content, the structural state of the muscle fibers (fresh or thawed raw material), the thickness of the product, the structure of the salt used and its degree of impurity. Under conventional salting conditions, the product loses some water. A surface rinse removes the brine and stops the salt from penetrating. Increasing the salting temperature increases the kinetics, but the risks associated with bacterial growth are also increased, and it is therefore generally accepted that a good compromise is to work at temperatures of +12°C or lower. For a fresh salmon fillet of about 1 kg, i.e. about 2.5 cm thick, with a lipid content of 14%, a period of 5 hours at +12°C for salting with fine salt will result in an average salt content of around 2.8%. Dry salt salting can also be mechanized by using a salt cannon.

The fish can also be salted in a previously prepared artificial brine (salt + water mixture). The fillets are immersed directly into the brine. The osmotic exchange mechanisms are quite similar to the case of salting with dry salt. However, the initial amount of water around the product is greater, which, for the same temperature conditions, has the effect of minimizing the extraction of water from the product. In addition, the brine must remain microbiologically clean so as not to "contaminate" the product. The agitation of the brine or the use of the brine shower, the brine flowing by gravity, increases the speed of penetration of the salt. A brine shower can reduce the treatment time by up to 40% for the characteristics previously mentioned.

It is possible to inject the brine directly into the product, which makes it possible to salt the product while having good processing yields. In a machine equipped with hollow needles, the brine is injected, under light pressure so as not to burst the structure or avoid forming "pockets" of brine, into the products passing through. The initial distribution of the salt in the product is therefore a function of the depth and the location of the needles. The size of the needles does not leave very perceptible marks on the surface of the product. The diffusion of salt continues afterwards and the product has therefore quickly a homogeneous spatial concentration of salt. However, the risk of contamination of the following fish by a contaminated piece must be taken into account. Frequent checks are therefore necessary. A product color (more pink) and a texture specific to this type of treatment are often observed. Needle-free injection has been successfully tested in the meat sector but has not been as successful in the smoked salmon sector. The concept is based on a technique of medical origin: the needle-free injection of liquids by means of micro nozzles fed by ultra high pressure ramps. It seems to be possible to inject brine into the entire muscle tissue without mechanical intrusion. Mixed salting, i.e. the coupling between injection salting and dry salt salting, is also used in order to benefit from certain advantages of injection and to limit certain disadvantages (surface effect and color).

The simultaneous salting – drying by DII (Dehydration – Impregnation by Immersion) or by DID (Dehydration – Impregnation by Douching) is close, in principle, to the salting by brining. The brine is then a ternary solution (mixture of water + salt + sugar). The sugar accelerates the dehydration of the fillets by increasing the osmotic pressure of the brine, which increases the osmotic imbalance between the flesh and the solution, causing an increase in the flow of water from the flesh to the treatment solution. High molecular weight sugar molecules (dextrose 21 equivalent) do not penetrate the flesh. However, the flow of salt to the flesh is slightly slowed down. This technique can also be applied as a dry salt–sugar mixture on the product. The regeneration of a solution, whose viscosity increases with the concentration of sugar, can pose certain technical problems, especially when used on a large scale.

8.4.2. *Drying*

The purpose of drying is to remove some of the water from the product. Other processes are likely to remove water, such as salting (as already mentioned), pressing and freeze-drying. But the term drying normally means desiccation by evaporation. The principle most used is drying by convection in which a current of "hot" air, or at least of a higher temperature than the product, and drier than the product establishes a difference in temperature and partial pressure of vapor between the fish flesh and the surrounding air. This results in heat transfer from the air to the fish and water transfer in the opposite direction, with the air serving as both an evaporating fluid and a transport vector for the removed water. An equilibrium is created between the partial pressure of the vapor in the air and the liquid phase of the fish. Drying was first carried out in the open air, in the sun or in the shade, and this is still practiced in many developing countries. But in this case, the ambient atmosphere cannot exceed a certain amount of water vapor depending on the temperature. The air is said to be saturated, it can no longer absorb water. Consequently, the climatic conditions govern these techniques. In hot and humid countries, biological contamination during this type of drying can be a major problem. In air-conditioned drying, the air circulation is accelerated by an adequate mechanical ventilation and by the regulation of temperature and hygrometry of this air. The atmosphere is never saturated and evaporation continues under optimized conditions. Salmon, like most fish, must be dried without denaturing the surface proteins. This is often referred to crusting when a noticeable change in the surface condition occurs. This crusting is particularly harmful during the following smoking phase because the adsorption of compounds will be strongly reduced. The sensory perception of the texture is also modified and the phenomena of "gaping" of the raw material are amplified. Gaping is the breaking of bonds between myotomes, which creates "holes" on the surface of the product. This denaturation of proteins during crusting is essentially associated with the surface temperature. It starts for temperatures slightly lower than 28°C and becomes very pronounced above 30°C. This value is, however, dependent on the characteristics of the fish and the surrounding hygrometry. It is therefore recommended to work at relative air humidities of about 65% for the generally light drying before smoking, the finished product after smoking should have a water content between 60% and 65%. The air velocity around the product should be as high as possible. An air speed around the fish of about 1 m/s is generally used because beyond that the gain in kinetics becomes low, especially with regard to the additional energy costs involved.

Electrofluidodynamic drying, which involves exploiting ionic wind to facilitate the drying conditions, is mainly aimed at increasing the air flow speed in the immediate vicinity of the surface of the product to be treated by creating an ionic flow. The energy cost of this additional speed is very low. A so-called discharge electrode with a specific geometry allowing the generation of ions by corona effect above an applied voltage threshold leads to the creation of a current. The ions are directed towards the collecting electrode, connected to the ground potential, on which the fish are placed. Secondary reactions, such as ozone formation, are associated with this phenomenon and may also play a role in microbial development. Although promising at the research level, this last technology has not been developed in practice.

Figure 8.3. *Air-conditioned drying-smoking chamber*

8.4.3. *Smoking*

Smoking is, as we have already mentioned, intimately associated with the incomplete and slow combustion of wood. There are many modes of smoke generation which we will briefly describe. All these modes present different conditions of thermodegradation and oxidation and generate different quantities of smoke and variable ratios between compounds. The initial composition of the wood (in France mainly beech and oak) also plays a considerable role. The amount of smoke deposited is a complex function of the concentration, physical characteristics, smoke composition, environmental conditions and the nature of the surface to be treated. Smoke is a mixture of a gaseous phase and a particulate phase. The particle size can vary with the residence time due to the coalescence phenomenon of the particles. The aqueous phase of the fish surface absorbs an important part of the compounds. Some compounds can partially bind to certain elements of the fish flesh. Some compounds react with the surface proteins to give a more or less intense browning. The volatile compounds in the smoke are divided between the gas phase and the particulate phase according to their boiling point. The temperature of the smoke has an influence on the equilibrium between phases and acts, as well as the speed of circulation, on the mechanisms of adsorption or deposition.

Smoking brings aromatic, coloring and preservative compounds to the product. The smoke compounds deposited during smoking provide a specific surface coloration (mainly carbonyls) and a notable modification of the flavor (via phenolic compounds, carbonyls, furans and acids). Some compounds, especially phenols, are also responsible for a slight antioxidant and antibacterial action (Maga 1988). Acids slightly modify the pH and thus can also have an impact on the preservation of the product. Some odorant molecules, such as benzaldehyde, are already naturally present in unsmoked salmon, and smoking does not significantly alter their concentrations (Varlet et al. 2006). Phenolic compounds are mainly derived from thermal degradation (depolymerization/oxidation) of the lignin in the wood used to generate the smoke. Among them, syringol, guaiacol, 4-methylguaiacol and cresols seem to be the main compounds responsible for smoky or even burnt or grilled odors. The other derivatives of guaiacol, thymol, eugenol, isoeugenols bring spicy notes such as vanilla, curry, clove or cinnamon. Note that the intensity of the perception of guaiacol by gas chromatography coupled with olfactometry is similar for the majority of smoked products regardless of their concentration which would seem to indicate that the concentrations commonly encountered for smoked salmon are above the saturation threshold of this odor. Furan compounds also show significant odor characteristics at high concentrations. These compounds are derived from the thermal degradation of cellulose and hemicellulose in wood, and their amount becomes more pronounced with increasing temperatures associated with smoke generation. Furfural does not appear to have a strong odor note. 5-methylfurfural, 2-acetylfuran or furfuryl alcohol show green or vegetal notes at high concentrations. Finally, enolones have also been detected by olfactometry (Varlet et al. 2007a) as being able to play a significant role in the "wood fire" odor. The production of PAHs in smoke depends on the modes of generation and the conditions implemented. Some aspects on these compounds are described in Varlet et al. (2007b).

Direct smoke and natural draft smokehouses are the oldest and simplest. The smoking chamber is directly above the fireplace. The products are generally suspended above the hearth. This is called vertical smoking. The temperature of the firebox depends on the natural draft. A fresh air dilution is generally useful to limit the progressive increase in ambient temperature around the product. Some of the fat migrates to the surface and flows out by gravity. The smoke density is generally low and the smoking times are quite long. However, there are many models with different geometries and multiple conditions of use depending on the "know-how", so the smoking times are very variable. For salmon, they reach at least 6–8 hours and can sometimes reach 24–48 hours. The control of smoking conditions, contamination and microbial development is very delicate with this type of smoker.

Industrial smoking is done under more standardized conditions. The production of smoke is independent of the smoking chamber. Smoking is horizontal and controlled by mechanical ventilation and by temperature and hygrometry regulations. The same air-conditioned chamber is generally used for preliminary drying and smoking. The drying is completed in a closed circuit, while the smoking is done with a continuous renewal of the smoke, except in the case of smoking by atomization, which we will describe in one of the following sections. The smoking of salmon is carried out at a temperature between 16 and 26°C (temperature control by a hot–cold system) and at a humidity of about 65%. Figure 8.4 shows an example of an air-conditioned drying-smoking room. Smoking times vary according to the practices, generally ranging from 2 to 4 hours. The drying phase continues during smoking, even though its intensity is lower.

The most commonly used smoke generator, shown in Figure 8.4, is the self-combustion generator. A heating ring is initially preheated to a temperature that triggers the combustion of the wood. Sawdust or wood shavings, depending on the model, which can be previously humidified, fall continuously or more often cyclically on this ring. At the same time, air is blown into the pyrolysis zone to provide the oxygen necessary for slow and incomplete combustion. The quantity of oxygen and wood, continuous or semi-continuous, makes it possible to maintain combustion without complementary energy contribution. In practice, self-combustion is not always easy to maintain and a complementary energy supply via the heating of the crown facilitates the task and is often available on this type of generator under the name of energy doser. The borderline between a self-combustion smoke generator and a smoke generator with a thermostatically controlled plate, where the temperature of the crown or the plate is regulated, is therefore sometimes tenuous.

Figure 8.4. *Self-combustion smoke generator*

Another type of generator sometimes used is the friction generator. This type of generator is illustrated by a photograph (Figure 8.5). In this case, a log of wood without knots if possible is strongly compressed on the surface of a hard steel rotor rotating at high speed. Under the effect of the friction, the temperature rises and causes pyrolysis. By adjusting the air intake, the pressure on the log, the speed of rotation, the friction and rest times (for cyclic operation), the smoke generation is regulated. Other types of generator exist in a rarer way as the generator of wet smoke and the generator of fluid smoke. In all cases, the generated smoke can be filtered before being transferred to the smoking chamber. There are numerous filtration possibilities, such as the use of baffles, metal chips, Raschig ring filters, water screens, electrofilters, bubbling filters, circulation in ice, cooling of the pipe walls, etc. The objective of filtration is to modify the characteristics of the smoke and, above all, to eliminate as many compounds, which are harmful to health and sensory perception, as possible.

In the techniques presented above, the smoke is always used in its "natural" form, either in its entirety or with certain compounds removed. The technology of smoke flavors consists, after filtration, of condensing the smoke. The smoke extracts are then mixed with stabilizers and possibly blended with other smoke extracts from other wood species. These extracts, called liquid smoke, are used to smoke fish by immersion, douching, spraying or vaporization. The atomization involves reconstituting the smoke in the form of very fine droplets (up to 1 micrometer in diameter, corresponding to the size of droplets for an auto-combustion smoke generation) and introducing it into an enclosed space where the fish are placed. Cyclic drying of the products is recommended in this case. The vaporization of the liquid smoke consists of pouring the extract on a very hot surface which restores almost instantaneously to the smoke its state "gas phase" which is immediately used in the smokehouse as in the traditional process. The use of liquid smoke seems to be rare in France for seafood products, but is a frequent practice in other European countries. The use of liquid smoke is under regulatory consideration.

Figure 8.5. *Frictional smoke generator*

A smoking process by electrostatic precipitation exists. As with electrofluidodynamic drying, the corona effect is used. Some of the charges of the generated ions are captured by the smoke particles flowing between the electrodes. The charged particles then experience an additional attractive force to the flow towards the collecting electrode (the Coulomb force). With the fish arranged or flowing on the collecting electrode (connected to ground), the particles are more easily deposited on the surface. The voltage on the discharge electrode can be positive or negative. The compounds deposited and adsorbed by this technology are not totally identical to those deposited and adsorbed by a traditional smoking process since the deposition rate of the particulate phase is increased, but the impregnation of the compounds of the gas phase is reduced due to a shorter treatment time. In this case, the smoking time can be reduced to a few minutes. This process does not allow a real drying of the product (short treatment time) and an additional drying before or after smoking is therefore desirable. This technology also seems to be applicable to the precipitation of atomized liquid smoke. This technology is not used a priori in the case of salmon smoking in France.

Whatever the technology used, industrial-scale production requires the sizing of installations to enable good control of heat and material transfers. The control of the aeraulics in a smoking chamber to allow a homogeneous heat exchange and circulation speeds is a tricky problem. It is common to observe average velocity ratios at different fixed points around the fish in a smoking chamber of up to 3 or 4. The minimization of these ratios and the homogenization for the treatment of the batches, in particular evaluated through the tools of simulations of flows and transfers (Computer Fluid Dynamics) led to geometrical adaptations and the use of modulated flows.

8.5. Conclusion

The smoking of seafood is an old process but is gradually evolving towards a more efficient technological environment contributing to increase the regularity of production and sanitary control, and to limit the energy and environmental impact while seeking to preserve the taste characteristics. The knowledge and control of the manufacturing conditions of smoked products make it easier to respond to consumer demand which, as shown by the European Eurosalmon program, is diversified in terms of characteristics (Cardinal et al. 2004). Innovation in the industry is still possible, as we have pointed out in this chapter, although in the agri-food sector in general it is still delicate and generally time-consuming because it is necessary to take into account many factors associated with the processes, the products and the process–product interactions.

8.6. Acknowledgements

The author would like to thank his colleagues or former colleagues, Mrs Cardinal, Mrs Chevalier-Gigout, Mrs Cornet, Mrs Leroi and Mrs Messrs. The author also thank Mr Joffraud, Mr Knockaert and Mr Vallet, who participated in the acquisition of the knowledge on smoking. The author would also like to thank his various research partners.

8.7. References

Brillet, A., Pilet, M.F., Prevost, H., Leroi, F. (2005). Effect of inoculation of *Carnobacterium divergens* V41, a biopreservative strain against *Listeria monocytogenes* risk, on the microbiological, chemical and sensory quality of cold-smoked salmon. *Int. J. Food Microbiol.*, 104(3), 309–324.

Cardinal, M., Berdague, J.L., Dinel, V., Knockaert, C., Vallet, J.L. (1997). Effet de différents techniques de fumage sur la nature des composés volatils et les caractéristiques sensorielles de la chair de saumon. *Sciences des aliments*, 17, 679–696.

Cardinal, M., Gunnlaugsdottir, H., Bjoernevik, M., Ouisse, A., Vallet, J.L., Leroi, F. (2004). Sensory characteristics of cold-smoked Atlantic salmon (*Salmo salar*) from European market and relationships with chemical, physical and microbiological measurements. *Food Res. Int.*, 37, 181–193.

Cardinal, M., Cornet, J., Sérot, T., Baron, R. (2006). Effects of the smoking process on odour characteristics of smoked herring (*Clupea harengus*) and relationships with phenolic compound content. *Food Chem.*, 96(1), 137–146.

Dalgaard, P., Murillo, E., Jørgensen, L.V. (2004). Modelling the effect of temperature on shelf-life and on the interaction between the spoilage microflora and *Listeria monocytogenes* in cold-smoked salmon. In *Seafood Quality and Safety. Advances in the New Millennium*, Shahidi, F. and Simpson, B.K. (eds). Science Tech. Publishing Company, St. John's, Newfoundland.

Gimenez, B. and Dalgaard, P. (2004). Modelling and predicting the simultaneous growth of *Listeria monocytogenes* and spoilage micro-organisms in cold-smoked salmon. *J. Appl. Microbiol.*, 96, 96–109.

Jorgensen, L.V. (2000). Spoilage and safety of cold-smoked salmon. PhD Thesis, Royal Veterinary and Agricultural University, Copenhagen.

Knockaert, C. (1990). *Le Fumage du poisson*. Édition Ifremer, Brest.

Leroi, F. (2002). La microbiologie du saumon fumé à froid : aspects hygiéniques et qualité. *Revue générale du froid*, 1028, 35–40.

Leroi, F., Joffraud, J.J., Chevalier, F. (2000). Effect of salt and smoke on the micro-biological quality of cold-smoked salmon during storage at 5°C as estimated by the factorial design method. *J. Food Prot.*, 63(4), 502–508.

Leroi, F., Joffraud, J.J., Chevalier, F., Cardinal, M. (2001). Research of quality indices for cold-smoked salmon using a stepwise multiple regression of microbiological counts and physico-chemical parameters. *J. Appl. Microbiol.*, 90, 578–587.

Maga, J. (1988). *Smoke in Food Processing*. CRC Press, Boca Raton.

Sérot, T., Baron, R., Knockaert, C., Vallet, J.L. (2003). Effect of smoking processes on the contents of ten major phenolic compounds in smoked fillets of herrings (*Clupea harengus*). *Food Chem.*, 85(1), 111–120.

Stohr, V., Joffraud, J.J., Cardinal, M., Leroi, F. (2001). Spoilage potential and sensory profile associated with bacteria isolated from cold-smoked salmon. *Food Res. Int.*, 34, 797–806.

Thurette, J., Membré, J.M., Han-Ching, L., Taillez, R., Catteau, M. (1998). Behavior of *Listeria* spp. in smoked fish products affected by liquid smoke, NaCl concentration and temperature. *J. Food Prot.*, 61(11), 1475–1479.

Varlet, V., Knockaert, C., Prost, C., Sérot, T. (2006). Comparison of odor-active volatile compounds of fresh and smoked salmon. *J. Agricult. Food Chem.*, 54(9), 3391–3401.

Varlet, V., Sérot, T., Cardinal, M., Knockaert, C., Prost, C. (2007a). Olfactometric determination of the most potent odor-active compounds in salmon muscle (*Salmo salar*) smoked by using four smoke generation techniques. *J. Agricult. Food Chem.*, 55(11), 4518–4525.

Varlet, V., Sérot, T., Monteau, F., Le Bizec, B., Prost, C. (2007b). Determination of PAH profiles by GC-MSMS in salmon processed by four cold-smoking techniques. *Food Addit. Cont.*, 24(7), 744–757.

9

Surimi and Derived Products

María Pilar MONTERO GARCÍA and Antonio Javier BORDERÍAS
Institute of Food Science, Technology and Nutrition (ICTAN), Spanish National Research Council (CSIC), Madrid, Spain

9.1. Introduction

Surimi is a myofibrillar fish protein concentrate processed in a series of steps in order to produce a tasteless material that retains the gel-forming ability of its proteins. It is considered an intermediate product that is further processed to make surimi-based products.

Frozen processing of surimi made from Alaska pollock (*Theragra chalcogramma*) was developed in Japan in the 1960s, and some years afterwards it became a very important industry in Western countries that introduced new surimi-based products. Surimi quality is affected by various factors, such as fish species, fish season and fish freshness, this last factor being of great importance, regardless of the species used.

9.2. Surimi sources

In recent years, there has been an increasing demand throughout the world, especially in Asian countries, for other fish products, such as fillets, headed and gutted, whole round, etc., which has diverted fish supply from surimi. Hence, surimi producers have shown an interest in using new species and lower-grade surimi in

Current Challenges for the Aquatic Products Processing Industry, coordinated by Véronique VERREZ-BAGNIS.

order to make different types of surimi-based products. Also, producers are implementing new processing techniques to increase yield in order to maintain profitability. On the other hand, consumption of surimi products has changed in recent years. First, Japanese consumers, especially young people, have changed their eating habits to Western diets and often consume various kinds of meat. Western countries have also started to consume different types of products, mainly analog products based on surimi.

The use of new technologies makes it possible to process new species as raw material for the production of surimi. For example, new washing and decanting techniques allow processing of fatty species such as mackerel. The isoelectric precipitation method has made it possible to process a protein concentrate similar to surimi, obtained from squid. Research on enzymes and enzyme inhibition in fish muscle has made it possible to use Pacific whiting captured on the west coast of the United States and Canada.

The use of Alaska pollock (*Theragra chalcogramma*) as the principal raw material for surimi processing has changed, and nowadays it represents less than 25% of world surimi production. Southeast Asia species now represent 60% of the raw material for global surimi processing. The species involved are threadfin bream (*Nemipteridae* spp.), lizardfish (*Synodontidae* spp.), bigeye snapper (*Priacanthidae* spp.), ribbon fish or hairtail (*Trichiurus lepturus*), croaker (*Sciaenidae* spp.), etc.

Since 2013, global surimi production has been stable at around 820,000 metric tons (MT) per year, and supply and demand are balanced (Guenneugues 2019).

9.2.1. *Alaska pollock*

Alaska pollock is a cold water whitefish caught in the Bering Sea ecosystem. It is captured with large pelagic trawl nets. The use of effective cryoprotectants that stabilize protein has made it possible for Japanese and US vessels to fish for long periods and store frozen surimi, enabling surimi product processors to receive raw material throughout the year. This species is also processed to produce fish fillet blocks suitable for making prepared food such as fish fingers and other breaded products. Its roe is also used for commercial production.

Before 1960, Alaska pollock was harvested in very small quantities. Catches increased during the late 1960s and reached a peak of 1.9 million MT in 1972. Now the quota of Alaska Pollock is high (1.55 million MT), but pollock surimi production in the United States may have decreased to 200,000 MT in 2019 owing to a high demand for fillets in frozen blocks (Guenneugues 2019).

9.2.2. *Pacific whiting*

Pacific whiting (*Merluccius productus*) is another cold-water whitefish much used for surimi production. It is a gadoid fish and is related to pollock. It is caught primarily for frozen fillet blocks (Alheit and Pitcher 1995). Like other gadoids, it has white flesh and a delicate flavor. In the case of this species harvested on the Pacific coast of the United States and Canada, the stock has a problem of soft flesh due to high levels of muscle proteases, which are associated with the endemic myxosporidian parasite. This parasite is microscopic and does not induce any human illness. To avoid the consequences of the presence of proteases in surimi gelation, even after washing, various food-grade protease inhibitors have been investigated. The best inhibitor was found to be the beef plasma protein (cathepsin inhibitors), but because of the associated problem of mad cow disease, its use has been banned by most regulatory organizations. For this reason, other protease inhibitors, such as egg white or whey protein, are used as ingredients in surimi production. Other inhibitors, such as a soy cystatin inhibitor, have shown activity (Kang and Lanier 1999), but most producers use egg white, which is more effective for serine proteases.

The resource is healthy (Total Allowable Catch: 70% for the United States and 30% for Canada). Pacific whiting surimi production has increased in recent years from 20,000 MT in 2016 to 40,000 MT in 2019 (Guenneugues 2019). At present, the main limitations are the processing capacity of shore plants and ships involved in the fishery, as well as market demand.

9.2.3. *Southern blue whiting and hoki*

Southern blue whiting (SBW) (*Micromesistius australis*) and hoki (*Macruronus novaezelandiae* and *Macruronus magellanicus*) are found in sub-Antarctic waters. Currently, they constitute large fisheries in New Zealand, Argentina and Chile, and the surimi from these species is of excellent quality (Okada 1978). In the 1980s, hoki surimi represented 3% of the world surimi supply, but afterwards there was a shift away from surimi and most of the hoki caught is used for frozen fillet blocks. SBW is frequently contaminated with parasites that make it useless for fillet production, so it is diverted to surimi production; this surimi has great strength and is very deformable, and it provides the basis for top products in Japan.

There were no exports from Chile in 2017, and in 2019, there was only one vessel in operation. The production of hoki and SBW in 2019 was about 6,000 MT (Guenneugues 2019).

9.2.4. *Northern blue whiting*

This species has traditionally been used as raw material for fishmeal and to some extent for H&G fish. After years of over-fishing, the resource is depleted. This species gives very high-quality surimi, as in the case of SBW, if it is processed very quickly. In 2003, shore factories were built in the Faroes to make surimi from fresh fish and in Russia to make surimi from frozen fish, although the quality of the latter is very poor. In 2018, there was only one French vessel, which had a production of 2,000 MT owing to mechanical problems. In 2019, production increased to about 4,000 MT (Guenneugues 2019). In the Faroes, after a big fire in 2017, Vardin built a new factory that included a surimi line in 2020.

9.2.5. *Tropical fish*

In Southeast Asia, the main species used for surimi processing are threadfin bream (*Nemipterus* spp.), bigeye snapper (*Priacanthus* spp.), croaker (*Pennahia* and *Johnius* spp.), lizardfish (*Saurida* spp.), goatfish/red mullet (*Upeneus* spp. and *Parupeneus* spp.), sea bream (*Evynnis* spp.) and ribbon fish (*Trichiurus* spp.).

Tropical fish surimi has been the main variable of adjustment for the global surimi market. When prices of surimi made from non-tropical species rise on the international market, processing plants in Southeast Asia increase in demand and consequently prices of tropical fish increase and fishing is more intensive until production exceeds the demand. Then, surimi prices decrease and fishing also decreases.

The processing temperature of surimi from tropical species is somewhat higher because it facilitates processing, giving high-quality surimi. An example of this is threadfin bream surimi, production of which has increased dramatically during the last 20 years (Yongsawatdigul et al. 2002). This species forms the greater part of the trawl catch landed in Thailand, Vietnam, the Philippines and Malaysia. The size of threadfin bream used for surimi is small (30 g per fish) and consequently the fish are manually headed and gutted, giving employment to a large number of people.

Lizardfish is considered a high-grade raw material for surimi, but its gel-forming ability decreases quickly over time. Bigeye snapper has a long shelf life in ice. A large bigeye snapper is combined with threadfin bream to improve gel strength. The smaller size is frequently used for low-grade surimi.

Other tropical species used for surimi production include croaker (*Sciaenidae* spp.), goatfish (*Upeneus* spp.), red mullet (*Parupeneus* spp.) and hairtail (*Trichiurus lepturus*).

Countries that process surimi from tropical fish are India, Thailand, Vietnam, China, Indonesia, Pakistan, Malaysia and Myanmar. India had a production peak in 2017–2018 (100,000 MT), but production decreased in 2019 because of poor fish landings and strikes. Production in Thailand reached a maximum in 2005 (145,000 MT), but, as a result of pressure from the European Union and the United States for various reasons, it decreased to 60,000 MT in 2019. Vietnam had a production peak in 2016, with a record of about 175,000 MT, but it decreased slightly to 170,000 MT in 2019.

After 20 years of continuous growth, surimi production in China decreased from around 240,000 MT in 2012 to 150,000 MT in 2018. Hairtail fish surimi processing has dropped by 40% since 2017. After growing in response to demand from Chinese surimi product processors, silver carp production reached a peak in 2016, although afterwards it declined slightly.

The other countries mentioned above have a smaller quantity of surimi production. Other countries, such as the Philippines, Sri Lanka and Bangladesh, also have resources and will probably enter the surimi business in the near future. With regard to the future of the surimi industry in Southeast Asia, Guenneugues (2019) suggested three possible developments: new countries will enter the surimi business; new resources will be used, especially pelagic fish; and diversification will bring increased use of aquaculture resources. In connection with this last possibility, China has demonstrated the economic possibility of processing surimi from silver carp.

9.3. Manufacture of surimi

As mentioned earlier, surimi is a concentrate of myofibrillar protein, purified to a greater or lesser extent by various stages of processing that, in their traditional form, include heading, gutting, deboning-mincing, washing, refining and dewatering. Once prepared, surimi can be used directly, but it is usual to stabilize it with cryoprotectants and antioxidants (the latter in muscle of fatty species) and market it in a frozen state.

For the manufacture of surimi the protein must be functional, so the muscle must have very evident freshness parameters, since otherwise deterioration would lead to a notable denaturation of protein, which would cause a loss of functionality (gel-forming ability, water retention, etc.).

The most suitable raw material for surimi manufacture consists of species with a large amount of white muscle and low fat content. This does not mean that they are the only possibility, and it may also be possible to obtain surimi from fatty species or species with a large amount of red muscle.

9.3.1. *Stages of surimi manufacture from lean species*

The first stage consists of **heading and gutting** the fish. After this step, the fillet should be washed with plenty of water in order to remove any material still adhering to it. If this step is carried out in factory trawlers, refrigerated sea water is normally used, whereas if it is performed on shore, fresh water is employed. The use of pressurized air (4–5.4 atm) to remove adhering waste can reduce freshwater needs and maximize yield (Toyoda et al. 1992).

Next, **deboning-mincing** takes place, usually performed by machinery with a belt that presses the fish through holes in a drum, allowing the meat to be separated on one side, while the bones, fins and skin remain on the other side, providing a residue of biological material with great potential as a by-product or co-product. The holes in the drum should not be excessively large (Ø ≤ 4 mm), in order to prevent impurities, such as pieces of cartilage, scales, skin debris, etc., from passing through with the meat.

The **washing stage, consisting of several washing cycles,** is perhaps the most decisive and critical in surimi manufacture, entailing as it does the removal of most of the muscle components except for myofibrillar protein. Prominent among the components eliminated are sarcoplasmic proteins, fat, blood and myoglobin. The elimination of these compounds produces considerable decoloring and deodorization of the flesh, while improving gelling capacity. The washing process consists of mixing the ground flesh in continuous stirring with cold water (≤5°C), preferably water with low mineral content (soft water) (Park and Morrissey 2000). Rotary screens are used to remove the free water after each washing cycle.

This process can be repeated several times. The most common and traditional way is to repeat it two or three times, so that about 10–12% solids are removed. The number of washing cycles and the water/fish mince ratio varies depending on a number of factors such as the species, the state of freshness and the washing equipment. The usual ratio reported for most commonly used commercial processes ranges from 5:1 to 10:1 water/fish mince (Park and Morrissey 2000). In the final washing stage (last cycle), salt (NaCl and/or $CaCl_2$) is often added in the range 0.1 to 0.3% to facilitate subsequent draining.

Involving several washes as it does, this stage requires large amounts of water, a considerable volume of which must be purified later on. It also produces a large quantity of solid waste (organic matter), mainly proteins, that can be recovered and used for other purposes, especially if it is to be an environmentally sustainable process with full recycling of waste. In addition, the volume of water per wash and the number of washes must be adjusted to the requirements of the particular species and the desired quality of the surimi. Other strategies, such as the use of air in washing, have been tried (Chen 2002), and washing with oxidizing agents (such as hydrogen peroxide and sodium hypochlorite), have been shown to improve texture in purple-spotted bigeye (*Priacanthus tayenus*) by disulfide bond formation, since they oxidize sulfhydryl groups (Phatcharat et al. 2006). However, none of these processes have been routinely implemented at an industrial level. Then again, great commercial success has been achieved in implementing static mixing devices at certain intervals in the pipeline (Toyoda et al. 1992), a method requiring less water and washing steps.

Before final draining, impurities must be removed by a **refiner**, which usually has a screen size between 1.7 and 1.5 mm. In this way, any remains of connective tissue, small particles of scales and bones, etc., are removed (Park and Lin 2005).

Subsequently, **draining** is carried out by means of a screw press or an industrial decanter centrifuge, to remove all excess water in the muscle. Special care should be taken to limit the salt added in the last washing cycle, as this may solubilize the myofibrillar protein, while any salts remaining in the surimi may facilitate denaturation during storage. At this stage, the water content of the minced fish should be similar to the original content of the fillet; this can vary according to the species, with in a variable range (around 81–85%) of humidity (Park and Morrissey 2000; Park and Lin 2005).

9.3.2. *Stages of surimi manufacture from fatty species*

In pursuit of sustainability and to avoid overfishing, there are a number of other fish species which can be used for surimi manufacture, such as pelagic fatty species. However, in this case, there are a number of characteristics that may influence surimi processing and functionality, mainly higher red muscle content (sarcoplasmic protein, hemopigments, etc.) and lipids. The presence of lipids can promote greater oxidation but at the same time provide significant nutritional benefits.

Although the manufacturing process of surimi from these species is basically similar to lean species, it differs in some aspects, briefly as follows. Firstly, the first washing cycle is usually performed in weak alkaline conditions (diluted saline solutions of sodium bicarbonate 0.2% and sodium chloride 0.15%) with occasional stirring. The meat is decanted, and the wash process is repeated once or twice more in highly diluted sodium chloride solution. This stage, entailing several washing cycles, is fundamental for surimi from fatty species, as it serves to eliminate large amounts of sarcoplasmic proteins and hematic pigments, and so the flesh has a weaker color and flavor (Gómez-Guillén et al. 1997; Hultin and Kellerher 2000a). It is important to note that the pH tends to decay faster post-mortem and the meat exhibits buffering capacity, so that the most suitable wash water conditions are close to neutrality (around 6.6–7.0) (Gómez-Guillén et al. 1997).

In addition, excess lipids can be readily skimmed and recovered for potential use in other applications (Shimizu et al. 1992). Industrially, fish mince can be washed under vacuum conditions, where the beating facilitates the elimination of unwanted compounds and particles and increases skimming output (Mendes et al. 1997).

Somjid et al. (2017) studied the differences produced by adding bicarbonate in the first washing cycle or in successive washes for the production of mackerel (*Auxis thazard*) surimi. They observed that the presence of bicarbonate is more suitable only in the first and second washes, since it is in these first two cycles that most elimination of lipids and of hematic proteins occurs, while muscle drip increases with further washing in water with bicarbonate.

9.3.3. *Surimi stabilization, use of additives for preservation*

9.3.3.1. *Cryoprotectants addition*

Surimi is essentially a myofibrillar protein, and therefore freezing produces denaturation and a loss of functionality. To ensure maximum techno-functional potential, cryoprotectants need to be added to prevent or minimize the effect caused by freezing and frozen storage. The most widely used commercial cryoprotectants are sucrose and sorbitol in proportions of up to 9%. In addition, polyphosphates (sodium tripolyphosphate (SPP) and sodium tripyrophosphate (STP), around 0.3%) are frequently used for their synergistic action with cryoprotectants (Reynolds et al. 2002). In recent years, polyphosphates have been studied for their dual role: in addition to their protective effect during freezing, they favor gelling by increasing elasticity when phosphates are added in a ratio of 2:3 (SPP:STP) or improving hardness, chewiness, etc. in ratios of 1:2 (SPP:STP) (Huang et al. 2019).

The substitution of these sugars for others such as trehalose which seems to exert a better stabilizing effect by inhibiting biological damage (Wu et al. 2014), has been found to be effective (at concentrations between 2.5–10 g/100 g wet material), while improving post-frozen storage (–18°C) surimi gelling capacity, as it favors a higher proportion of non-frozen water and hence greater protection against protein denaturation (Wu 2016).

In addition, with the growing concern to reduce the sugar content in foods, sugars as cryoprotectants are being replaced by other compounds that are more health-beneficial or have no negative implications, for example, for diabetics. One such case is protein hydrolysates, obtained from various sources (Wu et al. 2015; Lin et al. 2019), since their peptides favor delayed ice crystal formation and protein stabilization in frozen surimi (Zhou et al. 2019). Certain hydrocolloids such as konjac have also been tested as sugar substitutes (Xiong et al. 2009; Liu et al. 2019). This raises the possibility of hydrocolloids performing a dual function, as cryoprotectants during surimi storage and as coadjuvants of gelation, as reported further below.

Once these ingredients have been added, they must be mixed quickly to prevent a rise too rapid in the temperature of the fish (<10°C), since a too rapid temperature rise may partially damage the protein.

Blocks are then formed, usually of approximately 10 kg, put in plastic bags and placed on trays to freeze. During freezing, they should reach at least –25°C at the thermal center; this is usually done by ultra-freezing so as to achieve a high-quality surimi. They are then passed through a metal detector for inspection and kept in frozen storage, commonly in cardboard boxes with 10 kg blocks, which is how they are habitually marketed.

9.3.3.2. *Protease inhibitor addition*

Some commercial surimi are obtained from species such as Pacific whiting, the meat of which contains more proteolytic enzymes. In fact, in some species such as this, the abundance of proteolytic enzymes comes from infestation by parasites, in this case myxosporidia. Despite washing cycles, some proteases remain in the surimi. Protease inhibitors therefore must be added to this type of surimi, the most common being the beef plasma protein, egg white, whey protein concentrate or potato extract. In recent years, the use of beef plasma protein has been ruled out due to concerns about bovine spongiform encephalopathy. Some studies have been carried out with salmon plasma protein as a substitute. Fowler and Park (2015) found that salmon plasma protein very effectively inhibited endogenous proteases in Pacific whiting surimi at a concentration of 0.5%. This is important in that plasma proteins are more effective than other ingredients as inhibitors of L-captepsins, the main proteases that remain associated with myofibrillar proteins after washing in Pacific whiting surimi.

In this case, ingredients can be added simultaneously to cryoprotectants during surimi manufacture so as to promote inhibition from the outset, but this is not strictly necessary and such ingredients may be added later as part of the formulation in the gelation process.

9.3.3.3. *Addition of others stabilizers*

In the case of fatty species, in addition to cryoprotectants, it is important to add antioxidants to prevent oxidation of the remaining lipids, the most commonly used being tocopherol. Some authors have studied the incorporation of antioxidants not only to surimi as a final product but also during the washing process, in order to palliate the oxidative phenomena that occur in these stages due to the addition of salt. It was noted that during the production of horse mackerel surimi, the presence of caffeic acid or propyl gallate was effective in preventing oxidation (Eymard et al. 2010). Other authors have incorporated protein hydrolysates to act as both cryoprotectants and antioxidants (Lin et al. 2019).

9.4. pH shift process

9.4.1. *pH shift process in fish*

The pH shift process has been shown to overcome problems of surimi production from mainly some fatty fish species. Traditionally, the use of these resources has been limited due to their fat content; also, it is often complicated to obtain pure muscle tissue owing to associated tissues such as peritoneum and kidney tissue along the backbone. To overcome these problems, Hulting and Kelleher (1999, 2000b) patented an alternative method based on muscle solubilization and isoelectric precipitation. In this method, the fish muscle is broken down and homogenized with enough water solution at a suitable pH (acid or alkaline) to solubilize most of the protein. For that purpose, the muscle must be reduced to a very fine particle size, since complete solubilization cannot be achieved by simple grinding. Solubilization is performed at pH 2.5–3.5 (acid process) or at pH 10–11 (alkaline process) depending on the species. The volume ratio between the soluble and insoluble fractions should be greater than 7:1; in this way, the viscosity is not high and the separation of membranes and proteins is more efficient. It is important that the protein is not denatured at too low a pH (less than pH 2). It is recommended to lower the pH in two steps, first to 5–5.5 and then to around 3, to facilitate separation of the insoluble and dissolved protein (Hulting and Kelleher 1999).

The alkaline process can have advantages over the acid process depending on the species. It has been demonstrated that alkaline solubilization produces a whiter protein isolate compared to acid because in the latter process, more heme protein is retained (Kristinsson and Liang 2006; Marmon et al. 2012). On the other hand, lipid oxidation is generally less in the alkaline than in the acid process.

Protein denaturation is not a function of pH alone as well as of temperature and time. It is therefore necessary to keep the temperature at under 10°C and to proceed as quickly as possible. The low-pH protein solution can be treated by filtration or by centrifugation to separate impurities such as lipids, oils, bone, skin and membranes. If centrifugation is used to separate insoluble compounds, there are four potential and clearly differentiated phases post-centrifugation: the top phase contains neutral lipids; the second and liquid phase consists mostly of proteins in soluble state; a third pellet phase containing impurities such as small bones, skin and cell and lipid membranes; plus, a small fourth phase between the soluble protein and pellet phases in which the gel contains a small quantity of entrapped proteins that can be later recovered.

Once collected, the soluble protein phase is adjusted to the pH at which most of the proteins precipitate (isoelectric point); this will depend on the fish species but is generally about 5.3–5.5. The protein is then recovered, mainly by centrifugation (although filtration may also be possible), and this constitutes the precipitate. In the supernatant, other soluble muscle components such as nucleotides, amino acids and small peptides are discarded. This recovered protein isolate is composed of myofibrillar and sarcoplasmic proteins and is mostly washed out when surimi is processed in the normal way.

The moisture content of the protein isolate depends on pH and most importantly on the centrifugation method used. At isoelectric pH, water holding capacity is very poor, so that if possible, part of the water should be removed by applying pressure. Before mixing with cryoprotectants, the protein isolate should be adjusted from isoelectric pH to neutral pH. The type of base used for neutralization can also affect the surimi quality; KOH is thus preferable to NaOH since the K ion favors better myosin unfolding (Raghavan and Kristinsson 2008). This surimi will have good functional properties even although myosin rods tend to disassociate at low pH (acidic process), since these will be mostly recovered upon pH neutralization to pH 7–7.5.

The main advantage of this method is economy, in that it offers a high yield of muscle protein recovery and functional properties are preserved (Kristinsson and Liang 2006). Nowadays, yields of more than 85% can be obtained from fillets, well in excess of traditional processing, which achieve about 60%. It was previously supposed that sarcoplasmic proteins should be removed from the surimi to achieve good gelation, but in the pH shift method, there is no indication that sarcoplasmic proteins interfere with gelation, and under certain circumstances, the inclusion of these proteins could be advantageous (Kristinsson et al. 2014). Moreover, the process is faster than traditional surimi processing. Also, the lipid content of the protein obtained is very low, which should make for greater stability. In fact, some rancid odors and substances such as TBARS are effectively removed during pH shift processing.

The pH shift process offers some collateral advantages. When applying this method to Baltic herring, Marmon et al. (2009) found that not only was fat reduced in relation to the traditional method, but there was also a reduction of 70–80% in dioxin and polychlorinated biphenyls (PCBs). These contaminants were discarded together with the fat layer removed in the centrifugation process. Contamination by mercury can also be eliminated if it is in a lipid-soluble form such as methyl mercury. In the case of catfish as raw material, this surimi process has been found to reduce bacterial contamination by one order of magnitude (Kristinsson and Demir 2003). And also, the cost of pollution control is lower with the pH shift process since the processing water is poor in protein and salt, thus reducing the biological oxygen demand.

9.4.2. *pH shift process in cephalopods*

Cephalopod muscle, especially from some species like jumbo squid (*Dosidicus gigas*), has considerable potential for processing to make a surimi-like protein concentrate (the definition of surimi refers to fish protein concentrate, not cephalopod), given that the muscle is white, contains little fat and has a very bland flavor. Its muscle contains paramyosin, a non-fish protein. Particularly important for purposes of processing is that the muscle proteins solubilize at low ionic strength (Tsuchiya et al. 1978), which means that they may be leached out in the washing step during traditional surimi processing. To overcome this problem, a new procedure based on the pH shift process for fish has been devised to produce a functional concentrate from cephalopod muscle (Careche et al. 2004, 2005). The method entails (1) solubilization of the muscle in a mix of NaCl and $NaHCO_3$; (2) further filtration to eliminate remains of connective tissue, bringing the pH of the solution to 4.7–4.9 for precipitation (isoelectric point) to collect the precipitated protein by centrifugation and (3) further addition of cryoprotectants and freezing. For subsequent thermal gelation, in the final stage, the product must be neutralized with $Ca(OH)_2$. As in traditional surimi, a concentration of 1–3% NaCl is added for gelation (Sánchez-Alonso et al. 2007). Other very similar processes to the ones mentioned above have been published subsequently using an acid solution for muscle solubilization.

One problem for the production of "surimi" from jumbo squid is that the location of the resource often shifts from coastal to more distant waters due to the *El Niño* atmospheric phenomenon. Also, this resource migrates from south to north, disappearing from the north of Peru some years and from the north of Chile during others. The production of jumbo squid "surimi" reached 5,000–6,000 MT in Peru and Mexico in 2008 but decreased afterwards due to migration.

9.5. Mechanism of gel formation

As mentioned above, surimi is an intermediate product from which different products are created, most of them based on surimi gelation. In this gel, myosin plays a fundamental role, and its most outstanding characteristic is that it is an irreversible gel that traps water in its interior, constituting a hydrogel. This is usually a thermal process, and the structure of the gel is determined by the duration and rate of heating among other factors.

The first step is to achieve protein solubilization (sol step): surimi is partially thawed, chopped, ground and a viscous paste or sol is obtained with added salt by denaturing and dissociating the myofibrillar proteins, whereby the reactive surfaces of the molecules are exposed.

The proportion of salt can vary. Recent studies have thus tended to add the lowest salt (NaCl) dose necessary to dissolve the myofibrillar protein and facilitate polymerization, always bearing in mind that the final product must have a healthy salt content. Salt addition is normally around 1.5–2%. Various strategies have been considered to reduce the presence of sodium in formulations, such as the reduction of sodium chloride, or its partial replacement by potassium chloride (Lian et al. 2002; Nuñez-Flores et al. 2018).

Increasing chopping temperature enhances protein denaturation, up to a point (above the species' habitat temperature) at which the formation of disulfide bridges in the protein is favored by oxidation of SH groups (Poowakanjana and Park 2013). For the manufacture of some structures, it is especially important to eliminate any air inside the paste, for which a vacuum silent cutter is recommended (Poowakanjana and Park 2013).

Once the molecules are dissociated, interacting with the water and salt in the medium, and completely hydrated, the paste can be left to rest for a certain time at low temperatures (5–30°C), during which there is a reordering of the molecules to form an orderly, translucent and weak three-dimensional structure, which is called by the Japanese term "*suwari* gel". This setting can also occur at moderate temperatures of around 30–40°C, which differ from one species to another. Gels produced under controlled setting prior to definitive heating (discussed later) exhibit greater gel strength than those made without this setting step.

In a *suwari* gel, several myofibrillar protein bonds are formed simultaneously, producing the network that structures the gel. One of these interactions are hydrogen bonds, important as water binders for stabilization. Hydrogen bonds are broken by heat; according to Lanier et al. (2013), this causes the protein structure to break down between carbonyl and amide groups of each amino acid, favoring hydration of the structure. This water will then be retained in the final gel formation.

Ionic bonds are also formed, i.e. attraction between different sites charges in the protein. These saline or ionic bridges favor myofibrillar protein aggregation and are abundant in the myosin tail region (between glutamic acid or aspartic acid and the amino group of lysine or arginine) (Lanier et al. 2013). This aggregation promotes salt and water interactions, favoring some protein hydrophobic interactions.

In surimi still containing transglutaminase (protein-glutamine g-glutamyl-transferase) (TGase), which is found in the soluble fraction, this enzyme is significant in that it catalyzes the formation of ε-(ɣ-glutamyl)-lysine, a covalent interaction of bonding between glutamic acid and lysine. In this case, it may be useful to add calcium as a cofactor (Benjakul and Visessanguan 2003). Two phenomena are involved in setting: transglutaminase activity, which occurs mainly around 30°C, and actomyosin reordering/aggregation, which occurs mainly around 40°C (Nowsad et al. 1996). These temperatures vary slightly from one species to another. In industrial processing, this heating stage is commonly omitted, especially in products requiring very short heat treatment.

As temperature increases, protein thermal coagulation occurs and hydrophobic interactions increase, releasing water (Lanier et al. 2013). This is one of the causes of what is called *modori*. *Modori* is a phenomenon of structure degradation during heating in a temperature range of around 50–65°C. It varies according to the species, and the main changes are located in the myosin heavy chain (Shimizu et al. 1983). It is most evident if heating temperature is especially low.

There is a second and important mechanism associated with the presence of heat-active protease enzymes. In fact, *modori* may be less pronounced depending on the presence of certain inhibitors or other compounds which act on these enzymes (Morrissey et al. 1993; Jiang et al. 2000). For example, despite the washing cycles, pacific whiting surimi has been found to contain a large quantity of proteases, especially L-cathepsin, so that in this species *modori* in surimi gelation is very pronounced. These processing temperatures need to be reduced, and specific ingredients (especially protease inhibitors) should be considered to avoid this phenomenon.

A definitive gel is formed at 60–70°C. Firmness and elasticity increase as aggregation proceeds, and the gel acquires its final characteristic structure. While the myosin tail region has a predominant role at low temperatures, at these higher temperatures, heavy meromyosin (HMM) is the chief element in protein–protein interactions, favoring the formation of disulfide bridges and hydrophobic interactions, which seem to precede the gelation of myofibrillar proteins (Lanier et al. 2013).

Actin appears to play a synergic role in gelation, but it does not promote gel formation above 60–70°C. Above 40°C, actomyosin molecules become dissociated, and while myosin filaments combine, when the temperature is raised, single actin (G-actin) molecules combine to form an aggregate complex (F-actin) that is associated with the development of more elastic structures (Lanier et al. 2013).

9.5.1. *Heating methods for gel formation*

The most common heating methods are steaming, hot water immersion and contact heating, but there are other more recent methods worth mentioning.

9.5.1.1. *Ohmic heating*

Some industries have introduced ohmic heating. This consists of applying an electrical current to a pre-solubilized paste (sol) placed in molds; heating is thus very rapid and also homogeneous since the solid and liquid phases are affected simultaneously. Gelation is therefore very efficient, and in some species with high protease ratios, it improves significantly, even without inhibitors, while reducing myosin and actin degradation (Yongsawatdigul et al. 1995; Tadpitchayangkoon et al. 2012). Studies on Pacific whiting surimi where varying ohmic treatment conditions were applied along with addition of salmon plasma protein as an inhibitor have reported that a balance can be achieved between conditions favoring the TGase enzyme, inactivation of proteases and avoidance of excessive coloring or loss of elasticity, by performing ohmic heating in two steps at 25°C and 90°C (Fowler and Park 2015).

9.5.1.2. *Microwave heating*

In addition to being fast, this type of heating can promote protein solubilization, so that little or no added salt is required. Microwave heating in silver carp surimi, as well as inhibiting proteolysis, has been found to increase protein aggregation through covalent bonding (disulfide- and TGase-mediated) and diminish water retention (Fu et al. 2012). However, microwave heating is not uniform; it is more intense on the surface than at the center. In order to achieve uniform heating in microwave treatment (2,450 MHz), Cao et al. (2019) determined that the optimum thickness for surimi fiber should not exceed 2 cm, thus improving both rheological properties and water retention.

9.5.1.3. *High-pressure treatment (HPH)*

With high pressure, surimi gels can be made cold (<10°C) or by HPH-assisted heat treatment, i.e. by combined treatments. HPH is believed to induce muscle protein denaturation, facilitating transglutaminase access as temperature increases and also favoring the formation of disulfide bridges which stabilize the aggregates formed.

Cold gelation offers some interesting possibilities. The appearance of cold-set gels is different: they are more translucent and shinier, softer and more deformable, and have a manifestly different texture; hence, they are suitable for the manufacture of a wide range of products (Chung et al. 1994; Pérez-Mateos et al. 1997; Pérez-Mateos and Montero 2000; Tabilo-Munizaga and Barbosa-Cánovas 2004).

The results of combined HPH/heat treatment vary, probably in response to the HPH/T conditions applied and the particular species. In some cases, it has been found to improve texture, since it favors solubilization and TGase action and may or may not inhibit proteases depending on the treatment applied (Pérez-Mateos and Montero 1997; Zhu et al. 2015). There are also differences in the bonds that predominate in HPH-assisted cold gelation: Carlez et al. (1995) and Pérez-Mateos et al. (1997) reported abundant hydrophobic bonds in sea bream and blue whiting gels, while Gilleland et al. (1997) observed that hydrogen bridges predominated in Alaska pollack gels. All studies agree that > 300 MPa, myosin is polymerized by thermostable covalent bonds, and that myosin gelling capacity with HPH varies from species to species.

For instance, myosin from mammals and warm-water fish species is more stable to HPH than myosin from cold-water species (Ashie et al. 1999). Moreover,

Fernandez-Martín et al. (1998), using washed blue whiting mince, found that at non-denaturing temperatures (cold temperatures), HPH causes fish protein denaturation, with the greater the effect, the higher the pressure. However, at denaturing temperatures (higher temperatures), high pressure preserves proteins from thermal denaturation, and myosin is better preserved than actin.

9.5.2. *Gelation of fish protein isolate with pH shift processing*

The extreme pH values to which the protein is subjected during the shift process cause alterations in myosin, the main protein involved in the gelation process. At acidic pH, myosin is dissociated into several parts, but not at alkaline pH. In both cases, the globular part of the myosin molecule dissociates and most of the light chains are fragmented in either treatment. Consequently, when the pH is readjusted to neutral, myosin presents no ATPase activity but acquires a native structure. The folding and unfolding process during the acid or alkaline shift phase improves gelation properties, especially in the latter. In this connection, in a study on rockfish (*Sebastes flavidus*) surimi, Yongsawatdigul and Park (2004) reported that gelation was best, in terms of gel strength, using alkaline pH shift, and was worst (poorer than the traditional process) with acid pH shift. Shihka et al. (2006) reported that in the pH shift process, the neutralization of acidified Alaska pollack surimi improved gelation partially due to irreversible changes in the protein. After refolding, the protein contains more reactive thiol groups with either method, but the myosin is left in less of a native state with the alkaline procedure, which would explain the better gelation properties. Experiments with cod muscle have confirmed that processing at high pH (alkaline process) enhances myosin gelation, which correlated with myosin unfolding (Kristinsson et al. 2014).

Kim and Park (2008) reported inactivation of transglutaminase with both alkaline and acid processes, rendering the setting step less effective.

9.6. Waste management, food-grade coproducts and other materials

Various residues, both solid and liquid, are produced during surimi manufacture. Chief among the solid wastes are heads, viscera, spines, scales, skins, etc. Effluents include: 1) in the first stage, water used for washing fillets after gutting, which can be recovered; 2) the water used in the washing cycle is the most abundant effluent; the number of cycles and the amount of water used can vary according to the industry and species. These washing cycle effluents contain various solids, since it is in the first washing cycle that the largest amount of sarcoplasmic proteins and fat are accumulated. In the second cycle, smaller compounds such as polypeptides, blood components, minerals and some sarcoplasmic proteins are also removed, as well as fractions of myofibrillar proteins and other small particles.

To evaluate the most effective methods for solids removal and water purification, it is important to determine their potential uses as by-products and thus encourage implementation of the circular economy in surimi manufacture, so necessary today. Another important aspect of wash-water residue recovery is economic, in that if the solids are recovered before water purification, they can be put to use and effluent purification is rendered more economical.

Recovered solids may serve as food grade or not depending on whether or not they have been in clean waters. Recovery methods therefore vary, and also the functionality of the proteins and the properties of the compounds in these concentrated solids may or may not be maintained depending on the methods used.

Among the more common solids recovery methods are: filtration membranes, centrifugation (decanter centrifuges), isoelectric-point precipitation, coagulation (which can be carried out by thermal treatment, ohmic heating or electrocoagulation), etc., or a combination of these (Morrissey et al. 2005). With the first two, for example, solids can be recovered without heat or chemical treatment, and in the first case, some of them can even be separated sequentially by size (Montero and Gómez-Guillén 1998; Stine et al. 2012).

9.6.1. *Main types of by-products generated in surimi manufacture*

Fish processing waste after filleting can sometimes be as much as 60%. This waste can be differentiated by solid size. For example, collagen and gelatin can be obtained (Gómez-Guillén et al. 2011; López-Caballero et al. 2013) from skins, bones, scales and various enzymes and lipids can be obtained from viscera (López-Caballero et al. 2013), both examples constituting residues from first stage of processing.

However, solutes recovered from washing waters are a source of another variety of heterogeneous compounds as mentioned above, mainly fats (if the surimi is from a fatty species) and various proteins which have to be separated step by step if needed.

9.6.1.1. *Products derived from skins, bones, scales*

About 30% of waste consists of skin, bones and scales, all of which are rich in collagen. The degree of collagen aggregation is very variable, depending on the type of collagen, tissue, animal species, age, etc., and this may partially affect functionality. Gelatin is a product of collagen hydrolysis, and its properties depend on the extraction process and the properties of the protein, which also differ due these same factors. It is especially important to note that the hydroxyproline content of cold-water species is generally significantly lower than in warm-water species, whose amino acid profiles are closer to those of mammals and which therefore have different properties (Gómez-Guillén et al. 2011; López-Caballero et al. 2013). Both collagen and gelatin have multiple applications in the areas of medicine, cosmetics and food. In the case of spines, bones and scales, large amounts of minerals and other compounds could be recovered during collagen or gelatin extraction.

9.6.1.2. *Viscera derivates*

Viscera can be used for the production of polyunsaturated lipids (PUFAs) and are also a source of enzymes, including pepsin, trypsin, chymotrypsin and collagenase (Swapna et al. 2011). They can also be used as a substrate to obtain biologically active protein hydrolysates and peptides (López-Caballero et al. 2013).

9.6.1.3. *Protein concentrates from washing water*

The main compounds in solids from washing water are proteins. The protein concentrate obtained can be used for animal feeds such as fishmeal, fish silage or even fertilizers (Morrissey et al. 2005), but in many cases, they can also be used as functional ingredients, thus generating greater added value (Kim and Mendis 2006).

9.6.1.3.1. Protein hydrolysates

These concentrates are rich in sarcoplasmic proteins, and therefore contain enzymes; if the proteins are partially degraded, these concentrates may contain chiefly proteins, peptides and free amino acids, and also other compounds such as minerals and pigments. Protein hydrolysates have certain useful technological properties; they are highly soluble and can act as emulsifiers and foaming agents over a wide pH range (Thiansilakul et al. 2007; Giménez et al. 2009; Slyzite et al. 2009). Molecular size is important for these properties to manifest, and so the molecular weight of the hydrolysates should not be too small to exert these techno-functional properties. The greater the hydrolysis, the lesser is the emulsifying and foaming power, given that they are surface properties and require a minimum size to manifest their optimal conditions.

Once digested, the proteins consumed in the diet are hydrolyzed, releasing certain peptide sequences that express hypertensive, anticoagulant, immunomodulatory, hypoglycemic and other biological functions (López-Caballero et al. 2013). Many of these activities can be induced by hydrolyzing high-molecular-weight proteins or hydrolyzed concentrates to obtain the required short-sequence hydrolysates (Kim et al. 2008; López-Caballero et al. 2013). These hydrolysates are produced from the activity of species-specific endogenous enzymes and therefore constitute one more by-product; however, they may also be from exogenous sources. Ultrafiltration membranes are very useful for obtaining peptides from specific fractions, and they can be used for industrial purposes.

9.6.1.3.2. Skimming of washing waters

In surimi manufacture from fatty pelagic species, skimming during the washing cycles, and especially the first, retrieves a high percentage of fat (varying depending on the species and season of capture).

Low temperatures, frequently used in surimi manufacture to protect proteins, are also beneficial in preventing oxidation of extracted lipids. However, salt has a more harmful effect, as it renders them more susceptible to oxidation.

To improve processing, some authors (Mendes et al. 1997) have proposed a vacuum process and addition of antioxidants to prevent oxidation and maximize oil extraction. The lipids in this residue are highly unsaturated, rich in MUFAs and PUFAs. They are therefore very unstable but health-beneficial, and hence may offer a very useful food ingredient if recovered (Eymard et al. 2005, 2010).

9.7. Surimi seafood products

9.7.1. *Types of products*

9.7.1.1. *Kamaboko*

Kamaboko is the more traditional product. Known since 1672, it is made basically by kneading surimi with salt and steaming, to obtain a gel whose main characteristics are elasticity and a particular bland, fishy flavor. Nowadays, *kamaboko* is often sold on a wooden board.

Mass production of *kamaboko* started in the early 20th century with the invention of machinery and growing fish catches. The 1950s saw considerable growth in the production of *kamaboko* and fish sausage. In the late 1960s, surimi products were promoted by the development of new machinery and effective cryoprotectants. And the 1970s saw the invention of crabsticks, which very rapidly became popular.

Initially, only coastal species were used for surimi production as the fish had to be very fresh. But at the same time different products were developed through different local processing techniques with different fish. The growth of trawling in the East China Sea in the 1920s increased the supply of traditional and new fish species. With use of new cryoprotectants, frozen surimi production changed very rapidly in the 1960s and frozen blocks of Alaska pollock surimi became the premium material for surimi seafood development. That period saw a gradual shift in Japanese food preferences towards western products, and *kamaboko* production decreased.

There are several types of *kamaboko* in Japan depending on the heating method and form. After grinding with salt, the surimi paste can be steamed, boiled, roasted, fried, etc., to produce different textures and appearances. It can also be prepared on a skewer, in a roll or in a mold. Ooizumi (2014) reviewed different ways of processing *kamaboko* in Japan depending on geographical location.

9.7.1.2. *Other traditional products in Japan*

Other traditional Japanese surimi-based products include *chikuwa*, deep fried *kamaboko*, *hanpen* and others. The final product depends on the processing method and the raw material. *Chikuwa* is made by grinding surimi paste with salt and forming into a tubular shape; this is pierced by a skewer, which is then placed on a conveyor rack and passed through a gas flame. In some cases, surimi is mixed with finely chopped vegetables such as carrots and onions. The product usually comes in square, oval or even ball shapes. Deep-fried *kamaboko* is called *tempura* or *satsuma-age* depending on the geographical origin. *Hanpen* is a variety of boiled products whose final appearance and flavor vary depending on local traditions.

9.7.1.3. *Fish balls*

Other surimi products, widely consumed in China and Southeast Asia, are fish balls. There are different types of fish balls depending on the country and region, and they have different names, for example, *nga soke* in Myanmar, *bola-bola* in the Philippines, *bebola ikan* in Malaysia and Brunei, etc. Typical ingredients other than surimi are salt, sugar, starch and monosodium glutamate, but other ingredients can be added to accommodate local habits. The leading surimi product in China is the fish ball; it is primarily consumed in soup, although the range of products has diversified as well as the manner of consumption, so that products like barbecued fish balls have become popular as cold snacks. Chinese products very often include vegetables and garlic, as well as pork, chicken or beef. The Chinese conceive surimi as a nutrient to be combined with other proteins to confer a texture and a taste, but not necessarily to make seafood analogues. In Thailand, fish balls have a strong tradition, usually fried or grilled, and eaten as a snack or in soups.

9.7.1.4. *Fish sausages*

Fish sausages are another surimi product originating in Japan, although they have been widely consumed in South Korea since the 1960s. There are two types of sausage. The standard one is made with surimi and minced livestock meat along with other components, commonly vegetable proteins, starch, vegetable oil or lard and different spices. The other type also includes cheese and coarse ground meat. Different types of casing are used, mainly nylon film, but nowadays there are also sausages without a casing on the market. Functional compounds such as docosahexaenoic acid (DHA) and eicosapentaenoic acid (EPA), calcium and glucosamine are added to fish sausage for fortification.

9.7.1.5. *Crabsticks*

The invention of crabstick in the 1970s greatly impacted surimi product markets for the products described above, not only in Asian countries but also in American and European markets.

There are two ways to manufacture crabsticks: 1) extrusion in filaments and 2) forming with a mix of pre-formed filaments and surimi paste. In the first case, the blended surimi paste (surimi + salt + starch + etc.) is extruded as a thin (around 1.5 mm) sheet that is heated, either directly or using a transfer unit equipped either with a belt through a steam oven or a cooking drum. Fiberization is achieved by lengthwise cuts on the gelled sheet after heating; the fiberized sheet is pulled into a rope formation afterwards. Also, there are a variety of processing procedures for color application, cutting and wrapping.

In the second forming method, set and heated surimi seafood pastes are shredded and then blended with around 10% unheated surimi paste; the mixture is extruded in rope formation and then cooked in a steam oven and cooled down. The color is usually infused by co-extrusion.

New and different processing methods have been developed to achieve a juicier and more fibrous texture and a flavor closer to real crab meat

9.7.1.6. *Other analogs*

Other seafood analogs like scallop, shrimp, lobster and elver are processed by similar means. Scallops are usually made using method 1) as above, by direct extrusion of the blended surimi paste. In the case of shrimp and lobster analogs, method 2) is used, including pre-preparation of fibers, followed by mixing with a surimi paste and molding. There is also a product made by shredding only as in method 2), as crab meat analogs for salads, sandwiches or stuffing in pimentos or tomatoes. In Spain, there is a particular product, elver (juvenile eel) analog, which is formed by extrusion (about 2 mm) of a surimi (high-grade Alaska pollock) paste made by vacuum-blending surimi, salt and other components including vegetable oil and aroma; at the same time, a black surimi paste is co-extruded in a smaller quantity onto the other extrusion and the whole passed through a steam oven; the rope formation is then cut, packed and pasteurized. French markets offer other interesting products such as salmon sticks processed with salmon surimi, or products similar to crabsticks stuffed with cheese. A product very popular in France, and also known in Spain, is the crabmeat mosaic slab, made by packing 14–16 crabmeat sticks in a wide casing with an added gelatin solution to bind them together. And again, in Spain and France, surimi noodles are sold as in Japan. In Japan, there are numerous surimi products in different forms, many of them for children, or other party-style products in different shapes with cheese paste or other products. Again in Japan, there are products containing added functional compounds, such as docosahexaenoic acid (DHA) and eicosapentaenoic acid (EPA).

9.7.2. *Ingredients for surimi products*

Ziegler and Foegeding (1990) defined five gel models based on the manner of protein gelation. The first model was named "filled gels", being a single-phase gel containing soluble filler in the fluid within the gel network. Another model is where the filler is formed by dispersed particles or constitutes a secondary gel matrix unassociated with the main network, which in surimi is the myofibrillar network. The third model is when inter-component bonds are formed by physical interactions; thus, a non-gelling component may be associated with the myofibrillar network by nonspecific interactions. In the fourth model, several proteins may co-polymerize to form a continuous matrix. The fifth model is when an interpenetrating gel matrix is formed in combination with the main actomyosin network.

The most important ingredients for modification of texture characteristics in surimi gel-based products are water, starch, non-muscle proteins, hydrocolloids and phosphate, in addition to the salt already used to solubilize the protein.

9.7.2.1. *Water*

Myofibrillar proteins are dispersed in water to form an expanded network and facilitate a stable matrix by developing protein–protein bonds during heating. When salt is added, the myofibrillar proteins become soluble and are stabilized in a three-dimensional network through intramolecular forces and surface interactions with water. A minimum amount of water is required for this process, but too much water reduces shear stress. However, shear strain is maintained within certain limits of protein concentration.

There is a strong correlation between water content and color, such that with more water, the gels become whiter (increased L* value). The mineral content of the water is also important for texture development. Ca^{++} and Mg^{++}, especially the former, have a significant effect on setting. Other ions, such as Fe^{++} and Mn^{++}, affect color to some extent (Lee 1990).

9.7.2.2. *Starch*

Starch is important for the formation of good protein networks. High-amylose starches such as corn and wheat form brittle and opaque gels while high-amylopectin starches like tapioca or maize form cohesive, translucent gels (Hunt et al. 2009). Potato starch has the best swelling capacity. At low concentrations (< 5%), potato starch is a good ingredient for surimi products as its granules expand when heated, thus absorbing water, reinforcing the gel matrix and conferring greater gel strength; however, it should not be over-used to reduce the proportion of surimi. The influence of starch on surimi gelation has been reported in both Alaska pollack and whiting surimi by Yoon et al. (1997). Also, starch is used to improve the storage stability of refrigerated or frozen surimi products and to maintain textural characteristics in products intended for reheating.

Pre-gelatinized starches, which bind water without heating regardless of temperature, are also used in surimi gels. It is important not to use these starches in proportions higher than 3–5% in surimi products, since they absorb the water very quickly and inhibit gel formation. Hunt et al. (2010) propose the use of modified starches to improve the stability of surimi products during refrigeration and frozen storage.

The addition of starches alters gel color, so that fully swollen starch granules in the gel render this more translucent.

9.7.2.3. *Egg white protein*

Liquid egg white is one of the principal ingredients added to surimi products, although it can also be used in dry powder form. There are a number of reasons for adding egg white, for example, modification of texture, inhibition of "*modori*" (gel softening), achieving a whiter color, and prevention of freeze/thaw damage. The concentration is usually 2–5%, but up to 7% can be used to reduce the proportion of surimi. In surimi formulae, liquid and dried egg white are used to increase shear strength. Park et al. (2014) reported that in the US surimi industry, it is thought that pre-blending dried egg white with cryoprotectants for addition to surimi before freezing enhances performance as compared to the addition of egg white during gelation.

For the prevention of gel softening, egg white is used as a proteinase inhibitor in both surimi and surimi products, mainly since the banning of beef plasma protein. It contains various proteinase inhibitors, including serine, cysteine and metalloproteinases.

9.7.2.4. *Whey proteins*

Whey proteins, which contain mainly β-lactoglobulin and α-lactoalbumin, are a by-product of cheese processing. Whey proteins can differ depending on the type of cheese, so the choice of type is very important. There is a processed whey concentrate developed though concentration, refining, heating under special conditions and spray drying that has demonstrated improved functionality (Park et al. 2014). Also important is the manner of addition; for instance, significant differences in shear stress have been reported when whey proteins were added at the very outset of chopping before salt during gel processing, due to the negative influence of NaCl on whey protein gel formation.

9.7.2.5. *Soy proteins*

Soy proteins are very popular additives in Asia but not so much in Western countries. Protein concentrates and isolated soy proteins are both used. Soy protein gelation is highly dependent on temperature and ionic strength and is very sensitive to the presence of salt. Many producers use isolated soy proteins in the form of emulsion curd (protein isolate + water + vegetable oil) because such gels tend to possess higher cohesiveness (shear strength) if compared to gels in which powdered protein concentrates or isolates are added directly to the dough during gel processing. Additionally, the use of emulsion curd improves whiteness.

9.7.2.6. *Wheat gluten*

Gluten is the main source of wheat flour protein. Gluten contains two main proteins, gliadin and glutenin. Since the latter is not water soluble, it cannot readily be added by solubilizing in the surimi paste, and so the best way is by high-speed vacuum homogenizing. The literature reports varying results as regards gel strength after addition of wheat gluten. It has also been reported to improve freeze–thaw stability of surimi gel (Park et al. 2014).

9.7.2.7. *Hydrocolloids*

The hydrocolloids most commonly added to surimi paste as stabilizers and texturizers are carrageenan and konjac. There are three commercial carrageenans: kappa, lambda and iota. Surimi gels formed by addition of iota carrageenan are elastic and have good freeze–thaw stability. Also, the addition of this carrageenan increases water retention ability.

Konjac flour is rich in the polysaccharide glucomannan, which forms a thermostable gel when its acetyl groups are removed by raising the pH to around 11. Solo-de-Zaldivar et al. (2014) made suitable gels with a very high ratio of acetyl release (95%) at 3% glucomannan concentration and pH 10.7.

Wu and Suzuki (1991) made a shellfish meat analog using a formula with surimi + calcium hydroxide and konjac slurry. The dough was then frozen, thawed in cold water and then shredded into fibers. Park (1996) reported that addition of 5% konjac flour to surimi reinforced gel hardness, and moreover that this gel maintained consistent shear strain through repeated freeze–thaw abuse. Iglesias-Otero et al. (2010) found that gels were improved by the addition of 1% of a pre-blended dough with 10% glucomannan in water, at pH 10.4, observing increased shear strain and decrease of storage (G′) and loss moduli (G′′).

Other hydrocolloids like alginate, gellan and curdlan have also been studied, but they are hardly used by the industry.

9.7.2.8. *Phosphate*

The addition of a chemical blend of tetrasodium pyrophosphate and sodium tripolyphosphate (1:1) has been reported. Tripolyphosphate degrades to pyrophosphate during processing, and its chelating properties minimize protein freeze-denaturation. Park et al. (2014) reported that the addition of polyphosphates promoted better breaking deformation in surimi gel, but with no change in breaking force.

Phosphates are not recommended in combination with pyrophosphate calcium compounds as this forms a complex with Ca^{++} ions and reduces solubility.

9.7.2.9. *Transglutaminase*

Transglutaminase (protein glutamine g-carboxyamyltransferase) irreversibly catalyzes covalent cross-linking of proteins as mentioned above. The action of endogenous Transglutaminase (TGase) has been associated with setting in surimi gelation; it differs significantly from microbial transglutaminase (MTGase) secreted by *Streptoverticillium mobaraense* or *Streptomyces mobaraense* (Motoki and Seguro 1998). This MTGase, the one used as an additive and supplied mainly by the firm Ajinomoto, is totally independent of Ca^{++}, which is an advantage for use in surimi processing.

In an evaluation of the effect of MTGase concentration on surimi, Lanier and Carvajal (2000) reported that surimi gel was significantly affected by addition of 5 U/g of protein (1%), but that there was no appreciable effect when the concentration was higher. Duangmal and Taluengphol (2010) reported that the addition of 0.2% of MTGase to tilapia surimi increased the breaking force but reduced it when 0.3% was added, so it is important to find the right concentration for each type of surimi. Park et al. (2014) reported that the MTGase addition affected the gel breaking force more than the breaking deformation. These authors also found that MTGase affected surimi from warm-water species more than surimi from cold water species. Moreno et al. (2016) reported that low-quality surimi gel strength can be enhanced by the addition of MTGase, but that industries would need to adapt the methodology to include the setting time required for the enzyme to produce the desired effect.

9.8. References

Alheit, J. and Pitcher, T.J. (eds) (1995). *Hake: Fisheries, Ecology and Markets*. Chapman & Hall Pub., London.

Ashie, I.N.A., Lanier, T.C., MacDonald, G.A. (1999). Pressure induced denaturation of muscle proteins and its prevention by sugars and polyols. *J. Food Sci.*, 64(5), 818–822.

Benjakul, S. and Visessanguan, W. (2003). Transglutaminase-mediated setting in bigeye snapper Surimi. *Food Res. Int.*, 36, 253–266.

Cao, H., Fan, D., Jiao, X., Huan, J., Zhao, J., Yan, B., Zhou, W., Zhang, W., Ye, W., Zhang, H. (2019). Importance of thickness in electromagnetic properties and gel characteristics of surimi during microwave heating. *J. Food Eng.*, 246, 80–88.

Careche, M., Borderías A.J., Sánchez-Alonso, I. (2004). Method of producing a functional protein concentrate from cephalopod muscle and product thus obtained, which is used in the production of analog products and other novel products. Patent, PCT W2004/05117.

Careche, M., Sánchez-Alonso, I., Borderías A.J. (2005). Mejora del procedimiento de elaboración de un concentrado proteico funcional a partir de músculo de cefalópodos y producto así obtenido para ser utilizado en la elaboración de productos análogos y otros nuevos productos. Patent, ES 2 231 023.

Carlez, A., Borderías, J., Dumay, E., Cheftel, J.C. (1995). High pressure gelation of fish myofibrillar proteins. In *Food Macromolecules and Colloids*, Dickinson, E. and Lorient, D. (eds). Royal Society of Chemistry, Cambridge.

Chen, H.H. (2002). Decoloration and gel-forming ability of horse mackerel mince by air flotation washing. *J. Food Sci.*, 67(8), 2970–2975.

Chung, Y.C., Gebrehiwot, A., Farkas, D.F., Morrissey, M.T. (1994). Gelation of surimi by high hydrostatic pressure. *J. Food Sci.*, 59, 523–524.

Duangmal, K. and Taluengphol, A. (2010). Effect of protein additives, sodium ascorbate, and microbial transglutaminase on the texture and colour of red tilapia surimi gel. *Int. J. Food Sci. Tech.*, 45, 48–55.

Eymard, S., Carcouset, E., Rochet, M.J., Dumay, J., Chopin, C., Genot, C. (2005). All stages are critical for development of lipid oxidation during manufacturing of horse mackerel surimi. *J. Sci. Food Agric.*, 85, 1750–1756.

Eymard, S., Jacobsen, C., Baron, C.B. (2010). Assessment of washing with antioxidant on the oxidative stability of fatty fish mince during processing and storage. *J. Agric. Food Chem.*, 58(10), 6182–6189.

Fernández-Martín, F., Pérez-Mateos, M., Montero, P. (1998). Effect of pressure/heat combinations on blue whiting (*Micromesistius poutassou*) washed mince: Thermal and mechanical properties. *J. Agric. Food Chem.*, 46, 3257–3264.

Fowler, M. and Park, J.W. (2015). Effect of salmon plasma protein on Pacific whiting surimi gelation under various ohmic heating conditions. *LWT – Food Sci. and Tech.*, 61, 309–315.

Fu, X., Hayat, K., Li, Z., Lin, Q., Xu, S., Wang, S. (2012). Effect of microwave heating on the low-salt gel from silver carp (*Hypophthalmichthys molitrix*) surimi. *Food Hydrocoll.*, 27, 301–308.

Gilleland, G.M., Lanier, T.C., Hamann, D.C. (1997). Covalent bonding in pressure-induced fish protein gels. *J. Food Sci.*, 62, 713–716.

Giménez, B., Alemán, A., Montero, M.P., Gómez-Guillén, M.C. (2009). Antioxidant and functional properties of gelatin hydrolysates obtained from skin of sole and squid. *Food Chem.*, 114, 976–983.

Gómez-Guillén, C., Mendes, R., Montero, P. (1997). The effect of washing water parameters (pH, hardness and sodium pyrophosphate content) on the water-holding capacity and gelation characteristics of sardine (*Sardina pilchardus*) mince. *Z. Lebensm Unters Forsch A.*, 204, 13–20.

Gómez-Guillén, M.C., Giménez, B., López-Caballero, M.E., Montero, P. (2011). Functional and bioactive properties of collagen and gelatin from alternative sources: A review. *Food Hydrocoll.*, 25(8), 1813–1827

Guenneugues, P. (2019). Global surimi market update. In *The 11th Surimi School Europe. Proceedings*, Madrid.

Huang, J., Bakrya, A.M., Zengd, S., Xionga, S., Yina, T., You, J., Fan, M., Huang, Q. (2019). Effect of phosphates on gelling characteristics and water mobility of myofibrillar protein from grass carp (*Ctenopharyngodon idellus*). *Food Chem.*, 272, 84–92.

Hultin, H.O. and Kelleher, S.D. (1999). Process for isolating a protein composition from a muscle source and protein composition. Patent, US 6005073.

Hultin, H.O. and Kellerher, S.D. (2000a). Surimi processing from dark muscle fish. In *Surimi and Surimi Seafood*, Park, J.W. (ed.). Marcel Dekker Inc., New York.

Hultin, H.O. and Kelleher, S.D. (2000b). High efficiency alkaline protein extraction. Patent, US 6136959.

Hunt, A., Getty, K.J.K., Park J.W. (2009). Roles of starch in surimi seafood: A review. *Food Rev. Int.*, 24(4), 299–312.

Hunt, A., Getty, K.J.K., Park, J.W. (2010). Development of temperature tolerant surimi gels using starch-protein interactions. *J. Food Qual.*, 33, 119–136.

Iglesias-Otero, M.A., Borderías, J., Tovar, C.A. (2010). Use of Konjac glucomannan as additive to reinforce the gels from low quality squid surimi. *J. Food Eng.*, 101, 281–288.

Jiang, S.T., Hsieh, J.F., Ho, M.L., Chung, Y.C. (2000). Combination effects of microbial transglutaminase, reducing agent, and protease inhibitor on the quality of hairtail surimi. *J. Food Sci.*, 65(2), 241–245.

Kang, I.S. and Lanier, T.C. (1999). Bovine plasma protein functions in surimi gelation compared with cysteine protease inhibitors. *J. Food Sci.*, 64(5), 842–884.

Kim, A.-K. and Mendis, E. (2006). Bioactive compounds from marine processing by-products – A review. *Food Res. Int.*, 39, 383–393.

Kim, Y.S. and Park, J.W. (2008). Negative roles of salting gelation properties of fish protein isolate. *J. Food Sci.*, 73(8), C585–C588.

Kim, S.K., Mendis, E., Shahidi, F. (2008). Marine fisheries by-products as potential nutraceuticals: An overview. In *Marine Nutraceuticals and Functional Foods*, Barrow, C. and Shahidi, F. (eds). CRC Press/Taylor & Francis, Boca Raton.

Kristinsson, H.G. and Demir, N. (2003). Functional protein isolates from underutilized tropical and subtropical fish species and byproducts. In *Advances in Seafood Byproducts*, Bechtel P. (ed.). Alaska Sea Grant College Program, University of Alaska, Anchorage.

Kristinsson, H.G. and Liang, Y. (2006). Effect of pH shift processing and surimi processing on Atlantic Croaker (*Micropogonias undulates*) muscle proteins. *J. Food Sci.*, 71(5), 304–312.

Kristinsson, H.G., Lanier, T., Halldordottir, S.M., Geirsdottir, M., Park, J. (2014). Fish protein isolate by pH shift. In *Surimi and Surimi Seafood*, Park, J.W. (ed.). CRC Press/Taylor & Francis Group, Boca Raton.

Lanier, T.C. and Carvajal, P. (2000). Surimi gelation chemistry. In *Surimi and Surimi Seafood*, Park, J.W. (ed.). Marcel Decker Inc., New York.

Lanier, T.C., Carvajal, P., Yongsawatdigul, J. (2013). Surimi gelation chemistry. In *Surimi and Surimi Seafood*, Park J. (ed.). CRC Press/Taylor & Francis Group, Boca Raton.

Lee, C.M. (1990). Counter current and continuous washing systems. In *Engineered Seafood Including Surimi*, Martin, R.E. and Collette, R.I. (eds). Park Rigde, Noyes Data Corp., New York.

Lian, P.Z, Lee, C.M., Chung, K.H. (2002). Textural and physical properties of acid-induced and potassium-substituted low-sodium surimi gels. *J. Food Sci.*, 67(1), 109–112.

Lin, J., Honga, H., Zhanga, L., Zhanga, C., Luo, Y. (2019). Antioxidant and cryoprotective effects of hydrolysate from gill protein of bighead carp (*Hypophthalmichthys nobilis*) in preventing denaturation of frozen surimi. *Food Chem.*, 298, 124868, 1–10.

Liu, J., Fang, C.H., Luo, Y., Ding, Y., Liu, S. (2019). Effects of Konjac oligo-glucomannan on the physicochemical properties of frozen surimi from red gurnard (*Aspitrigla cuculus*). *Food Hydrocoll.*, 89, 668–673.

López-Caballero, M.E., Gimenez, B., Gómez-Guillén, M.C., Montero, P. (2013). Valorization and integral use of seafood by-products. In *Engineering Aspects of Food Biotechnology*, Texeira, J. and Vicente, A. (eds). CRC Press/Taylor & Francis, Boca Raton.

Marmon, S., Liljelind, P., Underland, I. (2009). Removal of lipids, dioxins and polychlorinated biphenils during production of protein isolates from Baltic herring (*Clupea arengus*) using pH shift processes. *J. Agric. Food Chem.*, 57, 7813–7825.

Marmon, S., Chaijan, M., Underland, I. (2012). Novel insights into pigmentation and salt solubility on pH-shift produced protein isolates. In *4th Trans-Atlantic Fisheries Technology Conference (TAFT)*, Tampa.

Mendes, R., Gómez-Guillén, C., Montero, P. (1997). Effect of a new vacuum leaching technology on the textural characteristics of sardine mince. *Z. Lebensm Unters Forsch A.*, 204, 113–120.

Montero, P. and Gómez-Guillén, M.C. (1998). Recovery and functionality of wash water protein from krill processing. *J. Agric. Food Chem.*, 46, 3300–3304.

Moreno, H.M., Herranz, B., Pérez-Mateos, M., Sánchez-Alonso, I., Borderías, A.J. (2016). New alternatives in seafood restructured products. *Crit. Rev. Food Sci. Nut.*, 56, 237–248.

Morrissey, M.T., Wu, J.W., Lin, D., An, H. (1993). Protease inhibitor effects on torsion measurements and autolysis of Pacific whiting surimi. *J. Food Sci.*, 58(5), 1050–1054.

Morrissey, M.T., Lin, J., Ismond, A. (2005). Waste management and by-product utilization. In *Surimi and Surimi Seafood*, 2nd edition, Park, J.W. (ed.). CRC Press/Taylor & Francis Group, Boca Raton.

Motoki, M. and Seguro, K. (1998). Transglutaminase and its use for food processing. *Trends Food Sci. Technol.*, 9(5), 204–210.

Nowsad, A., Kanoh, S., Niwa, E. (1996). Contribution of transglutaminase to the setting of fish pastes at various temperatures. *Fish. Sci.*, 62(1), 94–97.

Núñez-Flores, R., Cando, D., Borderías, A.J., Moreno, H.M. (2018). Importance of salt and temperature in myosin polymerization during surimi gelation. *Food Chem.*, 239, 1226–1234.

Okada, M. (1978). The gel forming capacity of some hake species from South America. Report, FAO, Technical Consultation on the Latin America Hake Industry.

Omjid, P., Panpipat, W., Chaijan, M. (2017). Carbonated water as a novel washing medium for mackerel (*Auxis thazard*) surimi production. *J. Food Sci. Technol.*, 54(12), 3979–3988.

Ooizumi, T. (2014). Manufacture of Kamaboko, Chikuwa, Tempura and Hanpen. In *Surimi and Surimi Seafood*, Park, J.W. (ed.). CRC Press/Taylor & Francis Group, Boca Raton.

Park, J.W. (1996). Temperature-tolerant fish proteins gels using Konjac flour. *J. Muscle Foods*, 7, 165–174.

Park, J.W. and Lin, T.M. (2005). Surimi: Manufacturing and evaluation. In *Surimi and Surimi Seafood*, 2nd edition, Park, J.W. (ed.). Taylor & Francis Group, Boca Raton.

Park, J.W. and Morrissey, M.T. (2000). Manufacturing of surimi from light muscle fish. In *Surimi and Surimi Sea Food*, Park, J.W. (ed.). Marcel Dekker Inc., New York.

Park, W.J., Ooizumi, T., Hunt, A.L. (2014). Ingredient technology for surimi and surimi seafood. In *Surimi and Surimi Seafood*, Park, J.W. (ed.). CRC Press/Taylor & Francis Group, Boca Raton.

Pérez-Mateos, M. and Montero, P. (1997). High-pressure-induced gel of sardine (*Sardina pilchardus*) washed mince as affected by pressure-time-temperature. *J. Food Sci.*, 62(6), 1183–1188.

Pérez-Mateos, M. and Montero, P. (2000). Response surface methodology multivaritate analysis of properties of high pressure induced fish mince gel. *Eur. Food Res. Technol.*, 211(2), 79–85.

Peréz-Mateos, M., Lorenço, H., Montero, P., Borderías, J. (1997). Rheological and chemical studies on high pressure-induced gels from blue whiting (*Micromesistius poutassou*) muscle. *J. Agric. Food Chem.*, 45, 44–49.

Phatcharat, S., Benjakul, S., Visessanguan, W. (2006). Effects of washing with oxidising agents on the gel-forming ability and physicochemical properties of surimi produced from bigeye snapper (*Priacanthus tayenus*). *Food Chem.*, 98, 431–439.

Poowakanjana, S. and Park, J.W. (2013). Biochemical characterization of Alaska pollock, Pacific whiting, and threadfin bream surimi as affected by comminution conditions. *Food Chem.*, 138, 200–207.

Raghavan, S. and Kristinsson, H.G. (2008). Conformational and rheological changes in catfish myosin during alkali-induced unfolding and refolding. *Food Chem.*, 107, 385–398.

Reynolds, J., Park, J.W., Choi, Y.J. (2002). Physicochemical properties of Pacific whiting surimi as affected by various freezing and storage conditions. *J. Food Sci.*, 67(6), 2072–2078.

Sánchez-Alonso, I., Careche, M., Borderías A.J. (2007). Method for producing a functional protein concentrates from giant squid (*Dosidicus gigas*) muscle. *Food Chem.*, 100, 48–54.

Shikha, F.H., Hossain, M.I., Morioka, K., Kubota, S., Itoh, Y. (2006). Effect of pH-shifting on the gel forming characteristics of salt-ground meat from walleye pollack. *Fish. Sci.*, 72, 870–876.

Shimizu, Y., Nishioka, F., Machida, R., Shien, C.M. (1983). Gelation characteristics of salt-added myosin sol. *Nippon Suisan Gakkaishi*, 49, 1239–1243.

Shimizu, Y., Toyohara, H., Lanier, T. (1992). Surimi production from fatty and dark-fleshed fish species. In *Surimi Technology*, Lanier, T.C. and Lee, C.M. (ed.). Marcel Dekker Inc., New York.

Slizyte, R., Mozuraityte, R., Martinez-Alvarez, O., Falch, E., Fouchereau-Peron, M., Rustad, T. (2009). Functional, bioactive and antioxidative properties of hydrolysates obtained from cod (*Gadus morh*ua) backbones. *Process Biochemistry*, 44, 668–677.

Solo-de Zaldivar, B., Tovar, C.A., Borderías, A.J., Herranz, B. (2014). Effect of deacetylation on the glucomannan gelation process for making restructured seafood products. *Food Hydrocoll.*, 35, 59–68.

Stine, J.J., Pedersen, L., Smiley, S., Bechtel, P.J. (2012). Recovery and utilization of protein derived from surimi wash-water. *J. Food Qual.*, 35, 43–50.

Swapna, C.H., Bijinu, B., Amit, K.R., Bhaskar, N. (2011). Simultaneous recovery of lipids and proteins by enzymatic hydrolysis of fish industry waste using different commercial proteases. *Appl. Biochem. Biotechnol.*, 164, 115–124.

Tabilo-Munizaga, G. and Barbosa-Canovas, G.V. (2004). Viscoelastic properties of pressurised and heat-induced surimi gels made from Alaska pollock (*Theragra chalcogramma*) and Pacific whiting (*Merluccius productus*). *Food Aust.*, 56(11), 543–548.

Tadpitchayangkoon, P., Park, J.W., Yongsawatdigul, J. (2012). Gelation characteristics of tropical surimi under water bath and ohmic heating. *Lwt – Food Sci. Tech.*, 46, 97–103.

Thiansilakul, Y., Benjakul, S., Shahidi, F. (2007). Compositions, functional properties and antioxidative activity of protein hydrolysates prepared from round scad (*Decapterus maruadsi*). *Food Chem.*, 103, 1385–1394.

Toyoda, K., Kimura, I., Fujita, T., Noguchi, S.F., Lee, C.M. (1992). The surimi manufacturing process. In *Surimi Technology*, Lanier, T.C. and Lee, C.M. (ed.). Marcel Dekker Inc., New York.

Tsuchiya, Y., Yamada, N., Matsumoto, J.J. (1978). Studies on squid mantle muscle proteins. Extraction and purification of squid myosin. *Bull. Japan Soc. Scient. Fish.*, 44(2), 175–179.

Wu, S. (2016). Effect of trehalose on the state of water, protein denaturation and gel-forming ability of weever Surimi. *Int. J. Food Prop.*, 19, 521–525.

Wu, M.C. and Suzuki, T. (1991). Preparation of konjac mannan gel. Patent US 50284445.

Wu, S.J., Pan, S.K., Wang, H.B. (2014). Effect of trehalose on *Lateolabrax Japonicus* myofibrillar protein during frozen storage. *Food Chem.*, 160, 281–285.

Wu, J.H., Rong, Y.Z., Wang, Z.W., Zhou, Y.F., Wang, S.Y., Zhao, B. (2015). Isolation and characterisation of sericin antifreeze peptides and molecular dynamics modelling of their ice-binding interaction. *Food Chem.*, 174, 621–629.

Xiong, G.Q., Wei, C., Ye, L.X., Du, X., Zhou, M., Lin, R.T., Geng, S., Chen, M., Corke, H., Cai, Y. (2009). Effects of Konjac glucomannan on physicochemical properties of myofibrillar protein and surimi gels from grass carp (*Ctenopharyngodon idella*). *Food Chem.*, 116(2), 413–418.

Yongsawatdigul, J. and Park, J.W. (2004). Effects of alkali and acid solubilization on gelation characteristics of rockfish muscle proteins. *J. Food Sci.*, 69(7), C499–C505

Yongsawatdigul, J., Park, J.W., Kolbe, E., Abu Dagga, Y., Morrissey, M.T. (1995). Ohmic heating maximizes gel functionality of Pacific Whiting surimi. *J. Food Sci.*, 60, 10–14.

Yongsawatdigul, J., Worratao, A., Park, J.W. (2002). Effect of endogenous transglutaminase on Threadfin Bream surimi gelation. *J. Food Sci.*, 67(9), 3258–3263.

Yoon, W.B., Park J.W., Kim, B.Y. (1997). Linear programming in blending various components of surimi seafood. *J. Food Sci.*, 62(3), 561–564.

Zhou, W., Wang, F., Yu, J., Li, X., Liu, Y. (2019). Cryoprotective effects of protein hydrolysates prepared from by-products of silver carp (*Hypophthalmichthys molitrix*) on freeze-thawed surimi. *Appl. Sci.*, 9, 563.

Zhu, Z., Lanier, T., Farkas, B. (2015). High pressure effects on heat-induced gelation of threadfin bream (*Nemipterus* spp.) surimi. *J. Food Eng.*, 146, 23–27.

Ziegler, G.R. and Foegedding, A.E. (1990). The gelation of proteins. In *Advances in Food and Nutrition Research*, Kinsella, J.E. (ed.). Academic Press, New York.

10

Little or Underutilized Marine Resources

Charles DELANNOY
PROCIDYS, Wimereux, France

10.1. What resources are not or hardly used in the fishing industry?

Marine co-products represent on average 50% of the weight of the raw material, i.e. the whole fish. Consisting of the parts of the animal not consumed by humans directly (heads, bones, skins, shells, etc.), they are rich in proteins, lipids, minerals and other molecules of interest and can be transformed into different finished products: meal, hydrolysates, oil, flesh, flavors, collagen, minerals, fertilizer, etc. For this, various processes can be implemented to stabilize, separate, purify, functionalize and concentrate the products of interest. While they have long been considered as waste, they are now the subject of an organized recovery to obtain an added value in several fields of application, ranging from the return to the soil to the niche industries (pharmacy, cosmetics, etc.) through animal feed.

This chapter presents the different types of marine co-products, the processes used to transform them and the applications of the products obtained.

For a color version of all figures in this chapter, see http://www.iste.co.uk/verrez-bagnis/aquaticproducts.zip.

10.1.1. *What are marine co-products?*

Marine co-products represent sources of proteins and lipids that have always been of interest (see Table 10.1).

At the regulatory level, marine co-products are part of animal by-products. The European Regulation (EC) No. 1069/2009 precisely defines the collection and treatment of different categories of animal co-products not intended for human consumption. This European regulation distinguishes three categories of animal by-products according to the level of risk to public health, animal health and the environment:

– Category 1 (C1): significant sanitary risk for public health (e.g. animals and therefore fish found dead in farms). They are destroyed by incineration or dumping.

– Category 2 (C2): non-contaminating disease risks to humans, classic microbiological risks, etc. The materials can be used in fertilization (organic fertilizer, biogas, composting).

– Category 3 (C3): healthy animal (fish) co-products suitable for human consumption. Fish co-products are generated by preparation activities, cutting plants or retailers. These perfectly healthy materials, from animals intended for human consumption, can be enhanced, after processing in an approved establishment, in food and feed, depending on their quality and the type of material, and are assimilated to co-products in the language of economic operators.

Marine co-products include: i) fish and aquaculture products that are not used for human consumption because of poor quality or lack of market, and these products tend to decrease, and ii) products resulting from the processing of products and not directly for human consumption (see Figure 10.1), i.e. heads, bones, skins and viscera for fish, and shells for shellfish.

The constraints to the enhancement of aquatic co-products are of a technical-economic nature (quantity, quality, seasonality, cost of collection and processing), regulatory (sanitary approval, traceability, maintenance of the cold chain necessary for certain enhancement) and human (co-products still considered as waste and not as noble raw materials).

The composition of these co-products is variable depending on the species of fish, the recoverable parts of the fish, their origins and their feeding. But globally, the main constituents of these co-products are proteins, lipids and minerals, as shown in Table 10.1.

	Type of compound	**Content**
Classical chemical analysis of marine co-products	Dry matter	18–25% for lean fish 25–35% for fatty fish
	Protein	13–16% for all species
	Minerals	4–7% for all species
	Lipids	1–5% for lean fish 5–15% for fatty fish

Table 10.1. *Average composition of fish co-products*

The volumes of co-products in France are significant, and the concentration of companies processing seafood products, for example, in fishing ports, favors the establishment of shared processing tools.

10.1.2. *What are the volumes of co-products?*

The tonnage of fish co-products generated in France is estimated at 150,000 tons, i.e. about 50% of the fish tonnage. These co-products come from the fish trade (more than 50% of the co-products generated), the sausage industry (31% of the volumes, especially salmon) and canning (18% of the volumes, especially blue fish). The regions of Hauts de France/Normandy/Brittany generate nearly 80% of the co-products of the fisheries and aquaculture sector in France (Ademe, Iterg, Actia, Aquimer Ifremer, CVG 2012).

French fisheries and aquaculture enhance almost all of their co-products, but towards uses of medium added value. Indeed, 96% of the co-products are the object of a mass valorization intended for animal feed (fishmeal and oils, protein hydrolysates, pulp and frozen minced meat). Only 4% are recovered in the form of other applications with higher added value (gelatin, collagen, chondroitin, aromatic extracts, refined oils, phospholipids, mineral supplements, chitin, protein hydrolysates).

The evolution of the volumes of co-products varies according to the species and product categories. For example, the volumes of salmonid co-products have increased (approximately 10 to 15%) due to the increase in salmon imports, while the volumes of white fish and cartilage have decreased.

10.2. The opportunities

The finished products produced from fish co-products can be destined for different markets:

– animal feed (aquaculture and dog and cat food) including:

- oil: the world market is estimated at 1 million tons,

- fishmeal: the world market is estimated at 4.5 million tons,

- protein hydrolysates: the world market for hydrolysates is estimated at more than 80,000 tons (including enzymatic hydrolysates, acids or silages),

- minerals;

– fertilization: organic fertilizers and amendments;

– fermentation: peptones as a source of nitrogen;

– human food:

- aromatic extracts,

- fish pulp,

- nutraceutical ingredients: peptides, mineral extracts, gelatin, DNA, etc.,

- oils: omega-3, vitamins, etc.;

– cosmetics:

- collagen and elastin,

- squalene,

- peptide extracts.

The demand for fishmeal and hydrolysates is increasing in particular due to the significant development of aquaculture and the dog and cat food market. Similarly, the markets for marine ingredients, aromatic extracts and pulp are also growing steadily thanks to the development of the dietary supplements and ready meals markets.

The volume distributions of co-products and finished products in France are presented in Table 10.2.

Breakdown by finished product	Raw material fish in tons	Finished products in tons
Fishmeal	55,000	11,000
Hydrolysates	40,000	6,000
Pulp	16,000	10,000
Petfood	36,000	36,000
Healthy ingredients	3,000	300
TOTAL	150,000	67,300

Table 10.2. *Presentation of co-product and finished product volumes in France by market and application*

10.3. Products and processes

10.3.1. *Production of fishmeal and fish oil*

This valorization was the first to be implemented in the field of marine co-product processing. It has been in existence for 60 to 70 years and represents the most traditional and developed valorization (Asche 2016).

Fishmeal is mainly used today in aquaculture (70%) and to a lesser extent in poultry and pig feed, as shown in Figure 10.1.

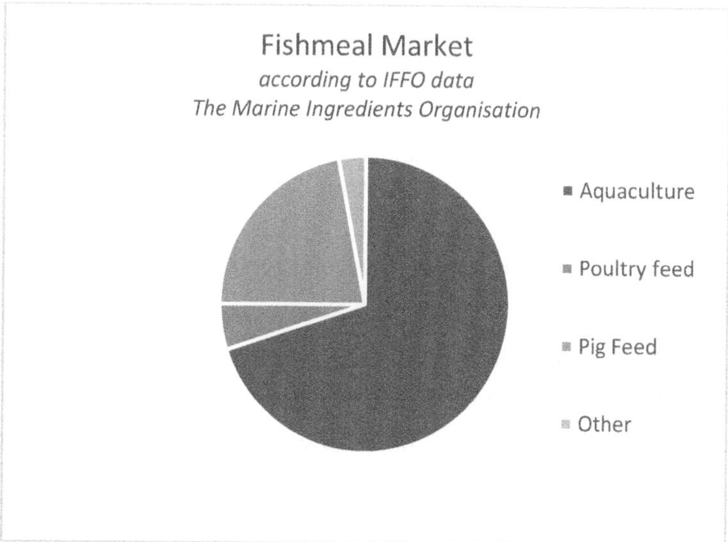

Figure 10.1. *Distribution of fishmeal use*

After a cooking stage of the co-products, the solid (proteins) and liquid (oil in particular) phases are separated before drying to give two finished products (fishmeal and oil). Figure 10.4 details the different steps of the process.

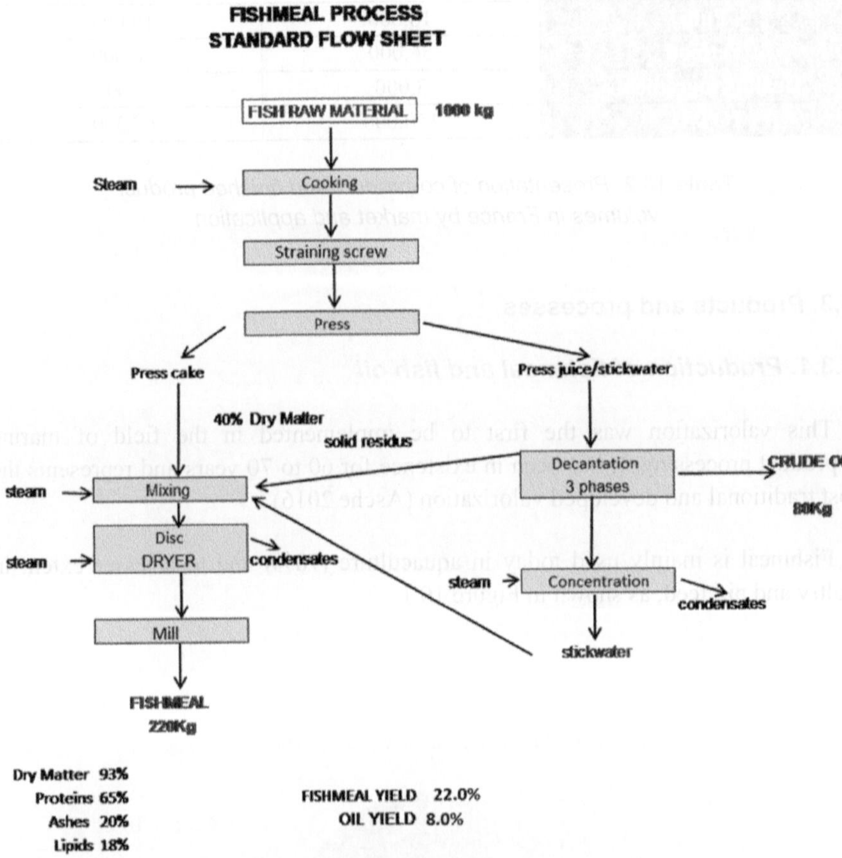

Figure 10.2. *Fishmeal production flowchart*

A conventional and optimized fishmeal production process can achieve the following yields, depending on the species used:

– production of 200 to 250 kg of fishmeal/ton of raw material, i.e. a yield of 20 to 25%;

– production of 60 to 100 kg of oil/ton of co-products, i.e. a yield of about 6 to 10% depending on the species of fish treated and their lipid content.

The main steps of this type of process are as follows:

– Continuous cooking in a disk cooker at a temperature of 100°C for 30–45 minutes.

– Pressing: after cooking, a screw press allows the majority of the protein juices and oil to be extracted. The other fraction is a press cake made of the fish flesh, which will be dried in a next step.

– Separation of oil phases and protein juice phases: the juice press or stickwater contains mainly oil and a diluted solution rich in protein and some solid residues of flesh passed through the mesh of the press. This separation is achieved by using centrifugal force to separate these two components, which have a different density.

– The classic equipment used are the horizontal bowl decanters, which operate continuously and allow the separation by centrifugation of either two phases (juice and solid residues) or three phases (protein juice, oil and solid residues).

– Concentration of stickwater (protein juices): after extraction of the oil, the residual protein juice is concentrated to 45% of dry extract by evaporation. The evaporation is carried out under vacuum in order to be able to work at lower temperatures and preserve the sensitive compounds of the products, and in particular, the proteins. The equipment used are single-effect or multiple-effect evaporators to save energy: in this case, several evaporation stages are installed in series and the evaporation vapor from the first effect is used as heating steam for the second effect, which operates at a lower temperature and pressure. Similarly, the evaporation vapor of the second effect is used as heating steam of the third effect and so on according to the number of effects. The vapor from the last effect is finally condensed on a tubular condenser.

– Drying and grinding: a disk dryer, heat with steam or a rotary dryer (Rotatube type) or a low-temperature dryer operating with steam enables homogeneous drying without overheating of the meal to be achieved. The residence time in this dryer and the drying temperature are very important parameters to obtain a good fishmeal with good digestibility and excellent nutritional characteristics: the drying process usually lasts two to three hours and is usually carried out at 100°C or at 60°C if the dryer works under vacuum. The dried product is then finely milled and sieved to obtain the fishmeal.

– Treatment of fish oils: the oil extracted from the stickwater (press juices) by centrifugal decantation can be treated and purified to improve its quality and commercial value. Depending on the market and the destination of the fish oils, they can be more or less clarified, neutralized, refined, detoxified and deodorized.

Figure 10.5 describes the refining steps that can be performed depending on the strategic choices and the quality of the oil required.

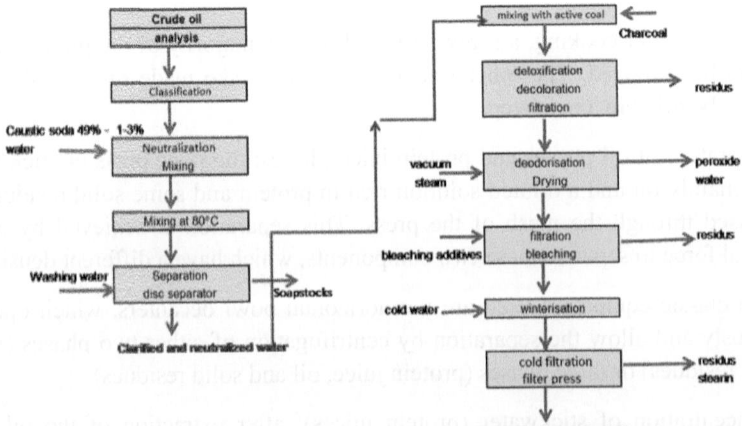

Figure 10.3. *Description of the process steps for refining and purifying fish oils*

10.3.2. *Production of hydrolysates*

Fish hydrolysates can be defined as essentially nitrogenous products resulting from the hydrolysis of fish proteins; they consist mostly of a mixture of polypeptides of varying sizes and free amino acids (He et al. 2015). The main objectives of enzymatic or chemical hydrolysis are to solubilize proteins, improve their digestibility and obtain polypeptides with functional properties.

The main characteristics of hydrolysates are as follows:

– A high protein content, which can vary from 75% to 90% of dry matter, or even 95% in the case of purified hydrolysates. These proteins are therefore soluble and have a high digestibility.

– Lipid content: the lipid content depends on the type of fish and its composition. Moreover, the residual lipid content of hydrolysates also depends on the centrifugal separation that allows the extraction of the oil and also on the expected applications (aquaculture or petfood). The usual contents vary between 10 and 25% of the dry matter of the hydrolysate.

– Low mineral content (5–6% of dry matter) compared to fishmeal (which generally contains two to five times more).

Moreover, depending on the hydrolysis, the molecular weight is often lower than 10,000 Da, with an average molecular weight between 500 and 4,000 Da. Depending on the application, it can be interesting to have very short polypeptides of four to five amino acids (from 500 to 1,500 Da) to have a better nutritional efficiency and interesting biological activities. The size of the peptides has a significant impact on the functional properties of the finished product.

Enzymatic hydrolysis consists of adding enzymes (proteases) to fish co-products, which hydrolyze the peptide bonds. Proteins under the effect of proteases will break down according to the following pattern:

Proteins → Polypeptides → Peptides → Amino acids

The proteases used to hydrolyze fish can be from different sources:

– animal (pancreas trypsin, pepsin);

– vegetable (papain, bromelain, ficin);

– bacterial (*Aspergillus* sp., *Bacillus* sp.).

This diversity leads to a wide range of pH (from 2 to 9) and activities being covered. The hydrolysis parameters to be set in order to optimize the enzymatic activity are the following:

– Temperature: the temperature exponentially increases the speed of the enzymatic reaction. However, the enzyme is a protein and it denatures around 70°C and loses its enzymatic activity. It should be noted that the inhibition of the enzyme at the end of the reaction occurs by heating the medium to 85°C.

– pH: an enzyme is only active in a certain pH range and has its maximum activity at a specific pH. This optimal pH varies from one enzyme to another.

– Duration: the duration of hydrolysis depends on the dose of enzyme, its activity and the degree of hydrolysis we want to obtain. On average, the duration of hydrolysis of fish is two to three hours.

– Enzyme/substrate ratio: the hydrolysis rate is proportional to the enzyme concentration. Because of the cost of the enzyme, the amount needed must be optimized. The enzyme/substrate ratio influences the hydrolysis speed: it is between 0.1 and 0.5% of enzyme compared to the protein.

– Stirring: a fairly effective stirring is usually necessary to improve the enzyme/substrate contact.

– Moreover, hydrolysis can be carried out in continuous or batch mode.

The enzymatic hydrolysis process of marine co-products is described in Figure 10.6.

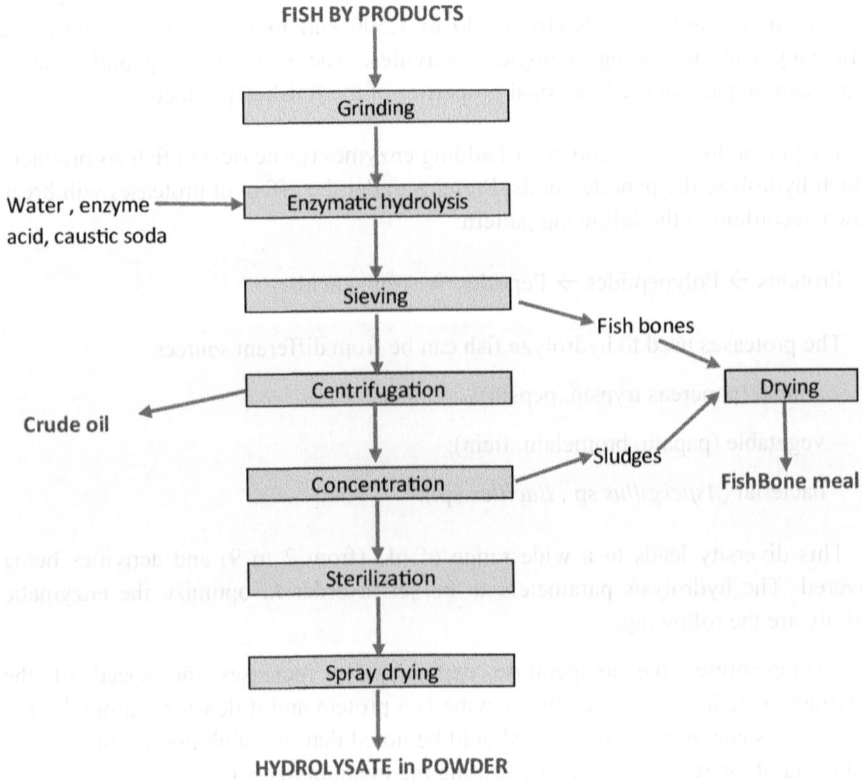

Figure 10.4. *Production diagram of fish enzymatic hydrolysates*

After a fine grinding of the raw material, this is introduced into a reactor to perform enzymatic hydrolysis in batch or continuous mode. This reactor is designed to allow a precise monitoring of the hydrolysis parameters, and is equipped with an efficient variable speed agitation and a temperature and pH regulation.

The hydrolysis reaction can be evaluated by different criteria:

– The study of the reaction kinetics by the pH-Stat method: the principle is to maintain the targeted pH by adding acid or base. The kinetics of the reaction is studied based on the consumption of the reagent used to maintain the pH of the medium at the optimum, as a function of time.

– Measurement of the solubilized dry matter: the follow-up of this measurement can be done by measuring the optical density or the refractive index (degree Brix), directly related to the solubilization of the protein compounds.

– The degree of hydrolysis: the number of hydrolyzed peptide bonds (h) is measured, by following the rate of soluble nitrogen as a function of time, by measuring the appearance of alpha-amino groups or by simple volumetric determination. The rate of hydrolysis (ΔH%) is expressed globally according to the following expression:

$$\Delta H\% = \text{total h/h} \times 100$$

– Peptide molar mass profile: by HPLC (High-Performance Liquid Chromatography on the column).

As shown in Figure 10.7, the elution of peptides from the column is carried out according to their sizes.

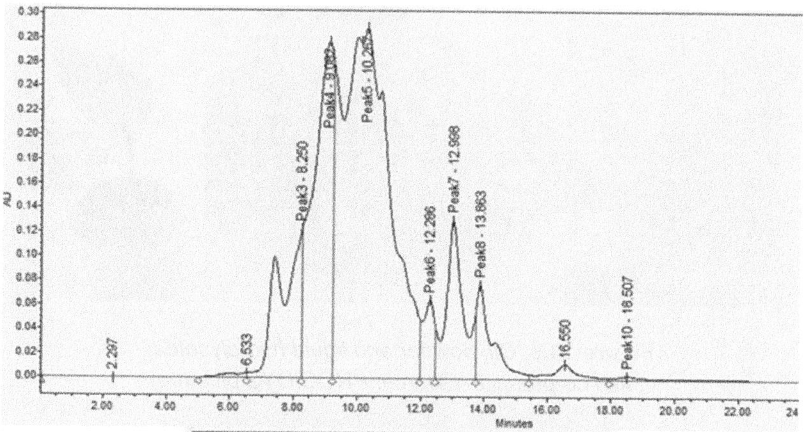

	Name	Retention Time (min)	Area (µV*sec)	% Area	Height (µV)	Int Type	Amount	Units	Peak Type	Peak Codes
1		2.297	26495	0.04	-183	BV			Unknown	
2	Peak1	3.686				Missing			Missing	
3	Peak2	5.560				Missing			Missing	
4		6.533	182858	0.31	4108	VV			Unknown	I06
5	Peak3	8.250	5690509	9.65	123216	VV			Found	I06 Q20
6	Peak4	9.082	11735562	19.89	277818	VV			Found	Q20
7	Peak5	10.267	31227185	52.94	288956	VV			Found	Q20
8	Peak6	12.286	1587867	2.69	64914	VV			Found	Q20
9	Peak7	12.998	4878521	8.27	130867	VV			Found	Q20
10	Peak8	13.863	2590358	4.39	77404	VV			Found	Q20
11		16.550	722831	1.23	12350	VV			Unknown	
12	Peak9	17.686				Missing			Missing	
13	Peak10	18.507	348808	0.59	3014	VB			Found	Q20

Figure 10.5. *Distribution curve of the molecular profiles of the hydrolysate by chromatography on the BioSep-SEC HPLC column (source: PROCIDYS)*

After hydrolysis, the fish hydrolysate solution may be sieved and then centrifuged to remove insoluble bone residues and oil. The product can then be concentrated by evaporation, as in the fishmeal production process. The product is then dried by spraying.

A process of enzymatic hydrolysis of fish co-products allows the following yields, depending on the species used, to be obtained:

– for the powdered hydrolysate, a yield of 15 to 18% compared to the raw material, depending on the composition of the hydrolysate and the raw material;

– for the oil, a yield of 5 to 9% depending on the species of fish and its lipid content;

– for the residual bones if dried, a yield of 5 to 8% compared to the raw material, depending on the structure of the co-products and the type of raw material.

Figure 10.6. *Oil, powder and liquid hydrolysates of fish co-products (source: PROCIDYS pictures)*

10.3.3. *Fish pulp*

Fish pulp is the product resulting from the mechanical separation of fish bones or gutted fish from the flesh. Fish pulp is to be distinguished from minced fish flesh or flesh consisting of particles of long muscle detached from the skeleton and essentially free of bones and skin (according to the Codex Alimentarius).

Color (light or red) is the most important quality and price criterion (other quality attributes are considered acceptable), but there is no recognized quality scale for the national and international markets. Several methods have been identified to improve pulp color, such as washing and the use of additives or processing additives.

The raw material used for the production of the pulp has an influence on the color of the pulp and the yield. The species used for the production of bone pulp are mainly the following marine fish: cod, haddock, whiting, Alaskan hake, and the following freshwater species: perch, roach, small carp and tilapia. In addition, salmon co-products are regularly used for pulp production.

10.3.3.1. *The process*

The mechanical separation of the bones and the freezing of the pulp are usually done on the same day as the filleting of the fish from which the bones originate. These two operations are often carried out continuously, but the bones can sometimes be kept in refrigeration for up to 24 hours before separation. This mechanical separation is done with a mechanical separator. Under the action of a high pressure through a pressing belt or a compression screw, the fish flesh is pressed through a die or a perforated drum. Skins, bones and ligaments are rejected.

Separation

The raw pulp is thus obtained by processing on this separator. There are two main types of separator:

– a belt separator and rotating drum;

– a mobile screw separator (and fixed drum or die).

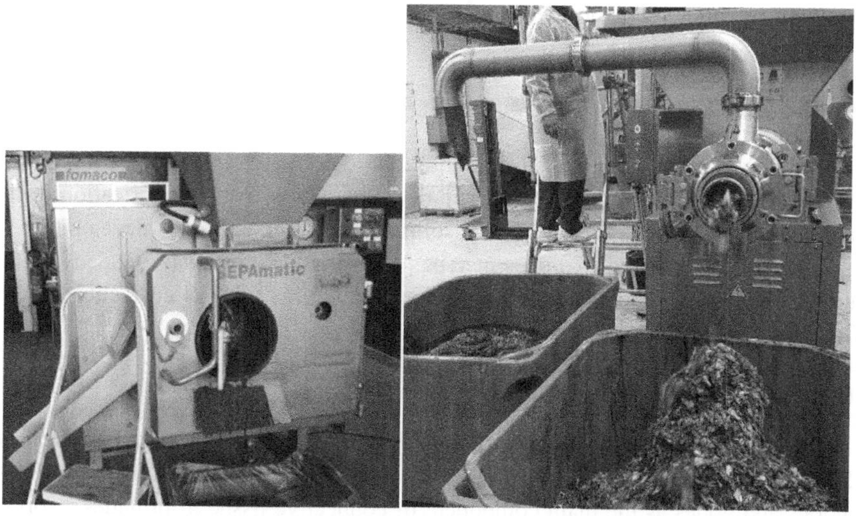

Figure 10.7. *Mechanical belt or screw separator for fish pulp extraction (source: PROCIDYS pictures)*

The raw material is introduced into a feed hopper. Conveyed by one or two feed screws, the raw material passes through a die with holes or slots. The meat is compressed by the screw in the die and expelled through it. The edges are removed at the end of the machine. The result is a raw pulp that can be used for human food if the necessary hygiene criteria are met. This pulp can then be refined by different steps:

– washing: this step improves the color of the pulp by eliminating the various impurities present in the flesh, as well as the proteins;

– freezing: the pulp is frozen in rectangular molds placed between the plates of a freezer. This is the most classical method for preservation and marketing.

10.3.3.2. *Applications*

Fish pulp is incorporated in a wide range of food products such as breaded fish products (mainly croquettes and sticks), products derived from surimi, seafood delicatessen products (terrines, rillettes, mousses, sausages, rolls), ready meals (gratins, brandades), spreadable products and butters, frozen or pre-cooked raw fish minced steaks, soups and aromatic extracts.

French production is estimated at 6,000 to 7,000 tons. The white fish pulp must be of grade A or B3, with few impurities. In 2021, it was mostly bought frozen and its price varied between 1.3 and 1.6 €/kg depending on the quality and grade, while the price of salmon pulp was between 1.2 and 1.3 €/kg.

10.3.4. *Ingredients for nutraceuticals and health nutrition*

Even if the production of fishmeal and, to a lesser extent, the production of fish enzymatic hydrolysate, are today the solutions that allow for the valorization of the majority of fish co-products, other solutions are increasingly developing, in particular the production of different types of marine ingredients intended for the human food, ready-to-eat, nutraceutical, health food, as well as cosmetic markets (Cashion et al. 2017).

10.3.4.1. *Gelatin and collagen*

Collagen is the most abundant protein in the human body. It represents 30 to 35% of the total proteins in the body. It is naturally present in the skin, muscles, tendons and bones as the main component of connective tissue. It provides a sort of framework for the human and animal body.

Fish co-products and in particular skins and cartilage represent a very important source of gelatin and collagen. It is therefore an interesting possibility to valorize these co-products, provided that they are collected under good conditions and that these skins are rich in collagen, which is variable from one species to another. The molecular weight of the collagen molecule is about 330 kDa. Gelatin is a form of fractionated collagen after hydrolysis or thermal denaturation of the triple helix characteristic of native collagen.

To date, 27 collagen variants have been identified (Gómez-Guillén et al. 2011) based on the nature and shape of their aggregates or their morphological differences. These collagen molecules consist of closely related but genetically distinct proteins, possessing a basic structure of three polypeptide chains, each with a glycine–X–Y repeat (where X is predominantly proline and Y is predominantly hydroxyproline), forming tightly linked triple helices (mainly stabilized by intra- and inter-chain hydrogen bonds), which subsequently aggregate to form various types of structures (Gómez-Guillén et al. 2011). The characteristic feature of collagen is the presence of certain amino acids in significant amounts, with glycine accounting for about 35% of the amino acids, proline 13% and hydroxyproline 8%.

Amino acids	Quantity (g/100 g of protein)	Amino acids	Quantity (g/100 g of protein)
Aspartic acid	4.6-5.5	**Leucine**	2.1-2.9
Glutamic acid	5.7-9.3	**Lysine**	2.7-3.2
Alanine	9.1-12	**Methionine**	1.0-1.6
Arginine	5.4-8.9	**Phenylalanine**	1.4-1.4
Cystine	0.0-0.5	**Proline**	9.3-13.3
Glycine	24.4-26.7	**Serine**	3.0-3.5
Histidine	0.4-0.5	**Threonine**	2.9-3.0
Hydroxylysine	0-5-0.7	**Tyrosine**	0.2-0.6
Hydroxyproline	7.3-7.7	**Valine**	1.9-2.3
Isoleucine	0.9-1.5		

Table 10.3. *Average composition of a fish collagen*

Fish collagen contains 20 to 30% less hydroxyproline (this amino acid plays an important role in mechanical strength) than porcine or bovine collagen. This explains the differences in texture between fish and meat. This is also particularity responsible for the lower denaturation temperature of fish collagens (Yata and Yoshida 2001).

The collagen extraction process essentially consists of preparing the fish skins by various acid and basic washes to remove degraded proteins, mucus and impurities often responsible for the smell. There are different process variants, but most often gelatin is extracted by solubilization in a hot aqueous process under agitation. The resulting gelatin is then purified, filtered, decolorized and deodorized. It is then concentrated and dried in the form of granules or a more or less fine powder.

This gelatin can therefore also be enzymatically hydrolyzed to give a collagen hydrolysate whose characteristics are very variable depending on how the enzymatic hydrolysis was carried out. The hydrolysis of collagen and gelatin can release interesting bioactive peptides for potential applications as functional foods in nutraceuticals. Thus, the beneficial effect of peptides obtained from collagen or gelatin hydrolysates on various pathologies has been demonstrated in animal or clinical studies. Currently, several ingredients, and in particular collagen peptides, have been patented and are currently marketed in the United States, Japan and Europe (Aleman and Martínez-Alvarez 2013).

Many articles and publications refer to certain bioactive activities attributed to collagen or gelatin peptides of marine origin. Thus, antioxidant, antihypertensive (ACE inhibitors), antimicrobial, opioid or joint regeneration peptides, peptides with calciotropic and secretagogic activities, as well as other interesting bioactive activities are often cited.

10.3.4.2. *Chondroitin sulfate*

Chondroitin sulfate is extracted from the cartilage of rays and sharks. This molecule, which is a mucopolysaccharide or glycosaminoglycan, is an integral part of cartilage, as well as of fish skin. Proteoglycans are made up of polysaccharide units (95%) and proteins (5%). They are covalently linked to the protein binding structure (fibrous proteins and glycoproteins) (Seeley et al. 2008).

It is also extracted more specifically from the fins and ligaments of tuna. Its effects on arthritic pain or on the flexibility of joints are known. Applications have existed for several decades in veterinary care and for some years in human medicine.

Numerous research projects have been carried out, and this molecule is used in a number of patented products.

10.3.4.3. *Chitin and chitosan*

Chitin is the most abundant polymer on Earth after cellulose. It is a nitrogenous polysaccharide composed of a chain of N-acetylglucosamine. Chitin is present in crustacean co-products as a chitin–protein–mineral complex (mainly calcium carbonate). Crustacean shells contain about 30% chitin, 25% protein and 45% minerals on a dry matter basis. Chitin is also present in the co-products of cephalopods (squid feathers).

It is usually extracted chemically in a few steps:

– demineralization by acid treatment to solubilize the minerals;

– deproteinization by basic treatment to eliminate proteins;

– elimination of soluble phases (mineral and protein);

– purification of residual chitin phases.

At the end of these treatments, there can remain pigments incompletely eliminated like carotenoids. We then use a treatment with oxidizing agents.

Chitin can be deacetylated to more than 60–70% to obtain chitosan either chemically at basic pH or enzymatically way. The chitosan thus obtained is soluble at acidic pH, whereas chitin is insoluble. These costly processes are used very little in Europe. Chitin and chitosan are mainly imported from Asia (China, India).

Chitin and chitosan have various applications:

– in cosmetics: moisturizing agents, anti-bacterial, dental gels;

– in the biomedical field: dressings, anticoagulants;

– in wastewater treatment: flocculants, coagulants, metal chelators;

– in agriculture: seed coating, biostimulants;

– in food processing as a technological auxiliary: emulsifiers, coatings, stabilizers, preservation (anti-bacterial, anti-fungal);

– in packaging: anti-bacterial additives.

10.3.4.4. *Bioactive peptides*

The use of seafood as a source of bioactive compounds is one of the ways to increase the value of marine co-products. It is indeed possible to industrially produce active peptides by enzymatic modification of proteins using selected proteolytic enzymes. A wide range of fish peptides has been developed in Europe, the United States and especially in Asia (Guérard et al. 2010). Peptides are resistant to pH variations and heat, with the advantage of being non-allergenic compared to proteins.

In addition to their technical properties, peptides are interesting molecules for their wide range of biological activities (Pérez-Gálvez et al. 2018). First, the nutritional value of marine products, and in particular fish, is generally recognized. Fish proteins are considered to have a higher nutritional value than those of plants, as they consist of essential amino acids, with a better balance. Added to their protein and amino acid quality, marine peptides are also the most bioavailable form of amino acid absorption. These peptides are generally composed of 2 to 14 amino acids and have hydrophobic amino acid residues, in addition to proline, lysine or arginine groups. Bioactive peptides are also resistant to digestion by the action of peptidases.

During enzymatic hydrolysis, two parameters influence the final bioactivity by selecting different peptide sequences, namely the type of enzymes used and the degree of hydrolysis. At the end of the hydrolysis, the peptides are isolated according to their molecular mass by filtration or ultrafiltration of the solubilized phase.

The main biological activities of marine peptides most often highlighted are anti-hypertensive effect/(ACE inhibition), antioxidant, anticoagulant, immuno-stimulant properties and anti-tumoral effects (Suarez-Jimenez et al. 2012).

We can also mention the development of peptides that act on the reduction of the glycemic index and the reinforcement of satiety, the reduction of triglycerides and cholesterol. Other peptides have gastrointestinal effects, anti-stress effects or interact for cardiovascular problems.

The development and production of marine peptides for health applications requires a significant investment in process development, validation of the health effect and biological activity through in vitro and in vivo studies, as well as marketing. This investment limits the marketing of new products. However, several products with anti-stress and satiety-enhancing properties are marketed in Europe and Asia.

Another constraint is the yield between the amount of peptides produced in relation to the fish raw material used, which is generally low. Therefore, it is important to find an outlet and added value for the residual products obtained after the different extraction and purification steps of the obtained peptides. These peptides and hydrolysates must be deodorized to eliminate the fishy odor and must also be quite pure and white or light in color if possible. Optimization of the production costs and standardization of the quality of the obtained product is necessary.

10.3.4.5. *Lipid extracts*

Many commercial lipid-based products exist on the market. Marine co-products are a very good source of lipids particularly rich in omega 3 and omega 6, phospholipids, vitamins D and A present in the oil phases, as well as alkylglycerols.

The role of omega-3 fatty acids in the prevention of cardiovascular disease has been the subject of much research, both from a clinical and basic perspective (Jacobsen 2010).

Alkylglycerols (AKGs) are naturally occurring ether lipids in two forms. It appears that the natural unsaturated AKGs 16:1 and 18:1 are by far the most potent antitumoral and antimetastatic agents. This oil is used in empirical medicine in the Nordic countries largely involved in fishing, such as the Scandinavian countries, or Japan, for healing properties or stimulation of the immune system. AKGs are predominant in shark liver oil (Deniau et al. 2010).

10.4. Synthesis of extracted products from marine co-products

The range of products extracted from marine co-products is therefore wide and the applications varied. Table 10.4 details the products obtained according to the type of co-products, including the species of fish.

Categories of finished products	Type of product	The most used species	Best known finished products
Fishmeal	Mixed fishmeal	Sardines, anchovies, blue whiting	Fishmeal for aquaculture and pig feed
	Mono-species flour	Sardines, tuna, mackerel, shrimp	Fishmeal for dog and cat food or aromatic fish powder
Hydrolysates	Hydrolysate Aquaculture	Sardines, anchovies, whiting, white fish, oily fish, salmon	Hydrolysate for fry feed or shrimp feed
	Hydrolysate Petfood	Tuna, salmon, white fish	Palatable products Flavors for wet food for dogs and cats Peptides with anti-stress and palatability properties
	Flavors for human food	Crustaceans, cod, tuna, pelagic fish	Flavors, markers, aromatic extracts
	Flavors for human food Peptides with biological activity	Crustaceans, cod, tuna, pelagic fish White fish	Flavors, markers, aromatic extracts Nutraceutical ingredients with properties: anti-hypertensive, antioxidant, anticoagulant, immuno-stimulant, reduction of the glycemic index, triglycerides and gastrointestinal cholesterol, anti-stress, satiety
Oils	Animal feed	Pelagic, various	Oil rich in omega-3 (EPA, DHA)
	Human food	Pelagic, various	Oil rich in omega-3 (EPA, DHA) Phospholipids
	Squalene Alkylglycerols	Shark, ray	Cosmetics: oil with anti-aging properties Nutraceuticals
Active molecule ingredients	Chondroitin	Shark, ray	Food supplements for cartilage preservation
	Collagen	Various fish	Cosmetics, nutrition
	Chitin chitosan	Shrimps and prawns	Cosmetics, nutraceuticals Water treatment
Miscellaneous	Inks Dyes Minerals	Cuttlefish Pelagic fish scales	Human food, ready meals

Table 10.4. *Presentation of possible products from co-products according to markets and applications*

10.5. Conclusion

Better valorization of the co-products generated by the various activities of production and processing of seafood products is a challenge for the actors of this sector. Indeed, to produce from by-products or marine raw materials not used, or even rejected at sea, allows a reasoned use of the whole biomass taken during the catching activities. It also allows us to improve the profitability of the sector by a financial valorization of these co-products, which were previously considered as waste, generating a charge.

In recent years, the valorization of co-products has been developed more and more. This trend has allowed a better financial valorization, and the development of new products, applications and processes, which has also contributed to changing their brand image.

Despite a real development of the valorization of co-products, several obstacles still exist, such as the variability and seasonality of products and raw materials (volumes, composition), the fragmentation of deposits and collection costs, and the sometimes restrictive and evolving regulations that may discourage some actors. In addition, the animal origin of products poses a problem for certain markets, such as cosmetics. Investment and development require research and development, marketing and materials, and it may be considered expensive to develop these processes and know-how.

10.6. References

Ademe, Iterg, Actia, Aquimer Ifremer, CVG (2012). Projet VAMACOPIA : valorisation matière des coproduits de l'industrie agro-alimentaire – Du co-produit à l'ingrédient, comment lever les verrous ?

Alemán, A. and Martínez-Alvarez, O. (2013). Marine collagen as a source of bioactive molecules: A review. *Nat. Prod. J.*, 3(2), 105–114.

Asche, F. (2016). Fishmeal and oil: Why bother? Opportunities and challenges. Presentation, Symposium on Future Perspectives of Fishmeal and Fish Oil, Hirtshals, 29–30 August.

Cashion, T., Le Manach, F., Zeller, D., Pauly, D. (2017). Most fish destined for fishmeal production are food-grade fish. *Fish Fish.*, 18(5), 837–844.

Deniau, A.L., Mosset, P., Pédrono, F., Mitre, R., Bot, D.L., Legrand, A.B. (2010). Multiple beneficial health effects of natural alkylglycerols from shark liver oil. *Mar. Drugs*, 8(7), 2175–2184.

Gómez-Guillén, M.C., Giménez, B., López-Caballero, M.A., Montero, M.P. (2011). Functional and bioactive properties of collagen and gelatin from alternative sources: A review. *Food Hydrocoll.*, 25(8), 1813–1827.

Guérard, F., Decourcelle, N., Sabourin, C., Floch-Laizet, C., Le Grel, L., Le Floc'H, P., Gourlay, F., Le Delezir, L., Jaouen, P., Bourseau, P. (2010). Recent developments of marine ingredients for food and nutraceutical applications: A review. *J. Sci. Halieut. Aquat.*, 2, 21–27.

He, S., Franco, C.M., Zhang, W. (2015). Economic feasibility analysis of the industrial production of fish protein hydrolysates using conceptual process simulation software. *J. Bioprocess. Biotech.*, 5(1), 1.

Jacobsen, C. (2010). Challenges when developing omega-3 enriched foods. *Oléagineux, Corps gras, Lipides*, 17(4), 251–258.

Pérez-Gálvez, R., Espejo-Carpio, F.J., Morales-Medina, R., García-Moreno, P.J., Guadix-Escobar, A., Guadix-Escobar, E. (2018). Fish discards as source of health-promoting biopeptides. In *Handbook of Food Bioengineering, Alternative and Replacement Foods*, Holban, A.M. and Mihai Grumezescu, A. (eds). Academic Press, Cambridge.

Seeley, R.R., Stephens, T.D., Tate, P. (2008). Articulations and movement. *Anat. Physiol.*, 252–277.

Suarez-Jimenez, G.M., Burgos-Hernandez, A., Ezquerra-Brauer, J.M. (2012). Bioactive peptides and depsipeptides with anticancer potential: Sources from marine animals. *Mar. Drugs*, 10(5), 963–986.

Yata, M., Yoshida, C., Fujisawa, S., Mizuta, S., Yoshinaka, R. (2001). Identification and characterization of molecular species of collagen in fish skin. *J. Food Sci.*, 66(2), 247–251.

11

Biorefinery of Underutilized Marine Resources Using the pH-Shift Technology

Mehdi ABDOLLAHI
Chalmers University of Technology, Gothenburg, Sweden

11.1. Introduction

Seafood contributes substantially to food security as a very important source of nutrients for billions of people worldwide. Fisheries also contribute globally to support the income of 660–820 million people and has a great role to play in gender equality and poverty (FAO 2020). Within EU, the fish processing industry alone has an overall turnover of nearly 28 million euros and employing more than 122,000 people (STECF 2014). However, the seafood production chain greatly suffers from unsustainability due to sometimes harmful fishing activities that has led to overexploited stocks and due to inefficiencies in the processing steps with up to 70% of the aquatic resources ending up as by-products (FAO 2020). Still, the latter are primarily used for low-value purposes or, in worse cases, wasted (Abdollahi and Undeland 2019). Therefore, better utilization of the valuable fisheries raw materials which are caught or harvested from the ocean is crucial to meet the global demand for seafood from the rising world population.

For a color version of all figures in this chapter, see: http://www.iste.co.uk/verrez-bagnis/aquaticproducts.zip.

Current Challenges for the Aquatic Products Processing Industry,

Most of fish processing by-products such as head, frame, tail and trimming are a rich source of high-value food-grade products like fish protein, n-3 PUFA, marine collagens, vitamins (e.g. vitamin B12 and D) and minerals such as selenium and iodine. Shrimp peeling by-products are also rich in chitin, astaxanthin, protein and biocalcium. However, these by-products normally have a complex structure and encompass a high amount of prooxidants such as heme proteins besides their sensitive unsaturated fats (Abdollahi et al. 2016). These challenges along with interest for more complete utilization of fisheries resources have led to continuous research to find technologies allowing successful fractionation of these by-products to multiple and higher-value products such as proteins, fish oil, collagen, peptides and chitin while dealing with their sensitive and complex nature. These efforts have resulted in the development of the concept of blue biorefinery where the so-called pH-shift technology can play a central role in fractionation of fisheries by-products to multiple high-value products for food, nutraceutical, biomedical and nutricosmetic application. This chapter will provide an overview of the most recent progresses in the application of the pH-shift technology as a tool for multiple-product blue biorefinery.

11.2. Multiple-product blue biorefinery

More than a decade ago, the concept of biorefinery was derived from the petroleum refinery in which a range of chemicals and fuels are manufactured from petroleum using different technologies (Chen et al. 2016). Biorefinery means integrated and sustainable processing of biomass into a spectrum of marketable chemicals, biomaterials and bioenergy (Cherubini 2010). This concept has been successfully implemented for woody biomass consisting primarily of cellulose, hemicellulose and lignin, which through an array of processes are fractionated and converted into a spectrum of products including transportation fuels and fine chemicals (Jin et al. 2018). By using fisheries biomasses such as underutilized fish, fish processing by-products, crustacean by-products and seaweeds as feedstock in biorefinery, the concept of blue biorefinery is developed. Blue biorefinery allows improved utilization of fisheries resources by generating value from all parts of particular biomass feedstock sources. As indicated in Figure 11.1, new biorefining and fractionation technologies such as pH-shift processing can help to decrease waste by making higher-value products such as fish oil, proteins, collagen, peptides and minerals from fish processing by-products for different applications and markets. Compared with the single-product approach conventionally applied in the valorization of fisheries biomasses, multiple-product blue biorefinery has several remarkable advantages, including (a) full utilization of fisheries biomasses, thus minimizing waste generation during processing; (b) diversification of the revenues by covering multiple markets/niches; (c) synergy effects of different technologies and (d) sharing manpower and equipment (Jin et al. 2018).

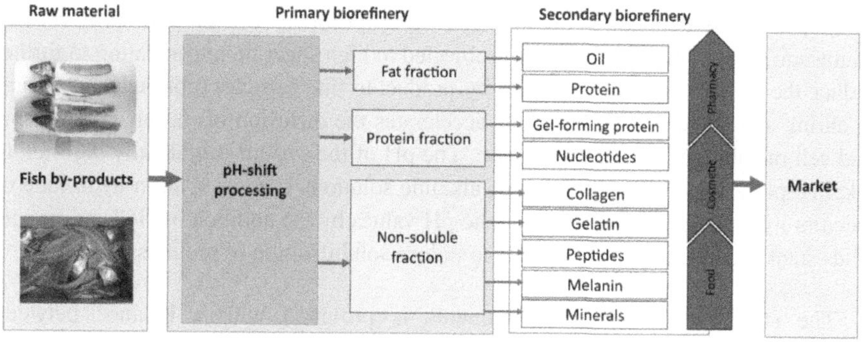

Figure 11.1. *Schematic overview of a potential blue biorefinery approach for fish processing by-products based on the pH-shift technology*

11.3. pH-shift technology and its fractionation principles as a biorefinery tool

The pH-shift process which is also called acid and/or alkaline solubilization and isoelectric precipitation has been for a long time industrially in use for isolation of plant proteins. However, following some pioneering works (Meinke et al. 1972; Meinke and Mattli 1973), the process was first adapted to fisheries raw materials by researchers at University of Massachusetts in the United States and patented (Hultin and Kelleher 2000 and 2001). As illustrated in Figure 11.2, the process, which is run at below ambient temperature, involves selectively extracting proteins and lipids from homogenized by-products in water using a high (> 10) or low (< 3.5) pH to solubilize the muscle proteins followed by centrifugation to separate the solubilized proteins from high- and low-density undissolved materials. The solubilized proteins are then recovered using isoelectric precipitation (usually pH 5.5) and dewatered by centrifugation or filtration (Nolsøe and Undeland 2009). It is important that the process is conducted at cold conditions to minimize the risk of both lipid oxidation and protein denaturation considering the sensitive nature of the marine raw materials.

The chemical principles behind the pH-shift process have been comprehensively described previously (Gehring et al. 2009, 2011). The process is based on the fact that protein solubility and insolubility in water can be induced by proper pH adjustment to extreme acid/alkaline pHs and to their isoelectric point (pI), respectively. The process is started with grinding the by-products to destructure the raw materials which facilitates their dispersion in water and subsequent solubilization of acid/alkaline soluble components of the by-products and their fractionation. The minced by-products are then mixed with 3–10 ratio of water (mostly 6 ratio) depending on the type of by-products and their protein content (Surasani 2018). The mixture is then subjected to high shear homogenization to further reduce the particle size of the minced by-product to fine particles (normally < 0.2 mm) (Gehring and Davenport 2009). This accelerates the diffusion of muscle components and cell membranes to the water phase. The pH of the system is gradually adjusted to alkaline pH values by the addition of alkaline solutions such as sodium hydroxide or calcium hydroxide (2–10 N) or to acidic pH values by the addition of HCL or organic acids such as acetic acid or citric acid to induce solubilization of proteins.

The stability of muscle components is provided with a balance between protein–protein hydrophobic interactions and electrostatic interactions of protein side chains with other cellular components and their surrounding environment. Depending on the selected process version, the protein side chains will be negatively or positively charged with the addition of base or acid to the homogenate, respectively. The accumulation of these introduced negative and positive charges results in protein–protein electrostatic repulsion and an increased hydrodynamic volume due to expansion and swelling (Kristinsson et al. 2005; Gehring et al. 2011). With further increasing or decreasing the pH, proteins with higher negative and positive surface net charge gradually start to interact with water. At the same time, the number of protein–protein hydrophobic interactions decreases. Finally, with increase in the surface charges of the proteins and association of more water to the protein's surface, they become water soluble and extractable (Nolsøe and Undeland 2009). At the same time, the homogenization process and the increased charge in the system induced by the pH adjustment facilitate the separation of both the neutral storage lipids and the more polar membrane phospholipids from proteins (Kristinsson et al. 2005). At optimum solubilization pH, which depends on the type of start material, ionic strength, water ratio, etc., the muscle proteins which are tightly enough bound to the water can be gravimetrically separated from non-soluble components of the by-products.

Centrifugation of the homogenate at an appropriate solubilization pH at 3,000–10,000×g for 10–20 min can result in the generation of two or three fractions depending on the type of start material (Surasani et al. 2017; Hinchcliffe et al. 2019). Fractionation of lean fish such as cod or crustacean by-products results in two fractions including solubilized proteins as supernatant and non-soluble collagenous or chitinous residue as precipitate, respectively (Khumallambam et al. 2011; Abdollahi and Undeland 2019; Khiari et al. 2020). Three fractions including a floating emulsion layer containing light neutral fats and fat-soluble compounds, a middle layer containing solubilized proteins and a precipitating layer encompassing collagenous residue such as bone, skin, scale and connective tissue are formed when using fatty fish or fisheries by-products with a high amount of neutral lipids such as herring, trout, carp and salmon as start material (Chen and Jaczynski 2007; Abdollahi et al. 2017; Abdollahi and Undeland 2019).

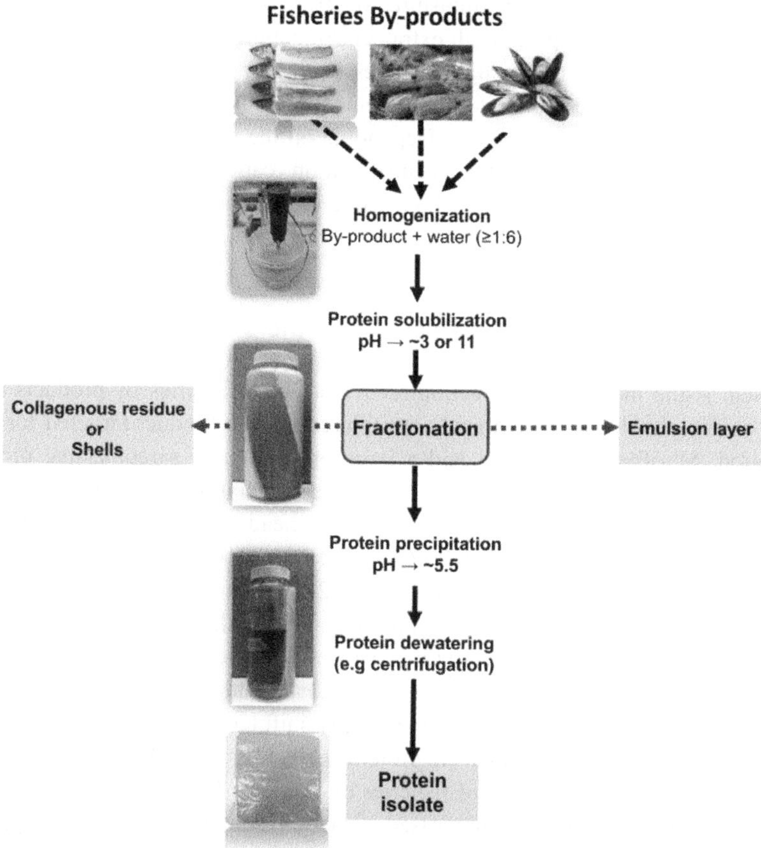

Figure 11.2. *Schematic overview of the classic pH-shift process for fractionation and protein isolation from fisheries by-products*

The solubilized functional proteins (both myofibrillar and sarcoplasmic) remaining in the middle layer can be then recovered by readjusting its pH to the pI of the proteins which is around 5.2–5.5 for fish proteins and 4–4.5 for crustacean proteins. At pI, proteins have zero net electrostatic charge and minimum water solubility. By adjusting pH to pI, protein–protein hydrophobic interactions take over protein electrostatic interactions with water which induce protein coagulation and precipitation (Gehring et al. 2011). The coagulated proteins can be then dewatered and recovered as a protein paste with filtration or centrifugation. The recovered proteins may still have high moisture content (> 90%) depending on the type of start material, pH, ionic strength, process version, centrifugation force, etc. Therefore, action for further removing the moisture to 80–85% should be applied, such as pressing the sample at the protein pI, extra centrifugation step or mixing the proteins with dried ingredients. The separated processing water will also contain different amounts of sarcoplasmic proteins such as heme pigments, salt, peptides, free amino acids, phospholipids and nucleotides. Depending on the type of start material, other fractions generated after the first centrifugation can be collected for further fractionation and extraction of other high-value compounds which are explained in sections 11.5 and 11.6.

11.4. Isolation of functional proteins using the pH-shift technology

Protein is the most studied fraction recovered from fisheries by-products using the pH-shift technology and is mostly considered as sole product targeted for food application. Myofibrillar proteins and a large proportion of sarcoplasmic proteins remaining on fisheries by-products can be recovered using the pH-shift technology. Different fisheries by-products such as fish frame, head, tail, roe, skin, dark muscle, viscera (Chen and Jaczynski 2007; Zhong et al. 2016; Shaviklo et al. 2017; Abdollahi and Undeland 2019; Yoon et al. 2019), shrimp peeling by-products (Khumallambam et al. 2011), sea cucumber guts (Du et al. 2019) and scallop gonad (Han et al. 2019) as well as underutilized resources such as invasive green crab (Khiari et al. 2020), krill (Tahergorabi et al. 2013) and empty crab have been studied as raw material for protein isolation using the pH-shift technology. As can be seen in Table 11.1, a large variation in total protein yield (30% to 90%) is found in the literature for pH-shift processing of fisheries by-product. This large variation can be related to differences in the processed fisheries source and species, type of by-products, pH-shift process version, solubilization/precipitation pH and even yield calculation method. For example, Abdollahi and Undeland (2019) found that the alkaline process version of the pH-shift process results in significantly higher total protein yield than its acid version when applied on by-product (head+frame) of three different species including salmon, cod and Atlantic herring. Solubilization pH resulting in maximum protein yield was also species-specific and needed a careful optimization as shown in that study. However, acid process version resulted in twice higher protein yield than the alkaline version when it was applied on anchovy's by-product mixture (viscera, head, fins, tail and spikes) (Rodrigues Freitas et al. 2016).

Source	By-product	Protein yield (%)		Protein content of isolate (% DW)	Gelation (breaking force in g)	Whiteness	Reference
		Acid version	Alkaline version				
Trout	Frame, skin, scales and fin	81–89	78–81	36–53	-	92–94	Chen and Jaczynski (2007)
Silver carp	Frame, dark muscle, cutoffs, viscera, skin, scales, bones and fins	60–70	50–70	76–82	-	61–68	Zhong et al. (2016)
Tilapia	Frame	15–16	16–19	87–90	220–320	-	Chomnawang and Yongsawatdigul (2013)
Grouper	Frame	81	87	-	97–254	52–54	Animal et al. (2020)
Skipjack tuna	Roe	7.8–7.9	7–9	71–76	-	48–53	Cha et al. (2020)
Cod	Head+frame	20–23	58–60	84–86	90–110	55–56	Abdollahi and Undeland (2019)
Salmon	Head+frame	40–45	63–79	67–69	30–56	64–67	Abdollahi and Undeland (2019)
Herring	Head+frame	32–42	42–65	86–89	60–90	49-51	Abdollahi and Undeland (2019)
Rohu	Head, skin, bone frame and viscera	31	31	-	-	41–52	Surasani et al. (2016)
Pangas	Frame	59	68	-	-	50–53	Surasani et al. (2017)
Argentine Anchovy	Viscera, head, fins, tail and spikes	64	35	-	-	49–55	Irene Rodrigues Freitas (2016)
Whitemouth Croaker	Viscera, head, fins, tail and spikes	56	64	-	-	47–56	Irene Rodrigues Freitas (2016)

a; The results have been reported as a range when the parameters have been reported for several pH values.

Table 11.1. *Overview of example by-products used for protein isolation using the pH-shift technology and their protein yield, protein content of the recovered proteins and their gel formation capacity and whiteness*

The pH-shift technology is in general very efficient in up-concentration and purification of myofibrillar proteins in the final protein isolate compared to the used start material via removal of high-density collagenous/chitinous residues. This results in a protein concentration and purity between 70% and 90% in final protein recovered using the pH-shift processing in most cases, as shown in Table 11.1. Part of this purification is also achieved by fat removal when the process is applied on the by-products of fatty species by removal of fat as a floating emulsion layer. However, the process may cause up-concentration of fat in the final protein isolate compared with the start material when it is applied on by-products of lean fish or crustacean. For example, an 87–119% increase in the fat content of cod protein isolate was found compared with its start material which was head+frame by-products (Abdollahi and Undeland 2019). From a nutritional perspective, this purification results in a higher amount of protein and essential amino acids per 100 g of final products developed from the protein isolates. However, the effect of this purification on other nutrients will be dependent on the processed species and water-soluble or fat-soluble nature of the nutrients as well as their binding affinity to proteins. For example, protein recovered from salmon and herring by-products contained lower content of n-3 PUFA and vitamin D compared with their frame while these nutrients were up-concentrated in cod protein isolate compared with its frame (Abdollahi et al. 2021). However, vitamin C, which is a water-soluble vitamin, was completely removed from the protein isolate of the three species compared with their original frame (Abdollahi et al. 2021).

The effect of pH-shift processing on the mineral content of proteins recovered from the fish frames was dependent on whether the minerals are located in the bone or muscle fraction of backbones as well as their binding affinity to proteins versus their water solubility. Minerals rich in bones, such as calcium and magnesium, were removed to a higher degree by the pH-shift process compared to mechanical separation, while a reverse trend was seen for minerals with high binding affinity to protein, for example, zinc. Highly water-soluble minerals, for example, potassium, were severely leached out by the pH-shift process (Lee et al. 2016; Abdollahi et al. 2021). Therefore, the effect of pH-shift processing on the content of nutrients in the recovered protein fraction from fisheries by-products varies a lot depending on the type of nutrient and the processed species. This should be carefully considered in application of the final protein isolate for food product development.

Proteins recovered from fish processing by-products such as head and frame or their combination can have gel-forming capacity, but the gel strength will depend on the processed species and processing condition. Proteins recovered from by-products will usually result in soft gels and their strength will be far from the strength of gels made of classic surimi from the fillet of white muscle fish such as Alaska Pollak.

However, their gel strength can be effectively improved with tuning the protein extraction process, some additives and optimizing gel production condition (Chen and Jaczynski 2007; Abdollahi et al. 2019). Therefore, this cannot be considered as a limitation for food development purposes. Other functional properties of proteins recovered from fisheries by-products such as solubility, emulsion and foaming capacity have shown a pH-dependent behavior, which normally improves by moving away from proteins pI and in this sense behaves more like soy protein (Panpipat and Chaijan 2017; Abdollahi and Undeland 2018; Shang et al. 2020).

Proteins recovered from fish processing by-products will have a brownish color. In case of crustacean, the protein will be more pinkish to orange color due to the astaxanthin residue coming from the crustacean shells. As can be seen in Table 11.1, the whiteness of proteins recovered from fisheries by-products is normally around 50, which is far from the whiteness of classic surimi from the fillet of white muscle fish and should be considered in their market development when whiteness is an important factor for a specific product and/or for a group of consumers. Abdollahi and Undeland (2018) showed that the application of proteins isolated from salmon and cod head+frame had no negative impact on the whiteness of mayonnaise-like emulsions developed with these proteins, but herring protein isolate reduced the whiteness of the mayonnaise. This means that the type of application of the proteins, their concentration and function in the product and color of the final product will define whether proteins from a specific fisheries by-product can suit that application or not. A new concept called cross-processing was suggested by Abdollahi et al. (2020) in which fish filleting by-products are cross-processed with agricultural side streams such as lingonberry, seaweed and shrimp that can produce proteins with a wide range of new colors for innovative applications.

11.5. Cold extraction of fish oil parallel with gel-forming proteins

Fish oil is the major commercially available source of long-chain n-3 polyunsaturated fatty acid (LC-n3 PUFA). The demand for high-quality fish oil is continuously increasing due to the growing health consciousness among consumers and awareness about the health benefits of LC n-3 PUFA, increasing incidence of chronic diseases and increasing demand for omega-3 in functional foods (Pike and Jackson 2010). Current global fish oil production is valued more than USD 2 billion (1.5 billion euros), in which omega-3 market is expected to reach USD 7,583.5 million by 2023, at a CAGR of 14.5% during the forecast period of 2018–2023 (ResearchAndMarkets.com 2018).

Currently, the industrial available process for production of fish oil from fish processing by-products is the classic combined fish meal/fish oil production process called wet reduction. In this technology, fish oil is produced as a co-product after cooking (90–94°C) of grinded raw materials and followed by decanting (Sobstad 1992). Unfortunately, the produced protein-rich fraction (which is called fishmeal after drying) is not suitable for human consumption due to its poor functionality and lower biological value. Also, sensory aspects are far from optimal. The heating involved in the wet reduction process causes lipid oxidation and possibly reduction of the LC-n3 PUFA content in the recovered fish oil too. Thus, the recovered fish oil will need subsequent costly refining steps to be used for human consumption (Okada and Morrissey 2007).

As mentioned before, the first fractionation step during the pH-shift processing of by-products of fatty fish such as salmon, herring, carp and trout results in the generation of a floating low-density emulsion layer, in parallel with the gel-forming protein-containing layer. This emulsion layer which is formed due to the high shear mechanical homogenization, high solubilization pHs and high-speed centrifugation involved in this process, encompasses mainly neutral lipids. It has been shown that the relative n-3 PUFA content in the emulsion layer recovered by pH-shift processing of trout by-product was similar to that of the original raw material (Chen et al. 2007). In addition, the emulsion layer formed during the pH-shift processing of gutted silver carp contained 50–80% moisture, but as much as 88–97% of total lipids on a dry weight basis (Taskaya et al. 2009). Recently, Abdollahi and Undeland, (2020b) patented a process for breaking the floating emulsion layer in cold condition (≤ 4°C) for extraction of high-quality fish oil from fish by-products using the pH-shift technology in parallel with gel-forming proteins. The process summarized in Figure 11.3 involves subjecting the emulsion layer to pH adjustment to proteins pI (~pH 5.5), or freeze/thaw and if needed to their combination. For example, pH adjustment or freeze/thawing alone can efficiently release the oil in the floating emulsion layer formed during pH-shift processing of salmon by-products at 4°C. However, for herring by-products, higher temperature (10°C) and a combination of the emulsion-breaking techniques was required for efficient oil release. The oil recovery based on the initial weight of salmon and herring by-products including head and frame was around 9% and 4.5%, respectively, which was lower than the oil recovery using the classic cooking at 90°C for 20 min. However, the alkaline version of the pH-shift process in combination with the emulsion breaking techniques resulted in substantially higher-quality fish oil, for example, with significantly less oxidation products (PV and TBARS) and FFA, as well as slightly higher amount of n-3 PUFA, particularly DHA for both species. The lower oxidation using the alkaline version of the pH-shift process was mainly related to conducting the process at cold condition (≤ 10°C), which inhibits most of oxidative reactions compared with the classic heating process. Oils extracted with the acid version of the pH-shift process, however, showed high levels of lipid oxidation in both species. This could be due to acid-induced met-Hb/met-Mb formation and denaturing of heme-pigments which, according to the above, could increase their pro-oxidative potential.

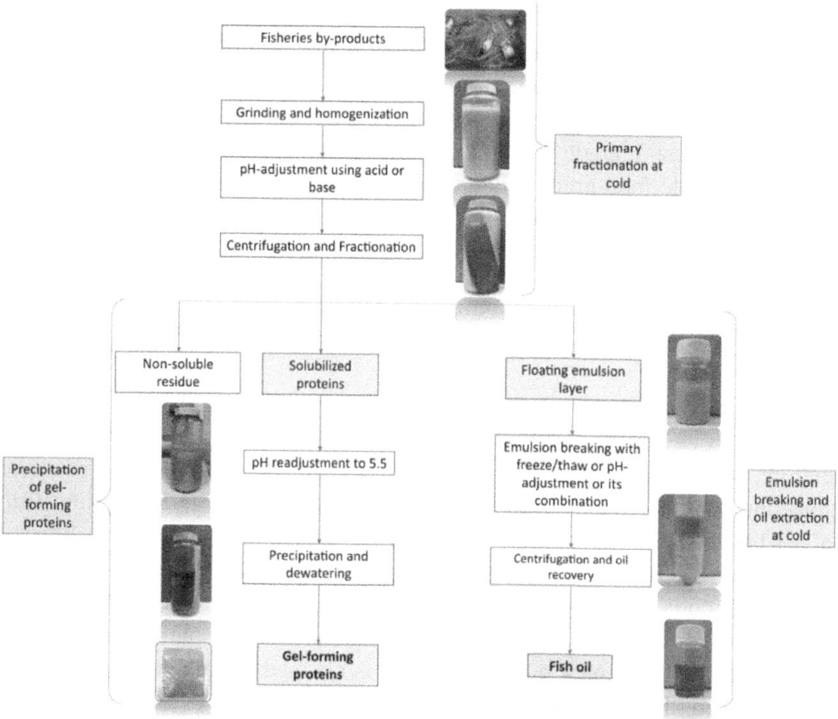

Figure 11.3. *A cold process for parallel isolation of fish oil and gel-forming proteins using the pH-shift technology. Adapted from Abdollahi and Undeland (2020a)*

The oil extracted from the by-product of both species using the pH-shift technology had a lower content of phospholipids compared with the oils extracted using the heating method. This can be explained by looking more deeply into the separation and removal of fats during the pH-shit processing. Centrifugation at extreme solubilization pH levels facilitates the separation of the charged proteins from both neutral storage lipids and the more polar membrane phospholipids. At this situation, it is expected that most of the storage lipids, which are usually less unsaturated than the membrane lipids and more hydrophobic, would float to the top emulsion layer. If the centrifugation force is high enough, a large portion of the more unsaturated phospholipids is expected to settle due to the relatively high density of membranes (Shi et al. 2017). However, this depends on the processed species, as a large proportion (70–75%) of the phospholipids present in gutted Atlantic menhaden and whole krill remained in the protein fraction and end up in the second supernatant upon pH-shift processing due to the amphiphilic nature of the phospholipids (Shi et al. 2017).

Extending the pH-shift process for the extraction of fish oil in parallel with gel-forming proteins in cold condition enables using the ignored or even wasted fractions generated during the process. This also allows for the production of high-quality omega-3 grade fish oil without heat or environmentally hazardous chemicals, multiple food-grade products generation from fish by-products, increase in the (whole process) profitability from a given amount of fish raw material and possible movement towards "zero waste" in the fish processing industry.

11.6. Collagen extraction using the pH-shift technology

Collagen is the most abundant protein in human body that plays very important structural roles as a main component of our skin, bone, joint and cellular skeleton. Therefore, native collagen has a wide range of applications in biomedical engineering such as wound healing, tissue engineering and 3D bioprinting (Silva et al. 2014). In addition, collagen hydrolysates or peptides is a unique bioactive nutrient which has gained great interest as both nutraceutical and nutricosmetic. As a nutraceutical, collagen is marketed with health claims such as promoting joint and bone health, antidiabetic activity, anti-inflammatory and antioxidant activity (Salvatore et al. 2020). As a nutricosmetic, it is claimed that collagen can promote skin, nail and hair health and beauty with anti-photoaging, anti-wrinkle, moisturizing, antioxidant and anti-inflammatory activities (Spiro and Lockyer 2018). Traditionally, animal by-products such as skin and bone from pig, bovine and chicken have been the main sources of commercial collagen-based products. However, some health concerns related to outbreak of bovine sponge encephalopathy, foot-and-mouth disease and avian influenza together with religious limitations for application of pig and bovine collagen in kosher and halal foods have caused great interest in finding alternative collagen resources (Silva et al. 2014). Within the health and beauty segment, marine collagen has gained significant interest as a healthy, environmentally sound and alternative to bovine and pork-derived products. A wide range of fisheries by-products including fish skin, head, bone, scale and swim bladder have shown a great potential as source of both native collagen and collagen hydrolysate (Salvatore et al. 2020). However, the focus has been always on a single-product approach.

It has been shown that sediments produced after the first centrifugation during the pH-shift processing of gutted silver carp (Taskaya et al. 2009) and trout by-products (Chen et al. 2007) still contain 42–67% of protein. As explained before, this sediment fraction mainly contains collagenous residues including bone, skin and connective tissue, which can be a great source for isolation of fish collagen, gelatin or collagen hydrolysate.

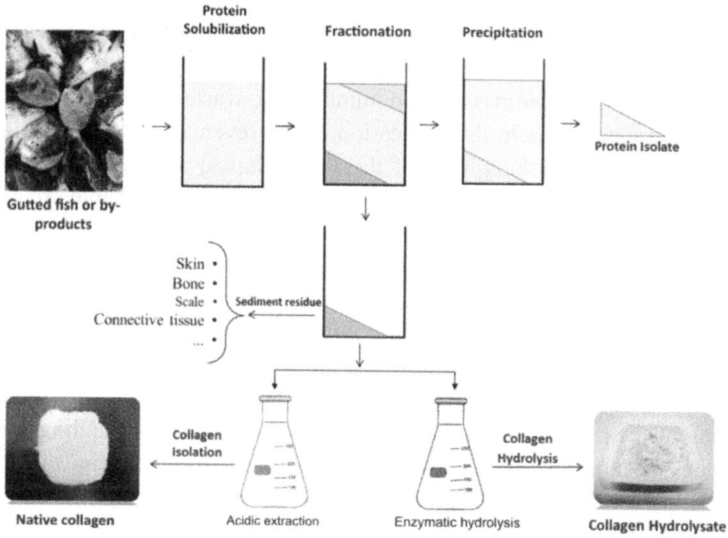

Figure 11.4. *Schematic overview of the process suggested for sequential isolation of gel-forming proteins and collagen from fish by-products. Adapted from Abdollahi et al. (2018)*

Application of these sediment residues emerging during isolation of gel-forming proteins can provide a more purified source for collagen since a large part of non-collagenous proteins are then removed as a main product. Based on this concept, as indicated in Figure 11.4, Abdollahi et al. (2018) used sediment residue emerging during protein isolation from gutted silver carp using the pH-shift processing for extraction of both native collagen and collagen hydrolysates. The study demonstrated that the sediment can be a very good source for isolation of high-quality type I native collagen and collagen peptides with large and small molecular weight. However, three to five times higher yield was obtained using the hydrolysis process compared with the isolation of native collagen. Addition of the extracted collagen peptides as nutraceutical ingredient to the protein isolated from the silver carp affected gel-forming properties of the protein isolate depending on their hydrolysis degree. Industrial implementation of this concept can enable resource-efficient biorefining of bone containing fish raw materials into two different products; functional proteins and collagen or collagen hydrolysate. This allows maximum utilization of marine resources and can create an additional stream of revenue for processors.

11.7. Conclusions and future prospects

The concept of biorefinery has great potential to enable more complete utilization of fisheries biomasses and minimizing waste generation during their valorization. It can also help diversification of the revenues for fisheries industries by covering multiple markets. One of the key technical aspects to succeed in the establishment of blue biorefinery platforms is finding suitable fractionation technologies which can deal with the complex and sensitive nature of fisheries raw materials. The pH-shift technology has a great potential in fractionation of fish and crustacean by-products in cold condition to high-quality and semi-intact gel-forming proteins, virgin n-3 PUFA-rich fish oil, collagen and chitin. The efficiency of the pH-shift technology in fractionation of the by-products to different valuable compounds depends, to a large extent, on the type of the fisheries by-product, and fine tuning of the process for each raw material is necessary. Expanding and testing the concept of multiple-product biorefinery on more fisheries raw materials and further efforts for the implementation of the concept in pilot and on an industrial scale is seen as a demand for future studies.

11.8. References

Abdollahi, M. and Undeland, I. (2018). Structural, functional, and sensorial properties of protein isolate produced from salmon, cod, and herring by-products. *Food and Bioprocess Technology*, 11, 1733–1749.

Abdollahi, M. and Undeland, I. (2019). Physicochemical and gel-forming properties of protein isolated from salmon, cod and herring by-products using the pH-shift method. *LWT*, 101, 678–684.

Abdollahi, M. and Undeland, I. (2020). Fish oil recovery from aquatic biomass. PCT/SE2020/050967.

Abdollahi, M., Marmon, S., Chaijan, M., Undeland, I. (2016). Tuning the pH-shift protein-isolation method for maximum hemoglobin-removal from blood rich fish muscle. *Food Chemistry*, 212, 213–224.

Abdollahi, M., Rezaei, M., Jafarpour, A., Undeland, I. (2017). Dynamic rheological, microstructural and physicochemical properties of blend fish protein recovered from kilka (*Clupeonella cultriventris*) and silver carp (*Hypophthalmichthys molitrix*) by the pH-shift process or washing-based technology. *Food Chemistry*, 229, 695–709.

Abdollahi, M., Rezaei, M., Jafarpour, A., Undeland, I. (2018). Sequential extraction of gel-forming proteins, collagen and collagen hydrolysate from gutted silver carp (*Hypophthalmichthys molitrix*), a biorefinery approach. *Food Chemistry*, 242, 568–578.

Abdollahi, M., Rezaei, M., Jafarpour, A., Undeland, I. (2019). Effect of microbial transglutaminase and setting condition on gel properties of blend fish protein isolate recovered by alkaline solubilisation/isoelectric precipitation. *International Journal of Food Science and Technology*, 54(3), 762–770.

Abdollahi, M., Olofsson, E., Zhang, J., Alminger, M., Undeland, I. (2020). Minimizing lipid oxidation during pH-shift processing of fish by-products by cross-processing with lingonberry press cake, shrimp shells or brown seaweed. *Food Chemistry*, 327, 127078.

Abdollahi, M., Wu, H., Undeland, I. (2021). Impact of processing technology on macro- and micronutrient profile of protein-enriched products from fish backbones. *Foods*, 10(5), 950.

Cha, J.W., Yoon, I.S., Lee, G.W., Kang, S.I., Park, S.Y., Kim, J.S., Heu, M.S. (2020). Food functionalities and bioactivities of protein isolates recovered from skipjack tuna roe by isoelectric solubilization and precipitation. *Food Science and Nutrition*, 8, 1874–1887.

Chen, Y.C. and Jaczynski, J. (2007). Protein recovery from rainbow trout (*Oncorhynchus mykiss*) processing byproducts via isoelectric solubilization/precipitation and its gelation properties as affected by functional additives. *Journal of Agricultural and Food Chemistry*, 55, 9079–9088.

Chen, Y.C., Tou, J.C., Jaczynski, J. (2007). Amino acid, fatty acid, and mineral profiles of materials recovered from rainbow trout (*Oncorhynchus mykiss*) processing by-products using isoelectric solubilization/precipitation. *Journal of Food Science*, 72(9), C527–C535.

Chen, X., Yang, H., Yan, N. (2016). Shell biorefinery: Dream or reality? *Chemistry – A European Journal*, 22, 13402–13421.

Cherubini, F. (2010). The biorefinery concept: Using biomass instead of oil for producing energy and chemicals. *Energy Conversion and Management*, 51, 1412–1421.

Chomnawang, C. and Yongsawatdigul, J. (2013). Protein recovery of Tilapia frame by-products by pH-shift method. *Journal of Aquatic Food Product Technology*, 22, 112–120.

Du, Y., Guo, X., Han, Y., Han, J., Yan, J., Shang, W., Wu, H. (2019). Physicochemical and functional properties of protein isolate from sea cucumber (*Stichopus japonicus*) guts. *Journal of Food Processing and Preservation*, 43(7), 1–11.

FAO (2020). The state of world fisheries and aquaculture 2020. Report, FAO, Rome.

Gehring, C.K., Davenport, M.P., Jaczynski, J.J. (2009). Functional and nutritional quality of protein and lipid recovered from fish processing by-products and underutilized aquatic species using isoelectric solubilization/precipitation. *Current Nutrition & Food Science*, 5, 17–39.

Gehring, C.K., Gigliotti, J.C., Moritz, J.S., Tou, J.C., Jaczynski, J. (2011). Functional and nutritional characteristics of proteins and lipids recovered by isoelectric processing of fish by-products and low-value fish: A review. *Food Chemistry*, 124, 422–431.

Han, J., Tang, Y., Li, Y., Shang, W., Yan, J., Du, Y., Wu, H., Zhu, B., Xiong, Y.L. (2019). Physiochemical properties and functional characteristics of protein isolates from the scallop (*Patinopecten yessoensis*) Gonad. *Journal of Food Science*, 1750–3841, 14598.

Hinchcliffe, J., Gunnar, N., Jönsson, E., Sundell, K., Undeland, I. (2019). Aquafeed ingredient production from herring (*Clupea harengus*) by-products using pH-shift processing: Effect from by-product combinations, protein solubilization-pH and centrifugation force. *Animal Feed Science and Technology*, 247, 273–284.

Hultin, H.O. and Kelleher, S.D. (2000). High efficiency alkaline protein extraction. U.S. Patent 6136959, University of Massachusetts, Boston.

Hultin, H.O. and Kelleher, S.D. (2001). Process for isolating a protein composition from a muscle source and protein composition. U.S. Patent 6136959A.

Jin, Q., Yang, L., Poe, N., Huang, H. (2018). Integrated processing of plant-derived waste to produce value-added products based on the biorefinery concept. *Trends in Food Science and Technology*, 74, 119–131.

Khiari, Z., Kelloway, S., Mason, B. (2020). Turning invasive green crab (*Carcinus maenas*) into opportunity: Recovery of chitin and protein isolate through isoelectric solubilization/precipitation. *Waste and Biomass Valorization*, 11, 133–142.

Khumallambam, M., Rama, S., Karuppannan, R.R., Manjabhat, S.N. (2011). Antioxidant activity of protein isolate obtained by the pH shift technique from hydrolyzed and unhydrolyzed shrimp processing discards. *Journal of Aquatic Food Product Technology*, 20, 209–221.

Kristinsson, H.G., Theodore, A.E., Demir, N., Ingadottir, B. (2005). A comparative study between acid- and alkali-aided processing and surimi processing for the recovery of proteins from channel catfish muscle. *Journal of Food Science*, 70, 298–306.

Lee, H.J., Lee, G.-W., Yoon, I.S., Park, S.H., Park, S.Y., Kim, J.-S., Heu, M.S. (2016). Preparation and characterization of protein isolate from Yellowfin tuna *Thunnus albacares* roe by isoelectric solubilization/precipitation process. *Fisheries and Aquatic Science*, 19, 1–10.

Meinke, W.W. and Mattli, K.F. (1973). Autolysis as a factor in the production of protein isolates from whole fish. *Journal of Food Science*, 38, 864–867.

Meinke, W.W., Rahman, M.A., Matil, K.F. (1972). Some factors influencing the production of protein isolates from whole fish. *Journal of Food Science*, 37, 195–198.

Nolsøe, H. and Undeland, I. (2009). The acid and alkaline solubilization process for the isolation of muscle proteins: State of the art. *Food and Bioprocess Technology*, 2, 1–27.

Okada, T. and Morrissey, M.T. (2007). Recovery and characterization of sardine oil extracted by pH adjustment. *Journal of Agricultural and Food Chemistry*, 55, 1808–1813.

Panpipat, W. and Chaijan, M. (2017). Functional properties of pH-shifted protein isolates from bigeye snapper (*Priacanthus tayenus*) head by-product. *International Journal of Food Properties*, 20, 596–610.

Pike, I.H. and Jackson, A. (2010). Fish oil: Production and use now and in the future. *Lipid Technology*, 22, 59–61.

ResearchAndMarkets.com (2018). Global omega-3 market trends, analysis & forecasts 2018–2023.

Rodrigues Freitas, I., Cortez-Vega, W.R.C., Prentice, C. (2016). Physicochemical and functional properties of protein recovered from fish waste. *Journal of Aquatic Food Product Technology*, 8850, 1–11.

Salvatore, L., Gallo, N., Natali, M.L., Campa, L., Lunetti, P., Madaghiele, M., Blasi, F.S., Corallo, A., Capobianco, L., Sannino, A. (2020). Marine collagen and its derivatives: Versatile and sustainable bio-resources for healthcare. *Materials Science and Engineering C*, 113, 110963.

Shafiq, U., Baraiya, K.G., Surasani, V.K.R., Joshi, S., Varadaraju, R.C. (2020). Extraction of protein isolates from grouper (*Epinephelus diacanthus*) filleting waste using pH shift method; influence of process variables on recoveries. 1–17.

Shang, W.H., Yan, J.N., Du, Y.N., Cui, X.F., Su, S.Y., Han, J.R., Xu, Y.S., Xue, C.F., Zhang, T.T., Wu, H.T. et al. (2020). Functional properties of gonad protein isolates from three species of sea urchin: A comparative study. *Journal of Food Science*, 85, 3679–3689.

Shaviklo, A.R., Rezapanah, S., Motamedzadegan, A., Damavandi-Kamali, N., Mozafari, H. (2017). Optimum conditions for protein extraction from tuna processing by-products using isoelectric solubilization and precipitation processes. *Iranian Journal of Fisheries Sciences*, 16, 774–792.

Shi, L., Beamer, S.K., Yin, T., Matak, K.E., Yang, H., Jaczynski, J. (2017). Mass balance for isoelectric solubilization/precipitation of carp, chicken, menhaden, and krill. *LWT – Food Science and Technology*, 81, 26–34.

Silva, T.H., Moreira-Silva, J., Marques, A.L.P., Domingues, A., Bayon, Y., Reis, R.L. (2014). Marine origin collagens and its potential applications. *Marine Drugs*, 12, 5881–5901.

Sobstad, G.E. (1992). Marine oil separation purification technology. *International News on Fats, Oils and Related Materials (USA)*.

Spiro, A. and Lockyer, S. (2018). Nutraceuticals and skin appearance: Is there any evidence to support this growing trend? *Nutrition Bulletin*, 43, 10–45.

STECF (2014). The Economic Performance Report on the EU Fish Processing (STECF-14-21). Report, Scientific Technical and Economic Committee for Fisheries.

Surasani, V.K.R. (2018). Acid and alkaline solubilization (pH shift) process: A better approach for the utilization of fish processing waste and by-products. *Environmental Science and Pollution Research*, 25, 18345–18363.

Surasani, V.K.R., Tyagi, A., Kudre, T. (2016). Recovery of proteins from rohu processing waste using pH shift method: Characterization of isolates. *Journal of Aquatic Food Product Technology*, 8850, 1–10.

Surasani, V.K.R., Khatkar, S.K., Singh, S. (2017). Effect of process variables on solubility and recovery yields of proteins from pangas (*Pangasius pangasius*) frames obtained by alkaline solubilization method: Characteristics of isolates. *Food and Bioproducts Processing*, 106, 137–146.

Tahergorabi, R., Beamer, S.K., Matak, K.E., Jaczynski, J. (2013). Chemical properties of ω-3 fortified gels made of protein isolate recovered with isoelectric solubilisation/precipitation from whole fish. *Food Chemistry*, 139, 777–785.

Taskaya, L., Chen, Y.-C., Beamer, S., Tou, J.C., Jaczynski, J. (2009). Compositional characteristics of materials recovered from whole gutted silver carp (*Hypophthalmichthys molitrix*) using isoelectric solubilization/precipitation. *Journal of Agricultural and Food Chemistry*, 57, 4259–4266.

Yoon, I.S., Lee, H.J., Kang, S.I., Park, S.Y., Kang, Y.M., Kim, J.S., Heu, M.S. (2019). Food functionality of protein isolates extracted from Yellowfin Tuna (*Thunnus albacares*) roe using alkaline solubilization and acid precipitation process. *Food Science and Nutrition*, 7, 412–424.

Zhong, S., Liu, S., Cao, J., Chen, S., Wang, W., Qin, X. (2016). Fish protein isolates recovered from Silver Carp (*Hypophthalmichthys molitrix*) by-products using alkaline pH solubilization and precipitation. *Journal of Aquatic Food Product Technology*, 25, 400–413.

List of Authors

Mehdi ABDOLLAHI
Chalmers University of Technology
Gothenburg
Sweden

Régis BARON
Ifremer
Nantes
France

Irineu BATISTA
Instituto Português do Mar
e da Atmosfera (IPMA)
Algés
Portugal

Antonio Javier BORDERÍAS
Institute of Food Science,
Technology and Nutrition (ICTAN)
Spanish National Research
Council (CSIC)
Madrid
Spain

Gervaise DEBUCQUET
Audencia Business School
Nantes
France

Alexandre DEHAUT
ANSES
Laboratoire de Sécurité des Aliments
Boulogne-sur-Mer
France

Charles DELANNOY
PROCIDYS
Wimereux
France

Guillaume DUFLOS
ANSES
Laboratoire de Sécurité des Aliments
Boulogne-sur-Mer
France

Mélanie GAY
ANSES
Boulogne-sur-Mer
France

Carmen GONZÁLEZ SOTELO
Instituto de Investigaciones
Marinas – CSIC
Vigo
Spain

Valur GUNNLAUGSSON
Matis
Reykjavík
Iceland

Grethe HYLDIG
Technical University of Denmark
Lyngby
Denmark

Kristina KAPPEL
Max Rubner-Institut
Kiel
Germany

Françoise LEROI
Ifremer
Nantes
France

Sabrina MACÉ
Ifremer
Nantes
France

Rogério MENDES
Instituto Português do Mar
e da Atmosfera (IPMA)
Algés
Portugal

María Pilar MONTERO GARCÍA
Institute of Food Science,
Technology and Nutrition (ICTAN)
Spanish National Research
Council (CSIC)
Madrid
Spain

Delphine PASSERINI
Ifremer
Nantes
France

Carla PIRES
Instituto Português do Mar
e da Atmosfera (IPMA)
Algés
Portugal

Ute SCHRÖDER
Max Rubner-Institut
Kiel
Germany

Helena SILVA
Instituto Português do Mar
e da Atmosfera (IPMA)
Algés
Portugal

Véronique VERREZ-BAGNIS
Ifremer
Nantes
France

Index

E, F

G, H

I, K

L, M

N, O

P, Q

R, S

T, V, W